Linear analysis of skeletal structures

Linear analysis of skeletal structures

David Johnson

Thomas Telford

Published by ICE Publishing, One Great George Street, Westminster, London SW1P 3AA

Full details of ICE Publishing sales representatives and distributors can be found at:
www.icevirtuallibrary.com/info/printbooksales

First published 2004
Reprinted 2016 (in paperback)
Other titles by ICE Publishing:
Elements of spatial structures: analysis and design. M. Y. H. Bangash and T. Bangash.
ISBN 978-0-7277-3149-1
Space structures 5. Edited by G. Parke. ISBN 978-0-7277-3173-4
Shell structures in civil and mechanical engineering. A. Zingoni. ISBN 978-0-7277-2574-2

www.icevirtuallibrary.com
A catalogue record for this book is available from the British Library

ISBN: 978-0-7277-3276-5
ISBN: 978-0-7277-6192-7 (paperback)

Typeset by Academic + Technical, Bristol
Printed and bound by CPI Group (UK) Ltd, Croydon, CR0 4YY

Contents

Preface

Since the introduction of computerised structural analysis in the 1960s, there has been a continuous debate as to the implications of this revolution for structural engineering education. Early changes to educational practice were based on the precept that an understanding of the appropriate theory was essential if the new technology was to be used reliably and intelligently. Accordingly, the stiffness and flexibility methods were introduced into curricula, it often being held that some experience of hand calculation of these techniques was essential if the basic theory was to be fully appreciated. There was also a tendency to consider that a complete knowledge of computer-based methods of analysis could only be obtained through a familiarity with programming and the coding of analysis programs.

Later, the need to incorporate finite element methodology and the dominance of the stiffness method, often resulted in the discard of flexibility based analysis. Concurrently with the move towards a stiffness-based orientation to curricula, concerns were expressed that the use of computer analysis and the concentration on teaching formal, matrix-based approaches had led to a lack of structural understanding.

The focus of the present text is on the promotion of a structural understanding of the linear behaviour of skeletal structures, especially when statically indeterminate. Traditionally, texts have sought to achieve this by describing both matrix and non-matrix solution techniques, which typically require substantial hand calculation on the part of the student. The expectation is that prolonged exposure to hand calculations will eventually promote structural understanding, as well as familiarity with the particular solution procedures adopted. In contrast, the present text avoids solution techniques almost completely and the reader is certainly not required to solve indeterminate structures by hand. Rather, solutions are to be obtained by the use of any convenient linear analysis computer package.

The elimination of solution techniques from the text has, it is hoped, led to a more direct promotion of structural understanding through a combination of descriptions of 'type' behaviour, qualitative behaviour investigations, and computer-based case studies. However, all of this is based on the assumption that the reader already has a sound knowledge of basic equilibrium and elasticity, which, it is considered, is best gained through the traditional hand calculation approach. The overall aim is to satisfy the demands of a typical prominent structural engineering educator, Professor Iain Macleod, that:

> *We need to teach students how to model, how to use computer packages in real contexts, to validate models, verify results and carry out parameter*

studies. Hand analysis is now only for very simple problems and for back-of-envelope checks.

The expected readership is twofold. Firstly, second and third year undergraduates, who are principally concerned with the study of statically indeterminate structures and, secondly, graduate engineers in practice who need to become familiar with computerised analysis and to be able to interpret the results of such analyses with a sound grasp of structural behaviour.

An initial chapter covers the fundamentals of linear analysis so that a student is aware of the basic principles underlying the material covered subsequently. Selective reading of this chapter, as and when necessary, is probably appropriate. However, if full benefit is to be obtained from the succeeding chapters, the reader should be aware of the limitations of linear analysis; the significance of statical determinacy; and the general features of computer systems based on the stiffness method. The stiffness method is described in outline only and some may well wish to amplify the material given by consulting the references provided.

Discarding hand solutions of indeterminate structures has allowed a complete range of skeletal structures to be considered, rather than a restricted consideration of plane structures, as is common in undergraduate programmes. Beams are considered first, since this allows a beam analogy to be used for the consideration of plane trusses. Arches are subsequently discussed before plane frames, so that arching principles can be related to frame action. Space trusses then provide an introduction to three-dimensional structural action. Grids offer the opportunity to explore torsional action, in particular, and space frames then provide a finale that incorporates all that has gone before.

Worksheets and case studies are provided throughout and answers are given, wherever feasible. For a small number of the case studies, unique solutions do not exist since they are dependent on reader derived data. In such instances solutions have either not been given, or are provided for typical data only.

In each chapter a general format has been followed of describing the general features of the behaviour of the structural form; the determination of its kinematic and statical indeterminacies; the application of computer analysis; and the derivation of influence lines. The chapters then conclude with relevant case studies.

Any author is indebted to advice from a wide range of colleagues and sources. I am especially grateful to Dr Angus Ramsay for his thorough reading of the draft text and for many valuable suggestions. My wife, Elizabeth, has vitally assisted by providing a number of the photographs and, even more importantly, by her continuous support and interest.

David Johnson
2004

Acknowledgements

Chapter 1: Title page (*Arch at Jervaux Abbey, England*): E. Johnson; C1.1b: D. Johnson. Chapter 2: Title page (*U-beam deck under construction*): Tarmac Precast Concrete; C2.1.1a: A. Johnson (*Sydney Opera House, Australia*). Chapter 3: Title page: E. Johnson *(Footbridge at Attenborough, Nottingham, England)*; 3.1: E. Johnson; 3.10: National Grid Transco; C3.1.1: D. Johnson. Chapter 4: Title page (*Arches at Manchester, England*): E. Johnson; C4.1.1 Corus Group plc Chapter 5: Title page *('A' Frame Bulk Store, Indonesia)*: Corus Group plc; C5.1.2: D. Johnson. Chapter 6: Title page *(Tower Cranes)*: D. Johnson; W6.3.1a: Babtie Group, *Independent Engineer for the London Eye*. Chapter 7: Title page *(Beam Grillage)*: D. Johnson. Chapter 8: Title page *(Eden Project Biomes, Cornwall, England)*: E. Johnson; C8.1.1 Anthony Hunt & partners; Chapter 9: Title page *(Design for a Sky Vault, East Midlands, England)*: Tom Hughes, *2hD*.

I Fundamentals

1 Fundamentals

1.1 PURPOSES AND TYPES OF ANALYSIS

Analysis objectives

The overall purpose of this text is to provide a course on the analysis of *skeletal* structures. Skeletal structures may be considered as assemblies of *line* elements, such as rods, beams and columns, the basic properties of which it is presumed have been studied previously. The major objective is to examine the *response* of skeletal structures to defined *causation* effects, which will generally be taken to be *static* (non-time dependent) loading, but which can also be temperature change, support settlement and other effects. The response which is sought will often be the distribution of the *stress resultants* that exist in the line elements and which can take four forms: *axial* (*direct or normal*) *force*; *shear* (*tangential*) *force*; *bending moment*; and *torque*. Occasionally the *displacement* response will also be of interest, and displacements will be interpreted in a general sense to include angular rotations as well as linear translations.

Mathematical model

To obtain a structural response from a given set of causation effects requires the creation of a mathematical model to link the two. The model to be used throughout this book is that of small deflection, linear elasticity. This is also the model most commonly used in analysis computer packages, although many systems will provide other mathematical models as options. Linear elasticity implies two things: first, that the constructional material is elastic, in the sense that removal of stress results in removal of strain, and, second, that stress and strain are linearly connected (Figure 1.1(a)). These two features do not necessarily co-exist, since non-linear, elastic behaviour (Figure 1.1(b)) is possible, as is linear, non-elastic behaviour (Figure 1.1(c)).

The small deflection assumption also has two principal implications. First, it allows equations of equilibrium to be formed in the undisplaced geometry rather than the displaced one. The moment reaction at the fixed support of the cantilever beam shown in Figure 1.2(a), for instance, will therefore be taken as WL and not WL'. Second, changes in the lengths of elements due to 'bowing' effects are neglected and axial elongation or contraction is therefore produced by axial load only. Thus, for the simply supported beam shown in Figure 1.2(b),

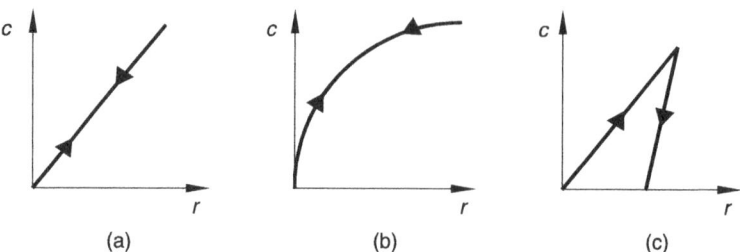

Figure 1.1 Cause (c)–response (r) plots: (a) linear elastic, (b) non-linear elastic, (c) linear non-elastic

the horizontal displacement at the right-hand side would be taken to be zero. This follows because the beam is not subjected to an axial load when the loading is applied in a direction normal to its axis, and therefore there is no axial extension or contraction.

The assumption of small displacements results in an inability to analyse effectively two main classes of structure. Flexible structures, principally those constructed from cables, clearly involve large displacements, in the sense that the movements are not negligible in comparison with the lengths of the structural elements. Such structures, which typically act in tension, therefore require a *large deflection* theory, in which equilibrium equations are formed in the displaced geometry (Figure 1.3(a)). Structures subject to buckling effects, notably compression structures, may well only experience 'small' displacements, but the effects of these displacements can be catastrophic. The column of Figure 1.3(b), for example, receives an additional bending moment due to the compression, P, if equilibrium is taken in the displaced geometry. This enhancement effect can then cause the element to become *unstable* (buckle) if the compression is sufficiently large. Structures liable to instability therefore also need a large-deflection (sometimes called buckling or $P - \Delta$ analysis) treatment.

If linear elasticity is coupled with a small displacement assumption, then the resulting analysis will be linear in an overall sense. This implies that not only will the deformation of the elements be directly proportional to the stress resultants they sustain, but also that the overall displacements will be proportional to the overall loading, as will the stress resultants themselves. Linearity has a major advantage in that the *principle of superposition* (Gere and Timoshenko, 1999) applies, whereby the response to a combination of causation effects is

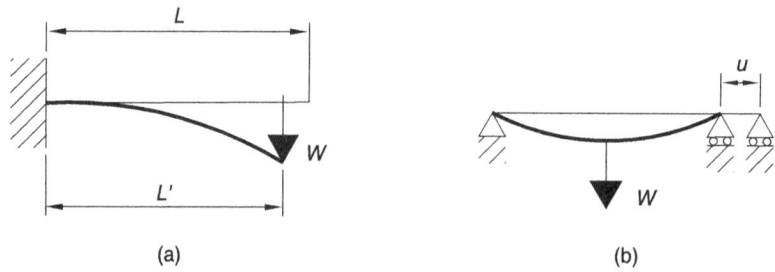

Figure 1.2 (a) Cantilever beam, (b) simply supported beam

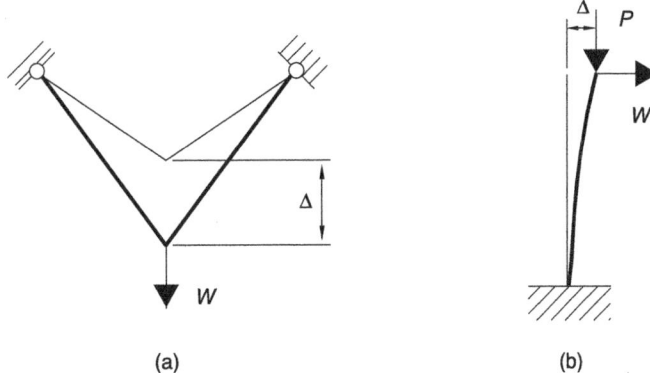

Figure 1.3   (a) Tension structure, (b) compression structure

simply the sum of the responses to the individual causes. A disadvantage is that the analysis will provide **no warning of failure**. The linearity of the material elasticity is presumed to continue indefinitely, and no yielding, fracture or other change in the material behaviour is modelled. Further, the small displacement assumption precludes the modelling of buckling failure.

In contrast to the small displacement formulation, large displacement, linear elastic analysis will result in a non-linear response. Tension structures will typically stiffen (Figure 1.4) under increasing load (as a clothes line under an increasing quantity of wet washing) and no warning of failure will be provided, since this is usually due to material failure. Compression structures (Figure 1.3(b)) will typically soften under increasing load and an indication of buckling failure can be provided, but this will ignore the effects that yield or other material degradation might produce.

With small displacement, linear elastic analysis, avoidance of failure therefore normally becomes a design issue, to be considered subsequent to the analysis. It might therefore be asked why the model is generally used if it offers no guidance on such a fundamental aspect of behaviour as failure? Several attributes of the model, in addition to the convenience of the linear nature, dictate its common adoption. It is easy to systemise and is therefore readily adapted to the production of general, robust computer systems. It also offers a generally acceptable estimate

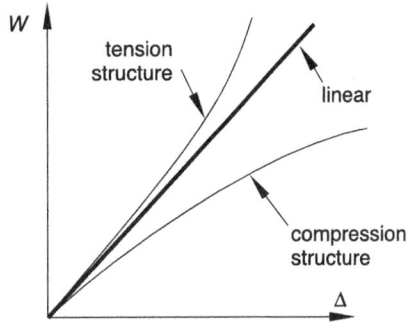

Figure 1.4   Load/deflection responses

of the likely response under *service* (working) conditions. Finally, a set of stress resultants will be derived which are in equilibrium and the '*lower bound*' (safe) theorem of plasticity (Moy, 1996) may therefore be invoked. This theorem states that provided **each** stress resultant of **any** equilibrium set of resultants is designed not to fail at less than the desired load factor, then the load factor at failure of the complete structure will be at least equal to the factor applied to the individual stress resultants. There is an important limitation of the theorem in that it applies to failure by ductile material yield. Structures in which instability effects are significant, or which contain brittle components, are therefore excluded and may require more specialised investigation.

1.2   KINEMATIC AND STATICAL DETERMINACY

Properties of statically determinate and indeterminate structures

The stress resultant response of some simple structures (Figure 1.5(a), (b)) may be found from the application of equilibrium conditions alone. Such structures are termed *statically determinate* and the establishment of a structure's determinacy, or otherwise, is of considerable importance. This is not particularly because determinate structures can be somewhat easier to analyse than indeterminate ones (Figure 1.5(c), (d)) (in fact it makes no difference if a computer package is to be used), but because the two types respond to causation effects in significantly different ways (Table 1.1).

Table 1.1 is worthy of considerable study because of the importance of its constructional implications. Avoiding stressing a structure by differential foundation settlement or extreme temperature changes (Properties 3 and 4, Table 1.1) can, for instance, encourage a determinate design. This, however, will lead to a structure with a 'single load path' (Property 2, Table 1.1). Again, the same penalty can be paid if a determinate structure is chosen because of its ability to tolerate errors in construction tolerances that lead to a lack of precise fit of the members.

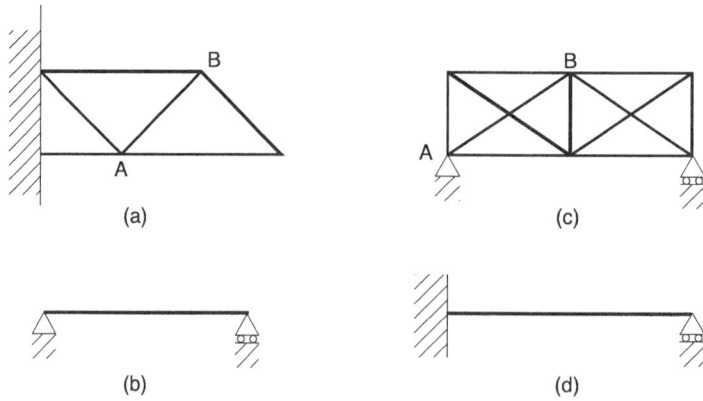

(a)	(c)
(b)	(d)

Table 1.1   Properties of statically determinate and indeterminate structures

	Determinate	Indeterminate
1.	Stress resultants and reactions may be found by statics.	Stress resultants and reactions may not be found fully by statics alone.
2.	Failure of a single stress resultant or reaction leads to at least partial collapse.	Failure of a single stress resultant or reaction does not necessarily lead to even partial collapse.
3.	Support movements do not produce stress resultants or reactions.	Support movements will generally produce stress resultants and reactions.
4.	Temperature and lack-of-fit effects do not produce stress resultants and reactions.	Temperature and lack-of-fit effects will generally produce stress resultants and reactions.
5.	Stress resultants and reactions are independent of member sizes and material properties.	Stress resultants and reactions are generally dependent on member sizes and material properties.
6.	Displacements are dependent on member sizes and material properties.	Displacements are dependent on member sizes and material properties.

For instance, if member AB in the truss shown in Figure 1.5(a) is fabricated too long, it will slightly alter the geometry of the truss when connected but will not induce loads in the structure. If the same were to occur with member AB of the indeterminate truss (Figure 1.5(c)), then the member would have to be forced into place. This process would certainly produce forces in at least some of the members of the truss, as well as altering the intended overall shape.

In addition to statical indeterminacy, structures may also be designated a degree of kinematic indeterminacy, k. This parameter has less significance for structural behaviour than the degree of statical indeterminacy, q. However, k is of importance for a number of theoretical reasons, one of which is that the determination of q is reasonably straightforward if k is found first. Kinematic and statical indeterminacy will therefore be examined next for the cases of trusses and beams, since these are basic structural forms and serve to illustrate the principles involved. The indeterminacy of other types of structure is considered in the appropriate chapters.

1.3   DETERMINACY OF PLANE TRUSSES

The degree of kinematic indeterminacy may be defined as the number of displacement degrees of freedom that are required to ensure the compatibility of the structure. Compatibility implies geometric integrity such that, for example, all the members of a truss remain connected when the truss displaces. This may be ensured by defining the deformations of the members in terms of the movements of the *nodes* (*joints*) of the structure. The nodal displacements are therefore taken to be the displacement degrees of freedom. Thus, for the simple truss shown in Figure 1.6(a), the nodal components of displacement in the horizontal and vertical

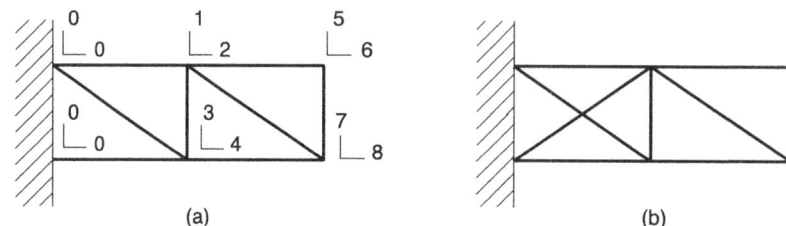

Figure 1.6   (a) Simple truss, (b) truss with extra member

directions (non-zero numbers in the figure) would be taken as the displacement degrees of freedom. Since there are eight of these in total, the truss of Figure 1.6(a) has a kinematic indeterminacy of eight. The kinematic indeterminacy of any truss may be found in this way by inspection or, alternatively, by noting that, in general, for a truss with a **total** of j joints (each with two potential degrees of displacement freedom) and d restrained **components** of displacement (indicated by zeros in Figure 1.6(a)), k will be given by

$$k = 2j - d \tag{1.1}$$

If the kinematic indeterminacy of the truss of Figure 1.6(a) is found from equation (1.1), therefore, since there are six joints in total and four restrained displacement components, it is again found that $k = (2 \times 6) - 4 = 8$.

The kinematic indeterminacy also represents the number of equilibrium equations which need to be solved for the chosen node set, since equilibrium must be maintained in each of the displacement freedom directions if accelerated movement (Newton's Second Law) is not to occur. This equilibrium equation interpretation of k is helpful in the calculation of the statical indeterminacy of the structure. The degree of statical indeterminacy may be defined as the number of stress resultants of the structure which cannot be determined by equilibrium (statics) alone. The number of equilibrium equations is, however, equal to k. Therefore, if the total number of stress resultants is s, the number that cannot be found by statics is given by

$$q = s - k \tag{1.2}$$

For a truss, there is just one stress resultant for each member since the members sustain axial loads (tension or compression) only. The number of stress resultants, s, is therefore the same as the number of members, m, and

$$q = m - k \tag{1.3}$$

The truss shown in Figure 1.6(a) has eight members so that $q = 8 - 8 = 0$. It therefore has no stress resultants that may not be determined by statics and is statically determinate. On the other hand, the introduction of an extra member, resulting in *cross-* (or *counter-*) *bracing* in one panel (Figure 1.6(b)), clearly increases q by one (without changing k) and indicates that the truss is statically indeterminate, having one *redundant* member, which may be removed without resulting in an unstable arrangement.

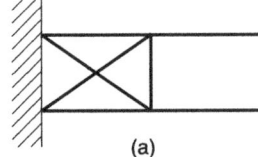 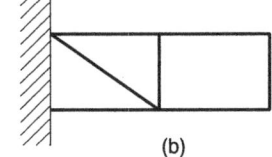

(a)   (b)

Figure 1.7   Mechanisms: (a) q = 0, (b) q < 0

The determinacy conclusion regarding the truss of Figure 1.6(a) should strictly have been potentially determinate. It is necessary to insert 'potentially', because of examples such as that shown in Figure 1.7(a). This truss also has $q = 0$ since k and m are unchanged from Figure 1.6(a), but it may be observed that the first bay is essentially indeterminate, while the second is a mechanism and would collapse. Satisfaction of $q = 0$ is therefore only a necessary condition for determinacy and the structure must also be shown to be stable. Unfortunately, the mathematical test for structural stability is too arithmetically demanding for hand calculations, even for simple structures. Reliance must therefore be placed on structural experience and computer analysis.

Structures that are deficient, regardless of possible member rearrangements, such as the truss shown in Figure 1.7(b), will produce a negative q value, which indicates the number of extra members needed for a stable arrangement. Thus, for the truss of Figure 1.7(b), $k = 8$ and, hence, $q = 7 - 8 = -1$. This indicates that the right-hand panel requires bracing if it is not to collapse in a shear mode due to the assumed pin-joints.

With practice, the determinacy predictions of equation (1.2) may be verified by inspection in reasonably straightforward cases, to which a more formal inspection procedure may also be applied. The formal approach is to establish 'stable' nodes in sequence. For a cantilever truss such as that of Figure 1.8(a), the starting point is a stable triangular arrangement of members linked to the support. This establishes three stable nodes at A, B and D. Node C may then be linked to A and D, thus becoming stable, and node E is finally linked to C and D. For a simply supported truss, such as that of Figure 1.8(b), a complete triangle is first established, and stable nodes built up from the initial three. Determinacy will require that the stability of the nodes can be established using all the members just once and it will also need a statically determinate support arrangement. Three non-concurrent support reaction

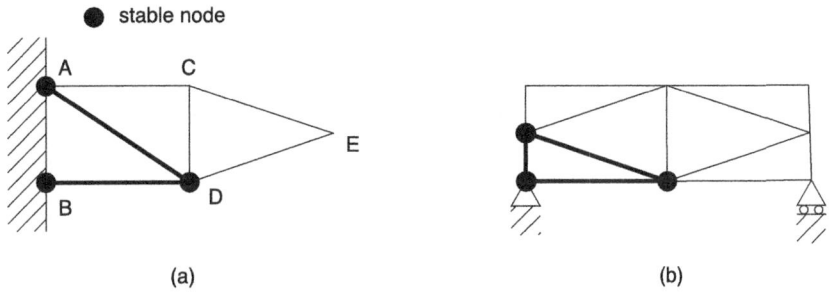

(a)   (b)

Figure 1.8   (a) Cantilever truss, (b) simply supported truss

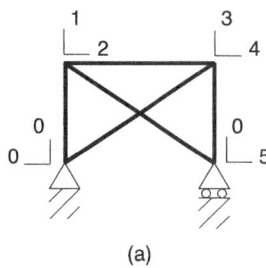
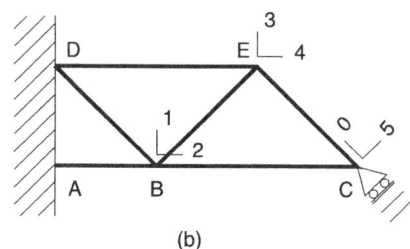

(a)   (b)

Figure 1.9   Trusses with roller supports

components can be evaluated from the three equations of planar equilibrium, so that the truss of Figure 1.8(b) is indeed determinate and stable. If the procedure is applied to the truss of Figure 1.6(b) it will be found that node stability can be achieved without involving one of the members in the first panel, indicating a redundancy. In the case of the truss of Figure 1.7(b), it proves impossible to establish the stability of the two rightmost nodes.

A roller support produces one displacement degree of freedom parallel to the direction of the support and a restrained component of displacement in the normal direction. This does **not** depend on the inclination of the roller (Figure 1.9(b)), because, even if horizontal and vertical displacement components are used, these are not independent of each other. The specification of one is sufficient, since the other is then defined by the condition that the roller remains on the support. The truss of Figure 1.9(a) has $k = (2 \times 4) - 3 = 5$ and $q = 5 - 5 = 0$. It is hence potentially determinate and may be shown to be stable by inspection. The truss of Figure 1.9(b) has $k = (2 \times 5) - 5 = 5$ and $q = 6 - 5 = 1$. It is therefore once redundant and can be made determinate by the removal of one appropriate member, such as *AB*. Member *BC* is not an appropriate member to remove as it would create a mechanism in which *EC* can rotate about *E*. Care therefore needs to be taken in the selection of members to remove and it is helpful to interpret member removal more generally as incorporating *releases* into structures. Releases can be formed by the elimination of stress resultants and/or components of reaction. Removing a member represents the elimination of the single direct force stress resultant carried by a truss member. Removal of the roller support in Figure 1.9(b) represents the elimination of a single reaction component and leaves a determinate truss, from which no member may be safely removed. This substantiates the finding that the structure is once redundant and is perhaps the most straightforward approach for this example.

Worksheet 1.1   Kinematic and statical determinacy of trusses

W1.1.1   Calculate the degrees of kinematic and statical indeterminacy for the trusses shown in Figure W1.1.1. State, also, whether the structures are (a) unstable or (b) statically determinate or indeterminate. If unstable, indicate a system of additional members and/or reaction components that will make the structure stable. If statically indeterminate, indicate a system of releases that will make the structure determinate and stable. It may be presumed that members are not connected at crossing points.

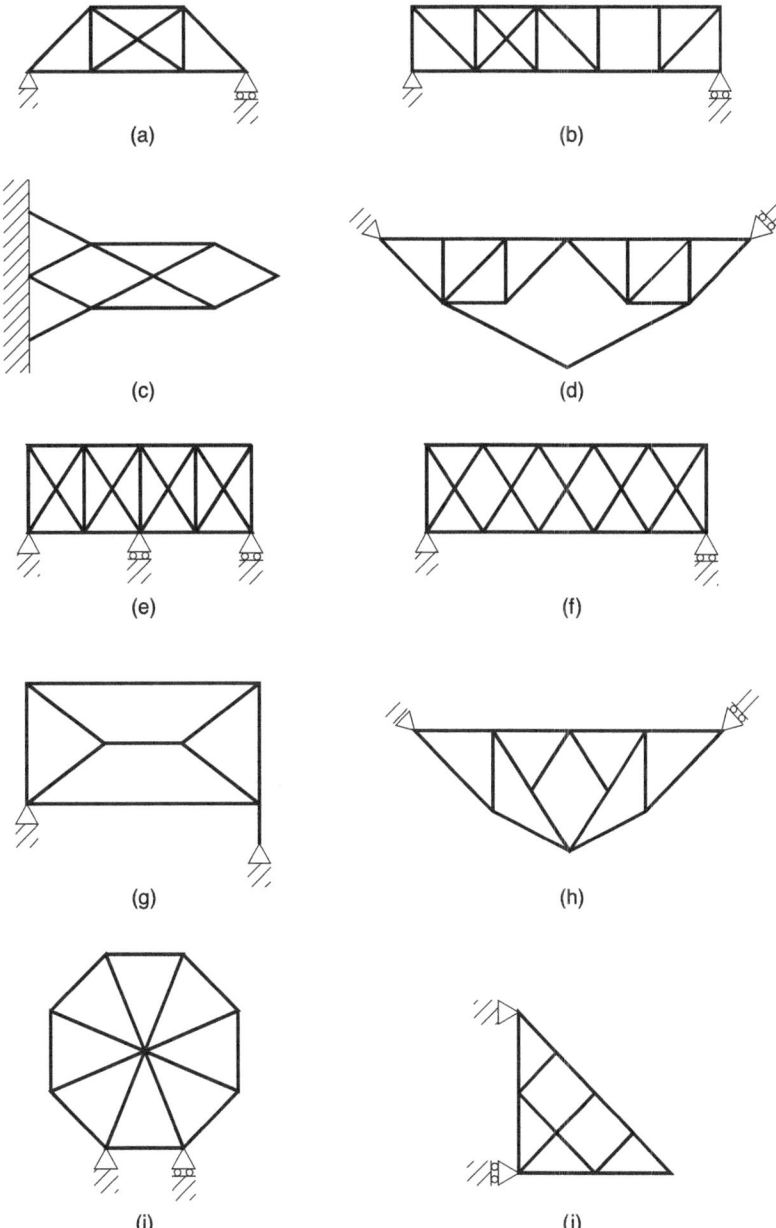

Figure W1.1.1   Worksheet 1.1 trusses

1.4   DETERMINACY OF PURE BEAMS

The kinematic and statical indeterminacy of beams may be evaluated by following a similar process to that adopted for trusses, although it is perhaps not so obvious as to what constitutes nodes and members in the case of a beam. For determinacy evaluation, and also for computerised analysis, it is generally sufficient to use support positions and free ends as nodes, designating the spans between these points as members. For trusses, the nodal displacement degrees of freedom

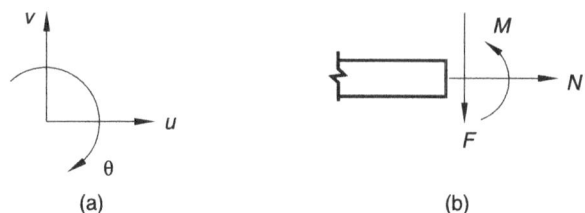

Figure 1.10   (a) Planar displacement components, (b) planar stress resultants

consisted of translation components only, since the assumed pin-joints allow freedom of rotation at the joints. Members therefore remain straight and do not sustain bending moments. Beams certainly transmit bending moments and the rigidity of their joints, over an internal support for example, requires that rotational degrees of freedom be considered. On the other hand, if a *pure* beam is defined as a bending element subjected to normal loading only, then no axial loads are sustained and no axial displacements are produced (Figure 1.2(b) *et seq.*). It follows that beams require consideration of two 'displacement' degrees of freedom (rotation and normal displacement) from the three components possible in a plane (Figure 1.10(a)). Equation (1.2) for kinematic indeterminacy therefore remains valid since the number (but not *type*) of nodal displacement components is the same as for a truss. A beam will also involve two of the possible three planar stress resultants (Figure 1.10(b)), shear force and bending moment, since axial load is absent, as just discussed. Shear force and bending moment will, in general, vary continuously along a beam and it might therefore be imagined that an infinite number of stress resultants would exist. However, if two stress resultant values are known within any given span, the complete span distribution may be evaluated by statics. The number of stress resultants to be determined in each span (member) is therefore two and equation (1.3) is replaced by

$$q = s - k = 2m - k \tag{1.4}$$

On the above basis, the two-span beam with an overhang shown in Figure 1.11 has a kinematic indeterminacy of four, by inspection. This may be verified by noting that $j = 4$; $d = 4$ and $m = 3$ so that, from equation (1.2), $k = (2 \times 4) - 4 = 4$. It then follows from equation (1.4) that $q = (2 \times 3) - 4 = 2$. The beam therefore has a degree of statical indeterminacy of two, and two releases may be made to achieve a stable, determinate structure. In the case of beams, reaction components are usually the most convenient releases and the two roller supports, for example, may be removed to leave a cantilever beam.

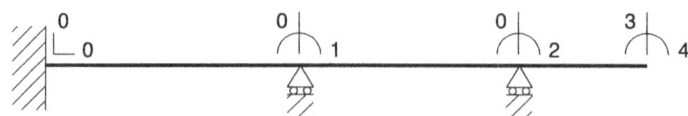

Figure 1.11   Continuous beam

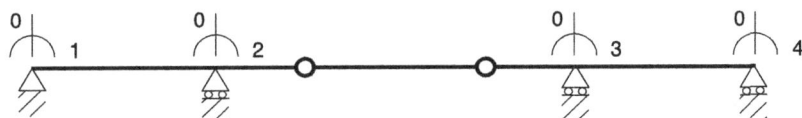

Figure 1.12   Beam with suspended span

A feature of beams that does not occur with trusses is that internal releases in the form of hinges are occasionally encountered. The bending moment at a frictionless hinge is known to be zero, hence producing one less unknown stress resultant. If there are h hinges in total, the number of unknown stress resultants becomes $s = 2m - h$, so that the statical indeterminacy is given by

$$q = s - k = (2m - h) - k \tag{1.5}$$

The beam shown in Figure 1.12 is of a 'suspended span' form that is sometimes used for highway bridges to facilitate construction, since the central, suspended span is independent of the side-spans and may be 'rolled-in' once they are complete. For this beam, k is found to be four from the figure, or, by calculation, as $k = (2 \times 4) - 4 = 4$. It follows from equation (1.5) that $q = ((2 \times 3) - 2) - 4 = 0$ and the beam is therefore determinate. It should be noted that internal hinges should not be taken as node points, so that a single central member exists, which happens to contain two internal hinges. The determinacy of this beam is a further factor that can influence its selection as a bridge form and the reader should verify that collapse would occur were any of the reaction components released.

A final property of the determinacy parameters, k and q, may be conveniently demonstrated by a consideration of the simply supported beam shown in Figure 1.13(a). First it should be emphasised that neither determinacy parameter is dependent upon loading, but solely on structural geometry and the support conditions. However, if point loads are applied to beams, it is sometimes convenient to place a node at the loading point, as shown in Figure 1.13(b), and the effect of doing so on the determinacy parameters is worthy of investigation. For Figure 1.13(a), $k = 2$, by inspection, and the determinacy of the beam is confirmed by $q = (2 \times 1) - 2 = 0$. With the intermediate node, $k = 4$, but the beam remains determinate since $q = (2 \times 2) - 4 = 0$. In general, kinematic indeterminacy depends on the modelling of a structure, specifically the selection of nodes and members, but statical indeterminacy is an intrinsic property and is not affected in this way.

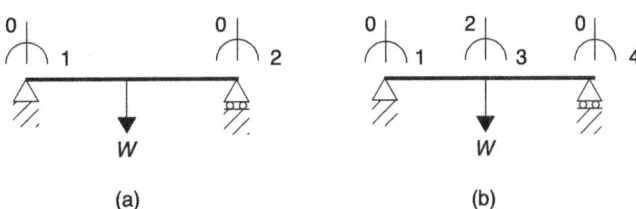

(a)   (b)

Figure 1.13   Simply supported beam

Worksheet 1.2   Kinematic and statical determinacy of beams

W1.2.1 Calculate the degrees of kinematic and statical indeterminacy for the beams shown in Figure W1.2.1. State, also, whether the structures are (i) unstable or (ii) statically determinate or indeterminate. If unstable, indicate a system of structural alterations that will make the structure stable. If statically indeterminate, indicate a system of releases that will make the structure determinate and stable.

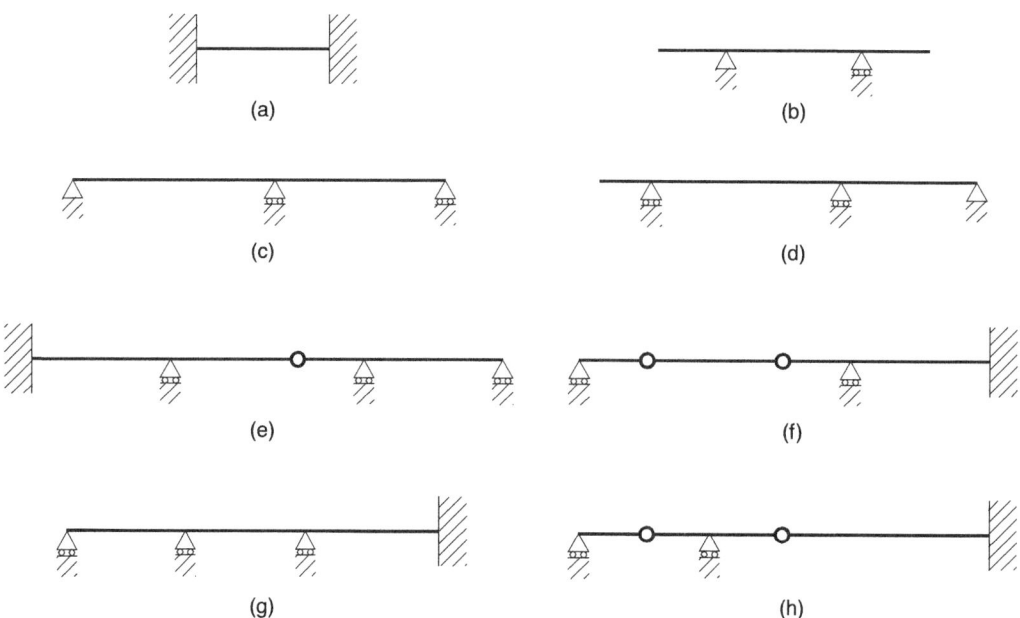

Figure W1.2.1   Worksheet 1.2 beams

1.5   STIFFNESS METHOD FOR LINEAR ELASTIC ANALYSIS

Basic requirements

In general, three sets of conditions need to be satisfied simultaneously for the analysis of a structure. Perhaps the most fundamental condition is that every part of the structure must be in equilibrium. For statically determinate structures, *equilibrium* alone will allow the internal stress resultants to be determined for a given set of loads, as has already been established (Table 1.1). If displacements are also of interest, however, *material laws* (assumed to be those of linear elasticity here) are needed, since these govern the deformation of the members in relation to their internal stress resultants. In turn, the member deformations produce the structural displacements. For statically indeterminate structures, the stress resultants themselves cannot be found unless the equilibrium and material law conditions are supplemented with *compatibility* statements. Compatibility ensures that geometric coherence is maintained such that no gaps or kinks, for instance, appear in a structure when it is loaded. In other words, its general form and topology (connectivity) are maintained.

Stiffness and flexibility methods

The equilibrium, material law and compatibility requirements may be combined together and solved in two main ways. The *stiffness* (*equilibrium*, *displacement*) method works by forming a set of equilibrium equations in terms of a compatible set of structural displacements, which, by definition, will be equal in number to the kinematic indeterminacy of the structure. The equations are then solved for the displacements, and stress resultants are determined subsequently. The *flexibility* (*compatibility*, *force*) method, on the other hand, forms a set of compatibility equations in terms of an equilibrium set of redundant forces (stress resultants). In this case, the number of equations to be solved is equal to the degree of statical indeterminacy and the solution of the equations will provide the stress resultant solution. Displacements may be found subsequently, if required.

For many structures, the kinematic indeterminacy is usually considerably greater than its statical counterpart. Prior to the introduction of computerised analysis, the flexibility method tended to be favoured, therefore, because of the smaller number of equations that needed to be solved by hand. The stiffness method was the first to be fully automated and is still somewhat more straightforward to automate than the flexibility method. For this reason, essentially all computer packages employ the stiffness method and an awareness of the method's basic properties is essential if packages are to be intelligently employed.

Stiffness method applied to plane trusses

The stiffness method starts by relating member internal forces to member nodal displacements. This is most conveniently done in relation to a set of *member* (*local*) axes. If the nodes of a typical planar member (Figure 1.14) are designated a and b, then direction a–b, along the axis of the member, will be taken as the first local axis, u. A second local axis, w, is then created in a direction normal to the member by rotating u in an anti-clockwise direction. The final axis, v, will then be taken perpendicular and 'into' the plane of u and w. Rotation about any one of the axes is taken positive if it acts in a clockwise sense when viewed in the positive direction of the axis. The axes u, v, w are said to be a *right-handed* set

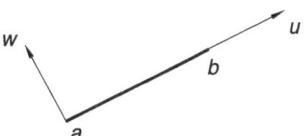

Figure 1.14   Axes for planar member

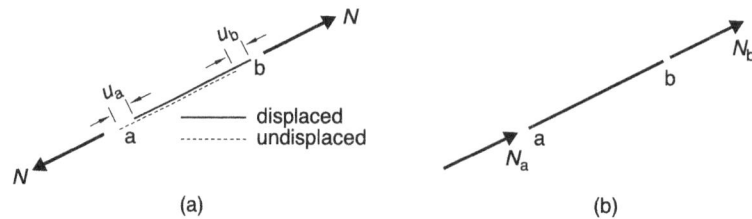

Figure 1.15   Typical truss member

since a quarter turn about u, in the positive direction, for example, causes v to move to the position of the subsequent axis, w.

For the typical truss member shown in Figure 1.15(a), linear elasticity (Gere and Timoshenko, 1999) may be invoked, in the form of Hooke's Law, to relate the axial load to the displacements of its end nodes by

$$N = \frac{EA}{L}(u_b - u_a) \tag{1.6}$$

However, in terms of member forces taken positive in the direction of the member axes (Figure 1.15(b)), we have $N_a = -N$ and $N_b = N$. So that

$$N_a = \frac{EA}{L}(u_a - u_b)$$
$$N_b = \frac{EA}{L}(-u_a + u_b) \tag{1.7}$$

Or, in matrix form

$$p = \left\{ \begin{array}{c} N_a \\ N_b \end{array} \right\} = \frac{EA}{L}\begin{bmatrix} 1 & -1 \\ -1 & 1 \end{bmatrix}\left\{ \begin{array}{c} u_a \\ u_b \end{array} \right\} = \mathbf{K_m}e \tag{1.8}$$

In equation (1.8), $\mathbf{K_m}$ is the *member* stiffness matrix and $_m$ indicates evaluation in local (member) axes.

If stiffness equations are to be eventually formed in terms of the nodal displacements relative to a set of global axes, x, y, z, it is necessary to relate the deformations of the member in the local axes, e, to displacements in the global (overall) axes. As may be derived from Figure 1.16(a), the desired relationships are

$$u_a = \cos\alpha x_a + \sin\alpha z_a \quad \text{and} \quad u_b = \cos\alpha x_b + \sin\alpha z_b \tag{1.9}$$

Or, in matrix form

$$e = \left\{ \begin{array}{c} u_a \\ u_b \end{array} \right\} = \begin{bmatrix} \cos\alpha & \sin\alpha & 0 & 0 \\ 0 & 0 & \cos\alpha & \sin\alpha \end{bmatrix}\left\{ \begin{array}{c} x_a \\ z_a \\ x_b \\ z_b \end{array} \right\} = \begin{bmatrix} \mathbf{T_0} & \mathbf{0} \\ \mathbf{0} & \mathbf{T_0} \end{bmatrix}d = \mathbf{T}d$$

$$(1.10)$$

In equation (1.10), $\mathbf{T}$ is known as the *transformation* matrix. A further transformation is needed if the member forces are to be combined together to form equations

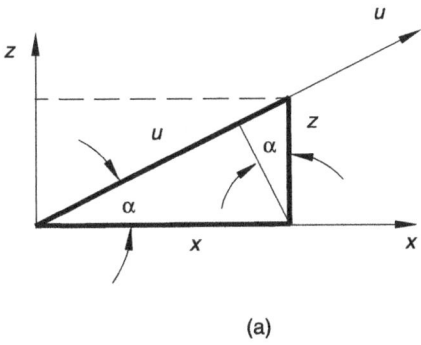 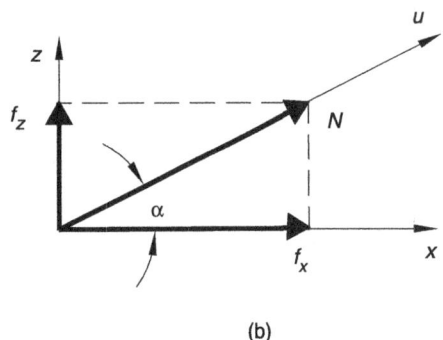

(a)   (b)

Figure 1.16   Transformation of axes

of joint equilibrium, since these need to be related to a common set of axes, rather than a collection of member axes which are local to each member. Using the global axes, x, y, z (Figure 1.16(b)), it is found, by resolution, that

$$\begin{Bmatrix} f_{xa} \\ f_{za} \end{Bmatrix} = \begin{Bmatrix} \cos \alpha \\ \sin \alpha \end{Bmatrix} N_a \quad \text{and} \quad \begin{Bmatrix} f_{xb} \\ f_{zb} \end{Bmatrix} = \begin{Bmatrix} \cos \alpha \\ \sin \alpha \end{Bmatrix} N_b \tag{1.11}$$

Or, in matrix form

$$f = \begin{Bmatrix} f_{xa} \\ f_{za} \\ f_{xb} \\ f_{zb} \end{Bmatrix} = \begin{bmatrix} \cos \alpha & 0 \\ \sin \alpha & 0 \\ 0 & \cos \alpha \\ 0 & \sin \alpha \end{bmatrix} \begin{Bmatrix} N_a \\ N_b \end{Bmatrix} = \mathbf{T}^t p \tag{1.12}$$

The occurrence of the transformation matrix, $\mathbf{T}$, and its transpose, $\mathbf{T}^t$, in equations (1.10) and (1.12) is an example of the *contragredient* nature of force and displacement transformations. Pre-multiplying equation (1.8) by $\mathbf{T}^t$ and then substituting from equations (1.10) and (1.12) gives

$$f = \mathbf{T}^t \mathbf{K}_m \mathbf{T} d = \mathbf{K}_g d \tag{1.13}$$

In equation (1.13), $\mathbf{K}_g$ is the member stiffness matrix referred to the global axes. It is now possible to combine contributions from the individual members together (Kennedy and Madugula, 1990), to produce a set of nodal equilibrium equations of the form

$$\mathbf{F} = \mathbf{K}_u \mathbf{D}_u \tag{1.14}$$

In equation (1.14), $\mathbf{K}_u$ is the unrestrained *structure* (*system, global*) stiffness matrix, and $\mathbf{D}_u$ is the vector of nodal displacement components (both restrained and unrestrained). The matrix $\mathbf{K}_u$ will be singular, and equation (1.14) is therefore impossible to solve, until the support conditions are enforced. This represents the ability of the structure to be given a rigid-body displacement in its unrestrained state. The support conditions are incorporated into equation (1.14) by constraining the restrained displacement components, d, to be zero. This reduces the equations represented by the matrix expression to those relevant to the unrestrained components of displacement, which will be equal in number to the kinematic indeterminacy of the truss. The vector, $\mathbf{F}$, will then represent the sum of the member end-force effects at the unrestrained nodes of the truss, which, for equilibrium, must be equivalent to the specified nodal loads, $\mathbf{W}$, so that

$$\mathbf{W} = \mathbf{KD} \tag{1.15}$$

Provided that $\mathbf{K}$ is non-singular, it can be inverted to produce the nodal displacements from

$$\mathbf{D} = \mathbf{K}^{-1}\mathbf{W} \tag{1.16}$$

Individual member displacements, d, may be extracted from $\mathbf{D}$ so that the forces in the members may be found by combining equations (1.8) and (1.10) to give

$$p = \mathbf{K}_{\mathrm{m}}\mathbf{T}d \tag{1.17}$$

Summary of stiffness method

The advantage of formulating the stiffness method in a matrix form is that it can then be readily applied to any form of structure. It is therefore worthwhile summarising the basic steps in a matrix format. These seven steps are:

Step 1   Form the member stiffness equation in local axes

$$p = \mathbf{K}_{\mathrm{m}}e \tag{1.18}$$

Step 2   Establish an appropriate transformation matrix such that

$$e = \mathbf{T}d \quad \text{and} \quad f = \mathbf{T}^{\mathrm{t}}p \tag{1.19}$$

Step 3   Form the member stiffness equations in the global axes

$$f = \mathbf{T}^{\mathrm{t}}\mathbf{K}_{\mathrm{m}}\mathbf{T}d = \mathbf{K}_{\mathrm{g}}d \tag{1.20}$$

Step 4   Form the unrestrained global stiffness equations by assembling member stiffness equations

$$\mathbf{F} = \mathbf{K}_{\mathrm{u}}\mathbf{D}_{\mathrm{u}} \tag{1.21}$$

Step 5   Apply the restraint conditions and apply equilibrium with the applied loads

$$\mathbf{W} = \mathbf{KD} \tag{1.22}$$

Step 6   Solve the stiffness equations for the nodal displacements

$$\mathbf{D} = \mathbf{K}^{-1}\mathbf{W} \tag{1.23}$$

Step 7   Determine the member forces (stress resultants)

$$p = \mathbf{K}_{\mathrm{m}}\mathbf{T}d \tag{1.24}$$

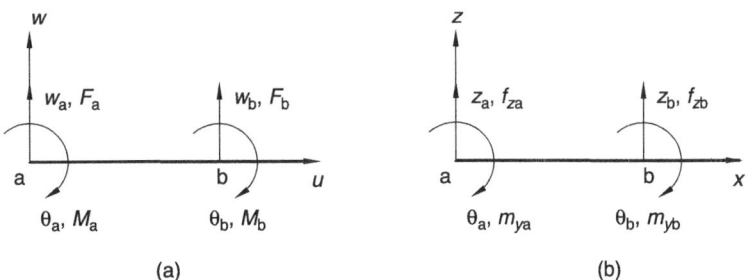

Figure 1.17   Pure beam member: (a) local axes, (b) global axes

Stiffness method applied to a pure beam

To apply the seven steps given by equations (1.18) to (1.24), it is simply necessary to define suitably the various vectors and matrices involved. Thus, for a pure beam, it follows from the determinacy discussion that the following definitions (Figure 1.17) are appropriate:

$$
p = \begin{Bmatrix} F_a \\ M_a \\ F_b \\ M_b \end{Bmatrix} \qquad
e = \begin{Bmatrix} w_a \\ \theta_a \\ w_b \\ \theta_b \end{Bmatrix} \qquad
f = \begin{Bmatrix} f_{za} \\ m_{ya} \\ f_{zb} \\ m_{yb} \end{Bmatrix} \qquad
d = \begin{Bmatrix} z_a \\ \theta_{ya} \\ z_b \\ \theta_{yb} \end{Bmatrix}
\qquad (1.25)
$$

Since, in this case, the local displacements/forces are in the same directions as the global displacements/forces, it follows that $\mathbf{T} = \mathbf{I}$, the identity matrix. The relationship between member forces and member displacements is established in standard texts (Kennedy and Madugula, 1990) to be as shown in Figure 1.18

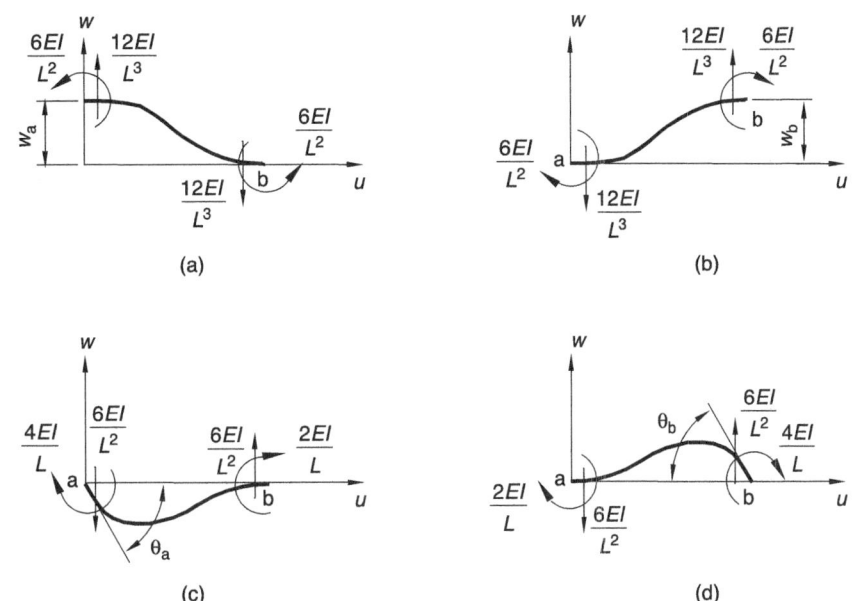

Figure 1.18   Member stiffness formation for a pure beam member

and equation (1.26):

$$p = \begin{Bmatrix} F_a \\ M_a \\ F_b \\ M_b \end{Bmatrix} = EI \begin{bmatrix} 12/L^3 & -6/L^2 & -12/L^3 & -6/L^2 \\ -6/L^2 & 4/L & 6/L^2 & 2/L \\ -12/L^3 & 6/L^2 & 12/L^3 & 6/L^2 \\ -6/L^2 & 2/L & 6/L^2 & 4/L \end{bmatrix} \begin{Bmatrix} w_a \\ \theta_a \\ w_b \\ \theta_b \end{Bmatrix} = \mathbf{K}_m e$$

(1.26)

1.6 COMPUTER IMPLEMENTATION OF THE STIFFNESS METHOD

Global and local axes

Almost all computer systems make use of a right-handed Cartesian set of global co-ordinate axes (Figure 1.19(a)). However, there are variations in respect of the sub-set that is used for planar structures, such as beams and frames. Some systems employ a x–z set of axes. This has the effect of making moments and rotations clockwise positive, according to the right-handed convention (Figure 1.19(b)). Other systems place two-dimensional structures in an x–y plane, which results in moments and rotations becoming anti-clockwise positive (Figure 1.19(c)), unless a departure from the right-handed convention is invoked. In this, as in other system dependent conventions, careful reference to the provider's user information is essential.

The local 'u' axis is invariably taken along the member axis, but its positive sense is variously defined. Sometimes, as here, it is simply from the first specified to the second specified end-node, but in other cases it may be defined as being from the smaller to the larger numbered end-node. Again, a number of conventions are used to define the remaining two local axes, some of which require the treatment of vertical members as special cases.

Member specification

The pair of connected nodes are used to specify members. Sometimes it is convenient to introduce intermediate nodes along members. This is used, for

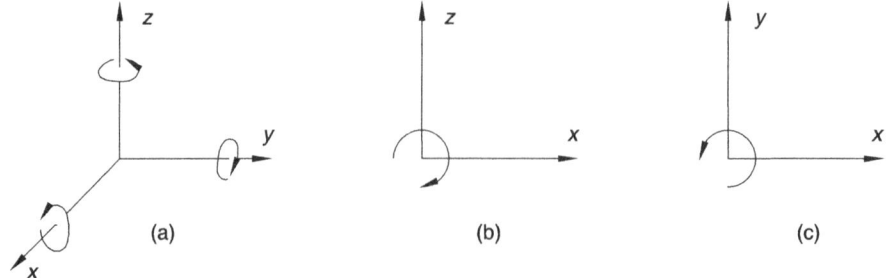

Figure 1.19   *Axes: (a) global, (b)* x–z *planar, (c)* x–y *planar*

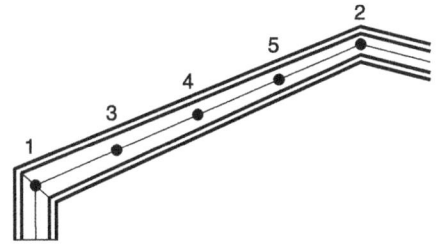

Figure 1.20   Member with intermediate nodes

example, to provide positions to which point loads might be applied or to ensure detailed output is provided at specific locations, should alternative facilities not be provided for the achievement of these results. It should be noted that it is vital that the sub-members are individually specified, otherwise the intermediate joints are structurally unconnected. Thus, in Figure 1.20, a series of members 1–3, 3–4, 4–5, 5–2 must be specified, not simply 1–2.

Members are generally assumed to be *prismatic* (constant cross-section along their length). Some computer systems offer facilities for variable section members, but, if this is not available, then intermediate nodes can be used to divide the member up into a number of approximating, prismatic sections. For linearly tapering members, for example, two or three intermediate nodes will generally provide an acceptable approximation (Steel Construction Institute, 1995).

A number of sectional properties will be required, according to the type of analysis being undertaken. Cross-sectional areas are normally needed since many systems will employ these to automatically determine self-weights, if specified. Bending analyses will additionally require second moment(s) of area (I values) and torsional constants (J values) will also be needed for grids or space frames. For metal structures, packages commonly incorporate database values for standard sections. Alternatively, reference may be made to standard tables (Owens and Knowles, 1992). For reinforced or prestressed concrete construction it is usual to employ the gross cross-sectional properties. For composite construction, use of a *modular ratio* (Gere and Timoshenko, 1999) may be made to convert the section into an equivalent homogeneous one. In the case of steel/concrete composite construction, care must be taken, however, to ensure that the contribution of the concrete is only included for portions of the beam over which the concrete acts in compression, and is therefore uncracked (Steel Construction Institute, 1995).

Restraints and releases

The displacement components which are restrained in equation (1.14) (either to be zero or to take prescribed values in the case of settlements) are related to the global axes. All software systems will therefore allow translations (x, y, z) and/ or rotations $(\theta_x, \theta_y, \theta_z)$ to be defined relative to the global directions. In terms of displacement component restraints, the specifications needed for common

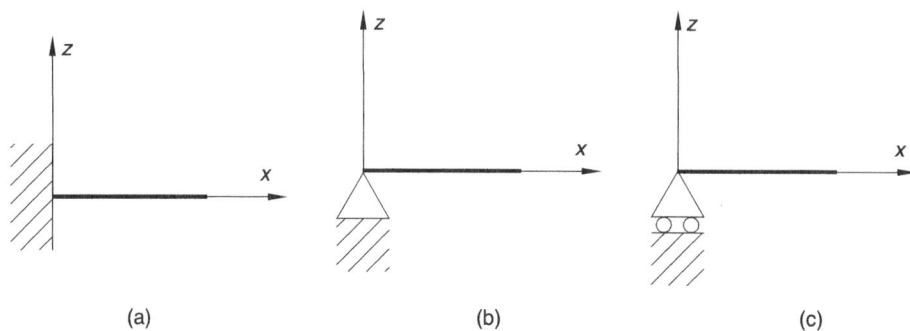

Figure 1.21   Common types of planar support: (a) fixed (x, z, θ_y), (b) pin (x, z), (c) roller (z)

planar forms of supports are as shown in Figure 1.21. Often packages will allow these support forms to be specified directly.

Some systems also allow restraint specification relative to *user defined axes*, so that inclined rollers, for example, can be described directly (Figure 1.22(a)). Should this not be available then a short, stiff pin-jointed member in the appropriate direction can be used to model the roller (Figure 1.22(b)).

Restraints, which relate to the displacement restrictions placed on the structure by the external environment, should not be confused with internal connection arrangements between members. Most computer systems assume that members are *rigidly* connected. This implies that the connection is fully effective for the transmission of internal forces (stress resultants). Rigid connection does **not** imply any restriction on displacements. If a member's end connection is not capable of transmitting a particular stress resultant, an internal *release* must be introduced at this point. Most computer packages only provide facilities for the release of bending moments in the form of internal hinges at the ends of members. If a joint is hinged, this may be achieved by introducing hinges at the ends of all the

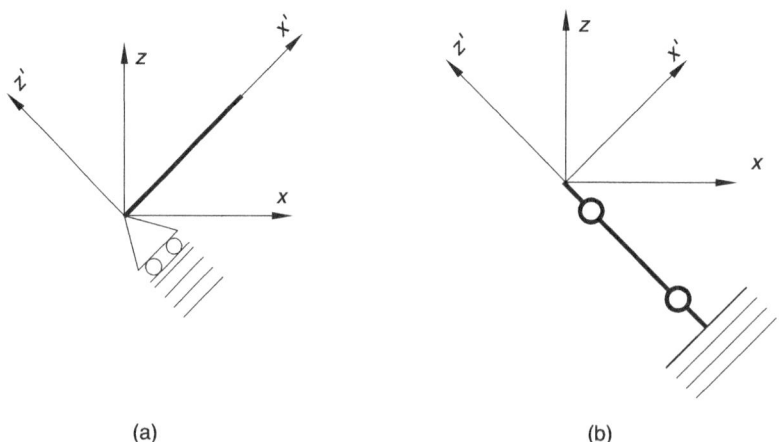

Figure 1.22   (a) Inclined roller, (b) modelling

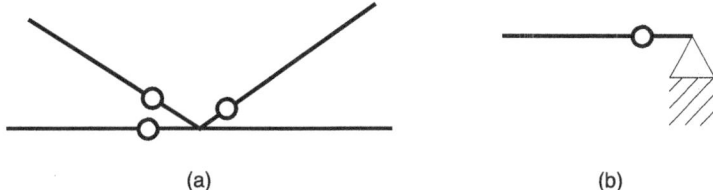

Figure 1.23   (a) Hinged joint, (b) erroneous pin description

inter-connected members but one (Figure 1.23(a)). If **all** the member ends were pinned the joint would be a local mechanism, since it would be free to rotate. The zero bending moment provided by the hinges, taken with the equilibrium requirement that there be no resultant moment at a joint, ensures that the bending moment in the member without a hinge is also free of bending moment.

Releases are not needed at restraint positions, since the appropriate freedoms can be incorporated into the restraint. The arrangement of Figure 1.23(b), for example, would result in a mechanism since the pin restraint already permits rotational freedom, and the hinge provision duplicates this.

Symmetry

Making use of symmetry can simplify the input preparation and output scrutiny for computer analyses of complex structures, although it may well be simpler to consider the full structure, if it is not too extensive. If symmetry is present, in respect of both structure and loading, then the symmetrical portion of the structure needs to be modelled such that the analysis is representative of the complete structure behaviour, rather than simply that of the portion. This is achieved by the enforcement of displacement and load conditions on, in general, planes of symmetry. In the case of a planar structure, the symmetry plane will be normal to the plane of the structure, and the plane therefore reduces to a line when the structure is viewed.

The basic concept is that translation normal to a plane of symmetry must be restrained to be zero (or this would destroy the symmetry) and, similarly, rotation about an axis contained by a plane of symmetry must be prevented. Forces applied within a line of symmetry should be halved, since only a half is resisted by the portion considered. Similarly, moments about an axis normal to a symmetry plane must be reduced by a half. If a structural member lies within a plane of symmetry then its member properties also need to be halved, and the output internal forces will relate to the half member and will therefore need to be doubled to give correct values.

Equation solution and mechanism detection

The restrained structure stiffness matrix, **K** (equation (1.15)), will be non-singular provided the structure is not a mechanism. Inversion of the stiffness matrix will be impossible if $\det \mathbf{K} = 0$, since (Stroud, 1995)

$$\mathbf{K}^{-1} = \frac{1}{\det \mathbf{K}} \operatorname{adj} \mathbf{K} \tag{1.27}$$

The singularity of $\mathbf{K}$ may therefore be checked from an evaluation of $\det \mathbf{K}$. Some computer packages will carry out a check on the determinant of the stiffness matrix and will produce errors or warnings should a zero, or small, value be found. Others will not do so and, due to the effects of arithmetic rounding errors, may be able to solve equation (1.15) even in the case of a mechanism. Such a solution will produce unrealistically large displacements, and structures for which computer analyses result in improbable displacements should be checked for possible mechanisms.

Member forces and sign conventions

Two quite distinct sign conventions are possible for member forces. Either a force direction-based convention may be used or a deformation-based convention. The force direction convention used here, and in most software packages, is to consider node-to-member forces as being positive if they act in the positive direction of the relevant member axis. This convention is generally used for printed output. The typical beam member of Figure 1.24 sustains a constant tension of 10 kN and a central, lateral point load of 8 kN. By standard theory (Gere and Timoshenko, 1999), the end reaction moments may be shown to be $WL/8 = 6\,\mathrm{kN\,m}$.

The node-to-member end forces on the two members that have been used to represent the beam are therefore as shown in Figure 1.25. With reference to the positive directions of the local axes in Figure 1.25, printed output for the member would therefore, typically, take the form shown in Table 1.2.

Figure 1.26 shows the positive senses of stress resultants that will be used in this text when a deformation-based convention is being used. The axial load convention is extension (tension) positive; the shear convention is related to a 'clockwise' deformation; and 'sagging' bending moments are taken positive. The bending moment convention needs to be referred to a designated side of the member, which is taken to be the lower one and is indicated by the broken lines in Figure 1.26. The broken line also establishes the 'base' side in the ambiguous case of a vertical member. For graphical purposes, positive

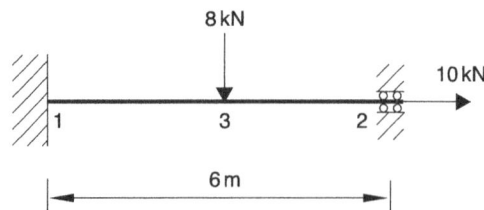

Figure 1.24   Example beam member

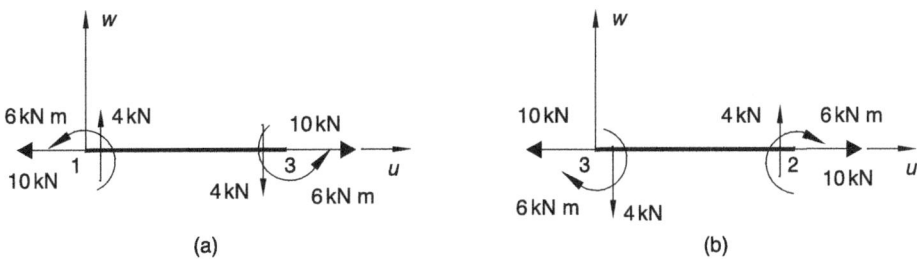

Figure 1.25   Node-to-member end forces: (a) member 1–3, (b) member 3–2

Table 1.2   Output results for typical beam member

Member	Axial load: kN	Shear force: kN	Bending moment: kN m
1–3	−10	+4	−6
	+10	−4	−6
3–2	−10	−4	+6
	+10	+4	+6

axial loads and shear forces will be drawn above the designated base line, but positive bending moments will be drawn **below** the base line. The anomalous treatment of bending moments is needed to ensure that bending moments are plotted on the tension-side of members and hence provide a visual reminder of the side of the member on which steel is required in the case of reinforced concrete construction.

A deformation-based convention is the norm for most purposes, including graphical output from computer packages. It is important to appreciate, therefore, that software results are presented differently in different forms of output. Graphical presentation of axial loads is usually based on a tension (or compression) positive convention, which may be helpfully reinforced by a colour code convention. Conventions for shear force vary considerably and the provisions of any given system need to be carefully checked. Many conventions are affected by the node ordering of members, since this can change the orientation of a

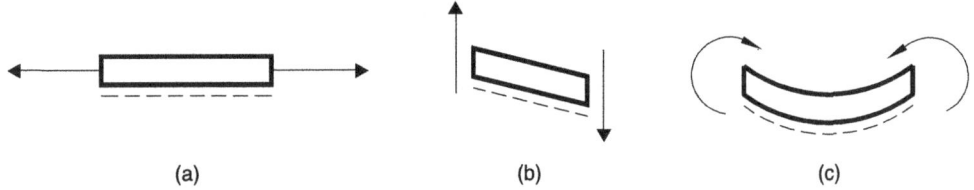

Figure 1.26   Deformation-based sign conventions: (a) axial load, (b) shear force, (c) bending moment

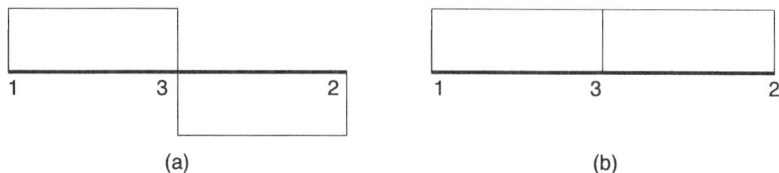

Figure 1.27   Possible shear force diagrams for example beam

member's local axes, which are normally used for plotting purposes. It is quite possible that the shear force diagram for the beam of Figure 1.24 will appear as shown in either Figure 1.27(a) or (b), according to whether the right-hand member is designated 3–2 or 2–3. United Kingdom-based software will normally be consistent in producing 'tension-side' bending moment diagrams.

1.7   INFLUENCE LINES

Introduction and definition

Shear force and bending moment diagrams are a familiar concept for the graphical depiction of the variation of an internal stress resultant with position on a structure, for example a beam. The diagrams are particularly convenient for the identification of maximal values, but relate to loading systems that remain fixed in position, such as dead loads. The maximal effects of live (imposed) loads can be more difficult to determine, since these loads can be variably positioned, or absent completely. To establish the critical positioning of an imposed loading system, it is often helpful to use an *influence line* that indicates the influence of a moving unit load on the variation of a particular stress resultant or reaction component at a fixed position. A shear force diagram, therefore, graphs shear force at **varying** position on the structure for a system of loading that is **fixed** in position. A shear force influence line, however, will plot shear force at a **fixed** point against the **varying** position of a unit load on the structure.

Reaction component influence lines

The general forms of influence lines are most readily obtained by application of Müller-Breslau's principle, the proof of which is given in standard texts (Ghali and Neville, 1997). The principle states that an influence line is given by the deflected shape of the structure produced by making a unit displacement in the direction of the quantity of interest. The displacement imposed will be a unit translation in the case of a force influence line but a unit rotation for a moment influence line. The sign convention that will be adopted here is to take an ordinate on an influence line (usually termed an *influence coefficient*) as positive if a unit load acts in the same direction as the ordinate. With this convention, influence lines for reaction components are produced by the deflected shape resulting

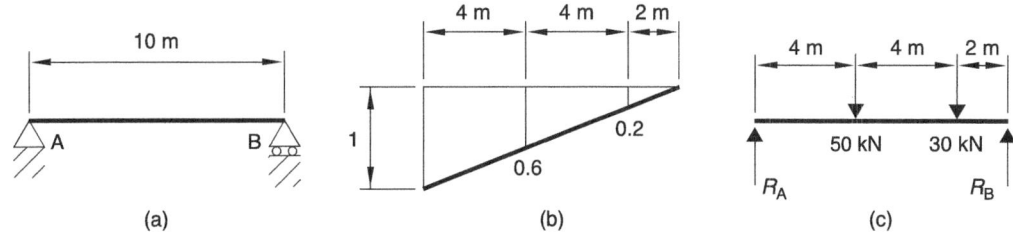

Figure 1.28 (a) Simply supported beam, (b) influence lines for vertical reaction, (c) point load system

from the application of a unit displacement acting in the opposite sense to that taken positive for the relevant component.

Simply supported beam

For the simply supported beam of Figure 1.28(a), the influence line for the vertical reaction component at the left-hand support (taken positive **upwards**) is obtained by the introduction of a unit **downwards** displacement at the support (Figure 1.28(b)). From the influence line, it follows that a unit load placed at a distance 2 m from the right-hand support will produce a reaction of 0.2 at the left-hand support. The reaction component will also act upwards, because the displacement is in the same direction as the load. The effects of multiple point loads may simply be summed, so that the upward reaction at the left-hand support due to the loads of Figure 1.28(c) is $(50 \times 0.6) + (30 \times 0.2) = 36\,\text{kN}$. Note that the ordinates of the influence line are dimensionless, so that the units derive from those of the applied loads. This result may, of course, be checked by statics by taking moments for the beam about the right-hand support to give

$$10R_A = (30 \times 2) + (50 \times 6) \text{ whence } R_A = 36\,\text{kN}$$

The effects of distributed loads may also be readily established. In Figure 1.29(a) the intensity of load over the length $\mathrm{d}x$ is w/length. Then, if $\mathrm{d}x$ is assumed negligibly small, variation of the influence coefficient i and of w over the length $\mathrm{d}x$ may be ignored. Hence, the effect of the load acting on the length $\mathrm{d}x$ is $(w\,\mathrm{d}x)i$. It follows that the effect, I, of a distributed load acting over the portion a–b (Figure 1.29(a)) of the structure is given by

$$I = \int_a^b iw\,\mathrm{d}x \tag{1.28}$$

If the load is uniformly distributed then

$$I = \int_a^b iw\,\mathrm{d}x = w\int_a^b i\,\mathrm{d}x = w \times \text{area under influence line between a and b}$$

$$\tag{1.29}$$

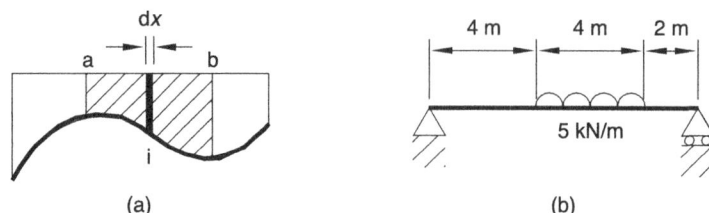

(a)   (b)

Figure 1.29   (a) Distributed load treatment, (b) uniformly distributed load example

If equation (1.29) is used to evaluate the left-hand vertical reaction for the beam shown in Figure 1.29(b), then, with reference to the influence line of Figure 1.28(b), the reaction is found to be $5 \times 4 \times (0.6 + 0.2/2 = 8\,\text{kN}$, as may be readily checked by statics.

Two-span beam

Using Müller-Breslau's principle, the influence line for the vertical, upward reaction, R_A, at support A of the two-span beam shown in Figure 1.30(a) is obtained by imposing a unit downward displacement at A. The resulting influence line is shown in Figure 1.30(b). It should be noted that, to comply with the fixed support condition, no rotation takes place during the vertical displacement. This constrains A–B to remain horizontal. The prevention of deflection at supports C, E and the articulation allowed by the pins B and D determine the remaining form of the influence line. For this statically determinate beam, all sections of the influence line will be linear. This property holds for all statically determinate structures, since the induced displacement may be regarded as a release. A

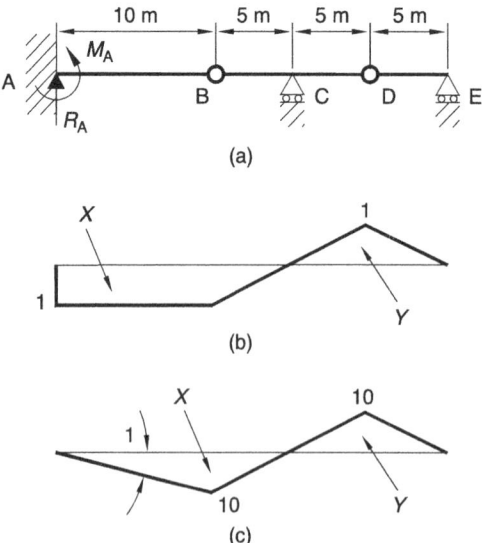

Figure 1.30   (a) Two span beam, (b) influence line for R_A, (c) influence line for M_A

single release applied to a determinate structure produces a mechanism so that the component parts move as rigid bodies and remain straight.

The influence line may also be checked by statics. The internal pins at B and D allow freedom of rotation and are unable to sustain a bending moment. Therefore, if the unit load is placed anywhere on A–B, the zero bending moments at D and B, respectively, show that the reactions at E and C must be zero. It follows from vertical equilibrium that the reaction at A is unity. If the unit load is placed at D, the reaction at E is still zero. The reaction at C may be determined from the zero bending moment condition at B as

$$5R_C = 10 \times 1 \text{ so that } R_C = 2$$

From vertical equilibrium, it follows that R_A is -1, as found previously. Taken with the zero R_A values arising from placing the unit load at the supports C and E, the linearity property of the influence line segments now allows the full influence line to be constructed.

From the influence line, it may be noted that downward load applied to span A–C will produce positive (upward) reaction at A. However, downward loading applied to C–D will produce negative (downward) reaction at A, since the displacement is opposite to the load. To find out if the support at A needs to be designed for uplift, it will be necessary to check whether the upward reaction effects of the dead load overcome the worst downward reaction effects of the live load. If it is presumed that the beam carries a dead load of 30 kN/m and a live load of 50 kN/m (any length) with a point load of 100 kN (both of which may be placed at any position), then the maximal values of R_A may be calculated as

$$\text{area } X \text{ (Figure 1.30(b))} = (10 \times 1) + (\tfrac{1}{2} \times 5 \times 1) = 12.5 \, \text{m}$$

$$\text{and area } Y = \tfrac{1}{2} \times 10 \times 1 = 5 \, \text{m}$$

The dead load acts across the complete beam, so that

$$R_A \text{ due to dead load} = 30 \times (12.5 - 5) = 225 \, \text{kN}$$

For maximum upward reaction at A, the distributed live load must be placed on A–C only and the point live load anywhere along A–B, hence

$$R_A \text{ due to live load} = (50 \times 12.5) + (100 \times 1) = 725 \, \text{kN}$$

It follows that the support at A must be designed for a maximal load of $225 + 725 = 950$ kN. To check for possible uplift, the distributed live load is now placed on C–E only and the live point load is placed at D. For this combination

$$R_A \text{ due to live load} = -[(50 \times 5) + (100 \times 1)] = -350 \, \text{kN}$$

Since the (constant) dead load reaction at A is only 225 kN it follows that the support needs to be designed for an uplift of 125 kN.

Maximal moment reactions at A may be found in a similar fashion from the influence line for M_A. In this case, a unit rotation is applied at A in the opposite sense to that assumed positive for the reaction moment (Figure 1.30(c)). The rotation is assumed to be small so that the small, natural angle assumption of

$\tan\theta \approx \theta$ can be made and

$$\tan 1 \approx 1 = \frac{BM_{\mathrm{B}}}{10} \quad \text{hence } BM_{\mathrm{B}} = 10\,\mathrm{m}$$

As before, the remainder of the influence line (Figure 1.30(c)) follows from the articulation allowed by the pins and the zero displacement at supports C and E.

The effects of the uniform loads can now be assessed from the areas contained by the influence line

$$\text{area } X \text{ (Figure 1.30(c))} = \tfrac{1}{2} \times 15 \times 10 = 75\,\mathrm{m}^2$$

$$\text{and} \quad \text{area } Y = \tfrac{1}{2} \times 10 \times 10 = 50\,\mathrm{m}^2$$

It follows that the dead load will create a reaction moment of $30 \times (75 - 50) = 750\,\mathrm{kN\,m}$. The maximal positive moment created by the live load will be $(50 \times 75) + (100 \times 10) = 4750\,\mathrm{kN\,m}$ and the maximal negative moment will be $-[(50 \times 50) + (100 \times 10)] = -3500\,\mathrm{kN\,m}$. It follows that the overall extreme moment values are $5500\,\mathrm{kN\,m}$ and $-2750\,\mathrm{kN\,m}$. Since positive reaction moment represents hogging curvature in the beam, it follows that the beam at the support must be designed to withstand hogging and sagging moments of $5500\,\mathrm{kN\,m}$ and $2750\,\mathrm{kN\,m}$, respectively.

Worksheet 1.3 Reaction component influence lines for beams

W1.3.1 Construct an influence line for the vertical reaction component at B for the simply supported beam with overhangs shown in Figure W1.3.1. Use the influence line to determine the vertical reaction at B due to the loading shown and check the result by statics.

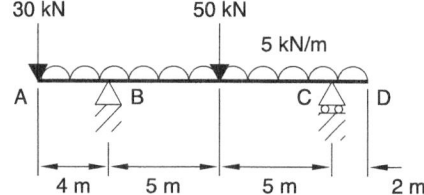

Figure W1.3.1   Worksheet 1.3.1

W1.3.2 Construct influence lines for the vertical force reaction at A (R_{A}) and the moment reaction at A (M_{A}) for the beam shown in Figure W1.3.2. Use the influence line to determine the magnitudes of R_{A} and M_{A} caused by the loading indicated, and check the solution by statics.

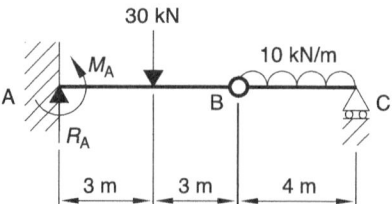

Figure W1.3.2   Worksheet 1.3.2

W1.3.3 Construct influence lines for the vertical force reactions at A and C for the three-span beam shown in Figure W1.3.3. If the structure carries a uniformly distributed dead load of 20 kN/m and a uniformly distributed live load of 40 kN/m (any length), determine the maximal values of the two reactions.

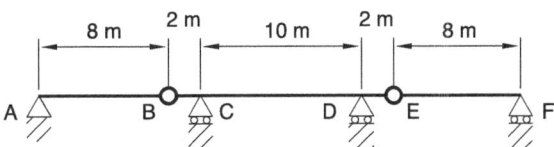

Figure W1.3.3   Worksheet 1.3.3

W1.3.4 Construct influence lines for the vertical force reactions at A and B for the suspended span structure shown in Figure W1.3.4(a). Determine the maximal values of the two reactions due to a vehicle crossing the structure. The vehicle has the axle loads shown in Figure W1.3.4(b) and may cross from either the left or the right.

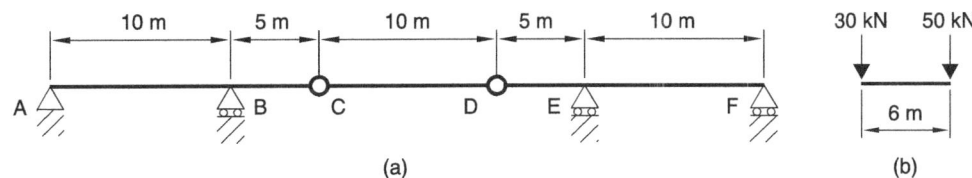

Figure W1.3.4   Worksheet 1.3.4

Stress resultant and displacement influence lines

Influence lines for internal stress resultants may be established by inducing unit displacements, which correspond to the internal action of interest. With the sign conventions adopted here, the unit displacements to be induced are as shown in Figure 1.31, and examples of the use of this technique are given in relation to differing structural forms in the relevant chapters.

Although perhaps less widely used, the shapes of displacement influence lines may also be rapidly generated by means of Müller-Breslau's principle. In this case, instead of applying a unit displacement at the point of interest, a unit load is applied and the resulting deflected shape provides the required influence line.

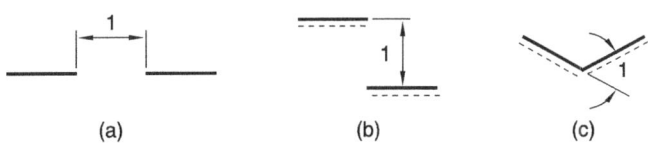

Figure 1.31   Positive unit displacements

Computer-generated influence lines

Müller-Breslau's principle may also be used in conjunction with computer analysis to generate influence lines. In this case the forces needed to produce the appropriate unit displacement (Figure 1.31) are applied at the point of the structure for which the influence line is sought. The forces needed to generate the unit displacement are calculated from equation (1.6) in the case of an axial displacement; from Figure 1.18(b) in the case of a shear displacement; and from Figure 1.18(c) or (d) in the case of a rotational displacement. Again applications of this technique are given in subsequent chapters.

Displacement influence lines generally present no real problem, it simply being necessary to run a load case consisting of the application of a unit load at the point of interest.

1.8 CASE STUDY

C1.1.1 In an area that had been extensively mined, it was considered appropriate to use statically determinate bridge designs. Why was this? One standard layout adopted was as shown in Figure C1.1.1. Show that this design is statically determinate.

The bridge was of reinforced concrete beam and slab construction. Making use of influence lines, and taking reasonable values for loading and dimensions, determine the maximal reactions for which an individual beam's bearings must be designed at the five supports of the bridge.

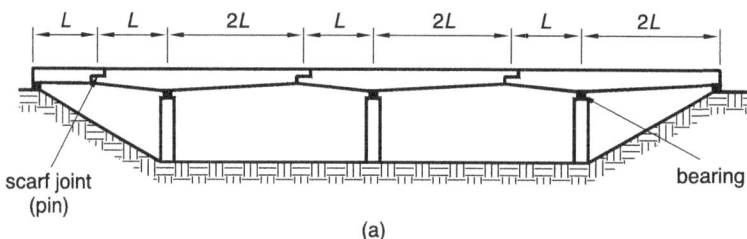

(a)

(b)

Figure C1.1.1 (a) Structural form for highway bridge, (b) prototype bridge

1.9 REFERENCES

Gere, J. M. and Timoshenko, S. P. (1999) *Mechanics of Materials*. 4th SI edn. London: Stanley Thornes – Chapter 2 of this reference covers the analysis of axial stress and strain; its Chapter 6 deals with composite construction, and its Chapter 10 includes fixed-ended beam results.

Ghali, A. and Neville, A. M. (1997) *Structural Analysis*. 4th edn. London: E & F N Spon – Chapter 12 of this reference includes a proof and applications of Müller-Breslau's principle for influence line purposes.

Kennedy, J. B. and Madugula, M. K. S. (1990) *Elastic Analysis of Structures*. New York: Harper and Row – Chapter 12 of this reference gives details of the application of the stiffness method to beam structures and its Chapter 14 deals with influence lines, including theorems for the determination of maximal effects.

Moy, S. S. J. (1996) *Plastic Methods for Steel and Concrete Structures*. 2nd edn. Basingstoke: Macmillan – Chapter 3 of this reference includes a statement of the fundamental theorems of plasticity.

Owens, G. A. and Knowles, P. R. (eds) (1992) *Steel Designers' Manual*. 5th edn. Oxford: Blackwell Scientific Publications – appendices provide section properties for standard British Structural Steel sections.

Steel Construction Institute (1995) *Modelling of Steel Structures for Computer Analysis*. Ascot: Steel Construction Institute – general information on computer analyses and detailed advice on modelling a variety of practical details.

Stroud, K. A. (1995) *Engineering Mathematics*. 3rd edn. Basingstoke: Macmillan – covers matrix theory including inversion by the adjoint (co-factor) method.

2 Beams

2 Beams

2.1 INTRODUCTION

In this chapter, beams will be taken to be linear structural elements that are subjected to loads normal to their axes only. In this case, resistance to loading relies on bending and shearing only, with axial loading being absent. If the element and its loading are not perpendicular, then compressive and/or tensile axial loads are induced and the system is more conveniently treated as a *plane frame*.

2.2 STATICALLY DETERMINATE BEAMS

Shear force, bending moment and deflection diagrams

Shear force, bending moment and deflection diagrams for the two simplest forms of statically determinate beam – simply supported and cantilevered – are shown in Figure 2.1. Diagrams for more complicated determinate beams may be derived by calculation, or, perhaps more readily, by use of a computer package. In either case, it is useful to be able to sketch at least the general form of the diagrams (and possibly to determine some 'spot' values), so that computed results can be validated and gross errors detected.

In sketching shear force diagrams, it is helpful to recall the following:

1. *On unloaded sections of beams the shear force is constant.*
2. *A point load causes a 'jump' in shear force that is of the same magnitude as the load and is in the same direction.*
3. *A point moment has **no** effect on a shear force diagram.*
4. *A uniformly distributed load causes a linear variation in shear force. The total change in shear force is equal to the total load traversed and is in the same direction.*
5. *The shear force is zero at a free end of a beam.*
6. *An internal hinge can transmit a shear force.*

Similarly, in sketching bending moment diagrams, one needs to be aware of the following:

1. *On unloaded sections of beams the bending moment diagram is linear.*

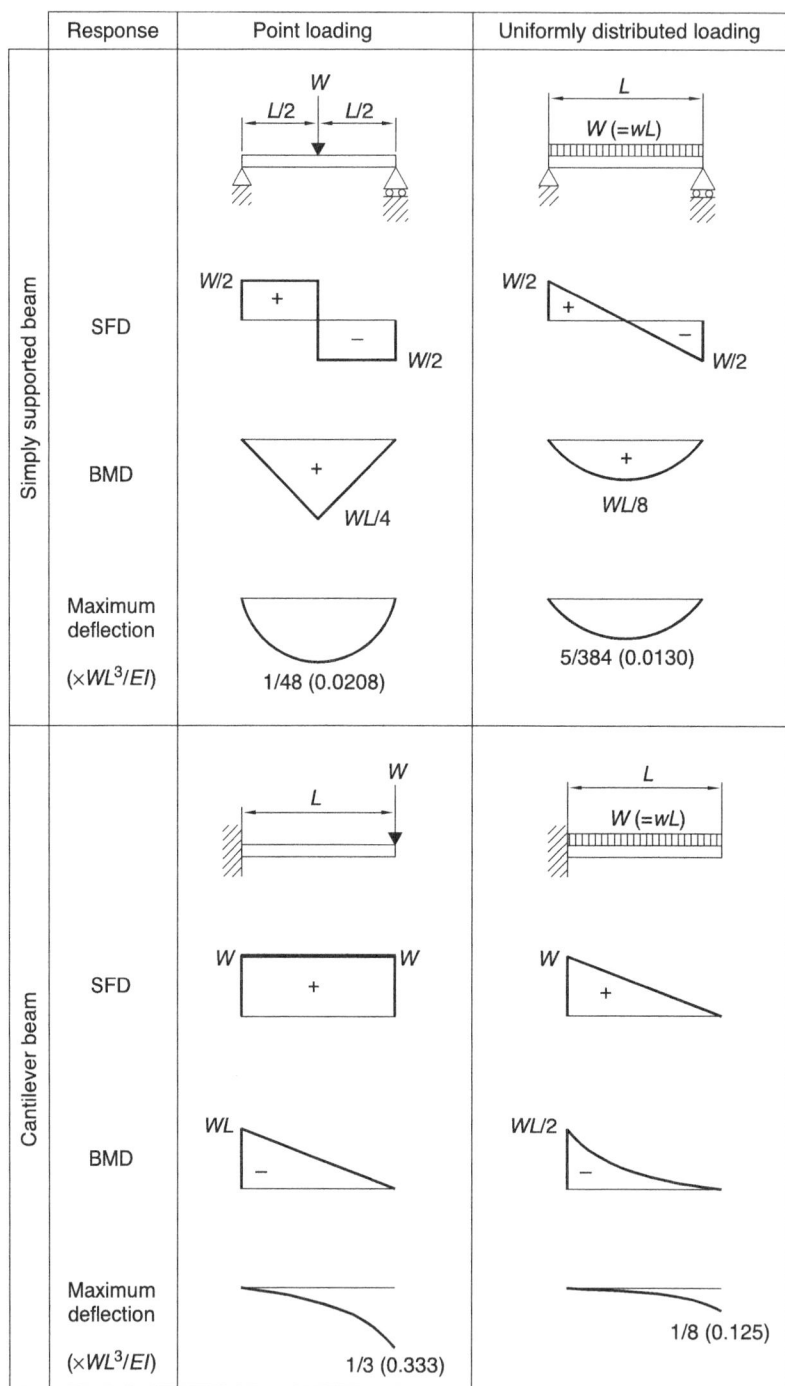

Figure 2.1   Simply supported and cantilever beam diagrams

2. A point load causes a 'kink' in a bending moment diagram.
3. A point moment causes a 'jump' in the bending moment diagram that is of the same magnitude as the moment and in a direction consistent with the sign convention.

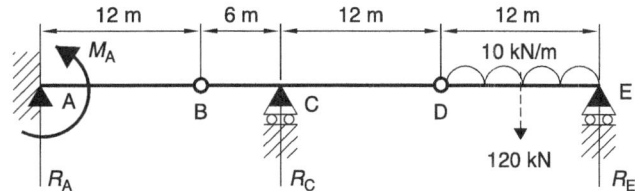

Figure 2.2   Example beam

4. *A uniformly distributed load causes a curved (parabolic) variation in bending moment.*
5. *The **change** in bending moment over a given length of beam is equal to the area under the shear force diagram over the same length.*
6. *The bending moment diagram has a turning point (maximum or minimum) at a point of zero shear force.*
7. *The bending moment is zero at the end of a beam that is simply supported or free.*
8. *An internal hinge **cannot** transmit a bending moment.*

Shear force and bending moment diagram sketching

Example 2.1   Shear force and bending moment diagrams

To sketch the shear force and bending moment diagrams for the composite beam shown in Figure 2.2, it is first helpful to evaluate the reactions at A, C and E. Thus, using the zero bending moment property at hinge D

$$12 \times R_E = 120 \times 6 \text{ whence } R_E = 60 \, \text{kN}$$

Then, using the zero bending moment property at hinge B

$$6 \times R_C = (120 \times 24) - (60 \times 30) \text{ whence } R_C = 180 \, \text{kN}$$

From vertical equilibrium, it then follows that $R_A = -120 \, \text{kN}$ and, using the zero bending moment property at B again

$$M_A = 12 \times R_A = -1440 \, \text{kN m}$$

The full set of reactions is therefore as shown in Figure 2.3(a). Using these reactions, the shear force diagram (Figure 2.3(b)) may be sketched by progressing from left to right along the beam and moving 'through' the reactions and loads as described in the guidelines given above. On the bending moment diagram, the point moment reaction at A causes an immediate, positive jump, as shown in Figure 2.3(c). The change in bending moment between A and C is then equal to the area under the shear force diagram between these points $(= 120 \times 18 = 2160 \, \text{kN m})$. It follows that the bending moment at C is $1440 - 2160 = -720 \, \text{kN m}$. The bending moment at hinge D is known to be

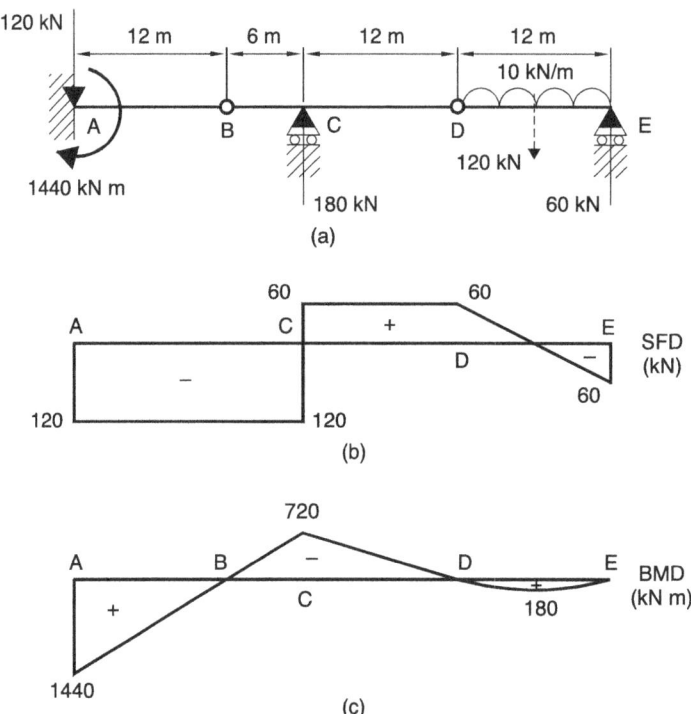

Figure 2.3   Example beam: (a) shear force diagram, (b) bending moment diagram

zero and the central moment between D and E is given by the area under the shear force diagram between D and the central point. The relevant bending moment is therefore $(1/2) \times 6 \times 60 = 180\,\text{kN m}$.

An alternative approach to this problem would be to divide the composite beam into its constituent parts, so producing a series of simpler problems. If the *method of sections* is used to 'cut' the beam at the hinge points B and D, then any internal forces cut at these points must be applied to the sections as external reactions. Since the bending moments at the hinges are zero, it follows that shear forces only need be applied at the cuts and these provide a pair of equal and opposite reactions at each location (Figure 2.4). By symmetry, it then follows that $R_D = R_E = 60\,\text{kN}$; by moments about B, that $R_C = 180\,\text{kN}$; and, by vertical equilibrium of BD, that $R_B = 120\,\text{kN}$. The reactions at A are then calculated as before. The shear force and bending moment diagrams may then be constructed by treating the three sections as individual components and subsequently assembling these. Thus, AB is a cantilever beam subjected to an upward, end point load. Section

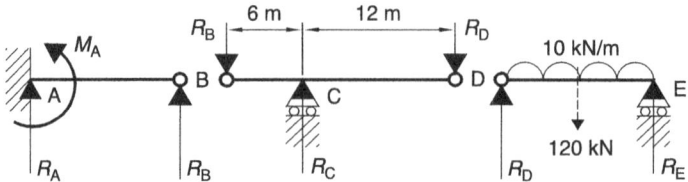

Figure 2.4   Method of sections approach to Example 2.1

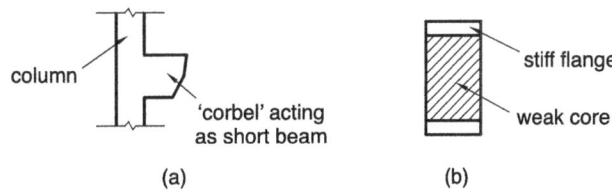

Figure 2.5   (a) Deep beam, (b) sandwich beam

BD is a balanced pair of cantilevers. Finally, DE is a simply supported beam subjected to a uniformly distributed load.

Bending moment and deflection diagram sketching

Both bending moments and shear forces cause beams to deflect, but in different ways. Pure bending causes a beam to become **curved**, while shear force induces **slope**. For normal solid beam proportions, the displacements due to the beam becoming curved are much greater than the displacements due to the slopes produced by shearing action. It is therefore usual to neglect displacements due to shear apart from the cases (Figure 2.5) of deep beams (span/depth ratio typically less than two) and beams which have particularly weak webs or cores, as is the case with some 'sandwich' beams.

The proportionality between curvature and bending moment is a key tool for sketching deflection diagrams. If a bending moment diagram is available, then the construction of a deflected shape sketch is normally reasonably straightforward. It is first necessary to ensure that the deflection restraints are complied with. Considering the beam of Example 2.1, the supports (Figure 2.6(b)) require that zero deflection **and slope** must be maintained at A and zero deflection at C and E. The bending moment diagram (Figure 2.6(a)) then shows that regions of sagging and hogging curvature alternate along the beam. The points at which

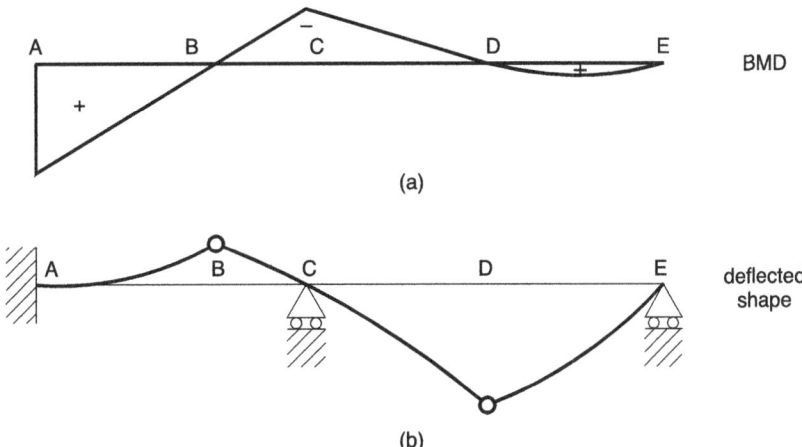

Figure 2.6   (a) Bending moment diagram, (b) deflected shape

the curvature changes (the hinge positions B and D in this case) are known as *points of contraflexure*, and the beam is instantaneously straight at such points since the curvature is zero. The sagging curvature between A and B shows that the beam must deflect upwards at B if it has an initial zero slope. On the other hand, the uniform load on DE will certainly cause downward displacement and the curvature must again be sagging. To complete the deflected shape, BD is then sketched with hogging curvature and continuity of slope at support C (as distinct from the hinges at B and D, which allow discontinuity of slope). Of course, especially with practice, the displaced shape could have been established directly by noting that the downward displacement of DE due to the load will cause BD to pivot around C, so causing AB to be forced upwards.

Example 2.2 Bending moment and deflected shape diagrams

Use of the connection between bending moment and curvature allows bending moment and deflection diagrams to be developed simultaneously, at least approximately, without prior calculation of bending moment values. Taking the beam of Figure 2.7(a) as an example, the overhang portion, CD, must be in negative bending moment and therefore be hogging. What happens in AC depends on the direction of the reaction at A. The overturning effects about C of the point moment at B and the point load at D are, however, certain to be larger than the restoring moment due to the relatively small, uniformly distributed load. It follows that the reaction at A will act downwards, which may be checked by calculation, if necessary. In AC, the bending moment must be negative initially, therefore, but the point moment will almost certainly change this to positive bending, leading to a possible bending moment diagram of the form shown in Figure 2.7(b).

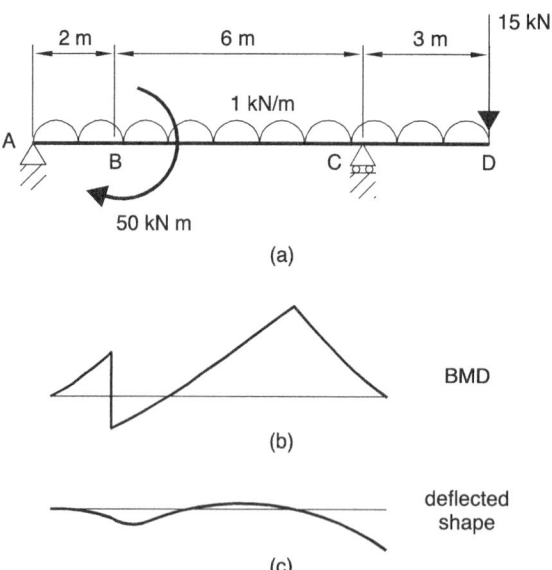

Figure 2.7   (a) Example beam, (b) bending moment sketch, (c) displaced shape sketch

To develop the displaced shape, it may be noted from the bending moment diagram that regions of hogging and sagging alternate. Clearly there must be no displacement at the supports A and C, but there are no constraints on the beam's slope in this case. At A, for example, it is therefore not known if the beam will initially slope upwards or downwards. In the sketch shown in Figure 2.7(c) a compromise zero slope has been used. After B the curvature sags for a short distance, before hogging over the support at C to produce the expected downward displacement under the point load at D. In this case, it has proved impossible to be precise in detail. It is not certain, for example, whether the displacements in AB are initially upwards or not. Nevertheless, the fact that it has proved possible to construct a feasible displaced shape gives reasonable confidence in the general form of the bending moment diagram.

Worksheet 2.1   Shear force, bending moment and deflection diagrams

W2.1.1 Sketch shear force and bending moment diagrams for the beam shown in Figure W2.1.1, indicating principal values. Sketch, also, the expected displaced shape of the beam.

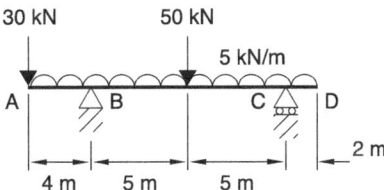

Figure W2.1.1   Worksheet 2.1.1

W2.1.2 Sketch shear force and bending moment diagrams for the beam shown in Figure W2.1.2, indicating principal values. Sketch, also, the expected displaced shape of the beam.

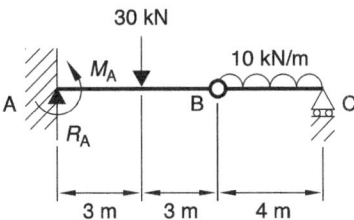

Figure W2.1.2   Worksheet 2.1.2

W2.1.3 Sketch the probable forms of the bending moment and deflection diagrams for the beam of Figure W2.1.3. Numerical values are not required.

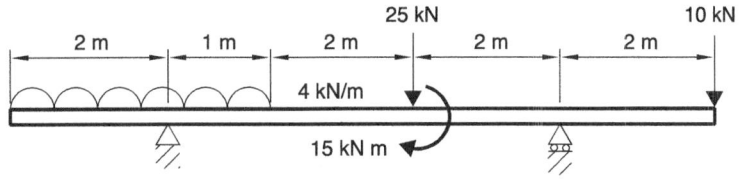

Figure W2.1.3   Worksheet 2.1.3

W2.1.4 Due to the effects of the uniformly distributed load shown, indicate the directions of the reactions at the supports of the beam shown in Figure W2.1.4. Sketch a displaced shape and a bending moment diagram for the beam. Numerical values are not required.

Figure W2.1.4   Worksheet 2.1.4

W2.1.5 Due to the effects of the uniformly distributed load shown, indicate the directions of the reactions at the supports of the beam shown in Figure W2.1.5. Sketch a displaced shape and a bending moment diagram for the beam. Numerical values are not required.

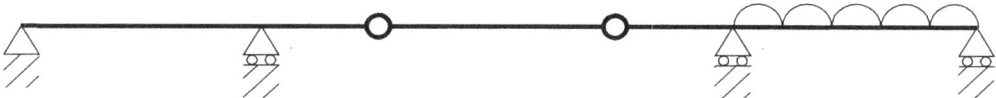

Figure W2.1.5   Worksheet 2.1.5

2.3   STATICALLY INDETERMINATE BEAMS

Shear force, bending moment and deflection diagrams

Shear force, bending moment and deflection diagrams for the two simplest forms of statically indeterminate beam – propped cantilever and fully fixed (encastré) – are shown in Figure 2.8. The degree of indeterminacy of the propped cantilever may be determined by noting that this type of beam has one more reaction component than a cantilever or simply supported beam (the prop force or rotational fixity respectively) and is therefore one degree indeterminate. Similarly, the fixed beam may be seen to be twice indeterminate since the moment reactions at either end are additional to the supports of the determinate simply supported beam. In assessing the determinacy, it should be noted that the structures are treated as 'pure' beams and that only a single horizontal restraint should therefore be provided. The supports of the fixed end beam, for example, should allow horizontal movement at one end only, although it is common practice to sketch fixed beams as shown in Figure 2.8.

By comparison of Figure 2.1 with Figure 2.8, it may be seen that the introduction of a prop to the free end of a cantilever obviously results in considerable stiffening due to the curvature reversal produced by the prop. Maximal deflections, shear forces and bending moments are therefore all reduced. Similarly, the introduction of fixed ends to a simply supported beam results in reversed curvature at both ends, with reductions in maximal deflections and bending moments, but not in shear force.

Perhaps the most common indeterminate beam is the *continuous beam* (Figure 2.9(a)), which may be considered to be a simply supported beam with intermediate

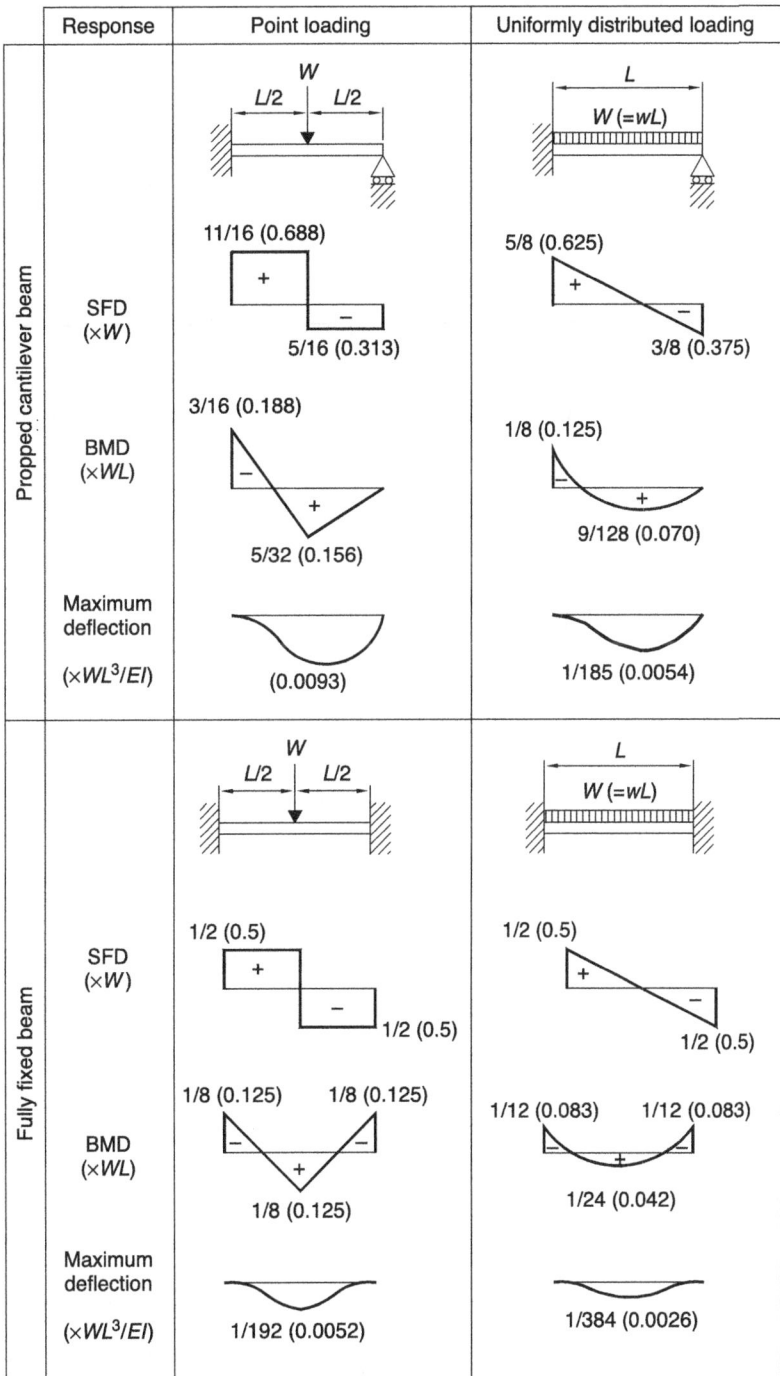

Figure 2.8   Propped cantilever and fully fixed beam diagrams

supports over which the beam is continuous. The degree of statical indeterminacy is simply equal to the number of intermediate supports that are introduced. Multi-span bridges are often of this form and beams in reinforced concrete buildings (Figure 2.9(b)) can be treated as continuous if the rotational resistance of the supporting columns is neglected.

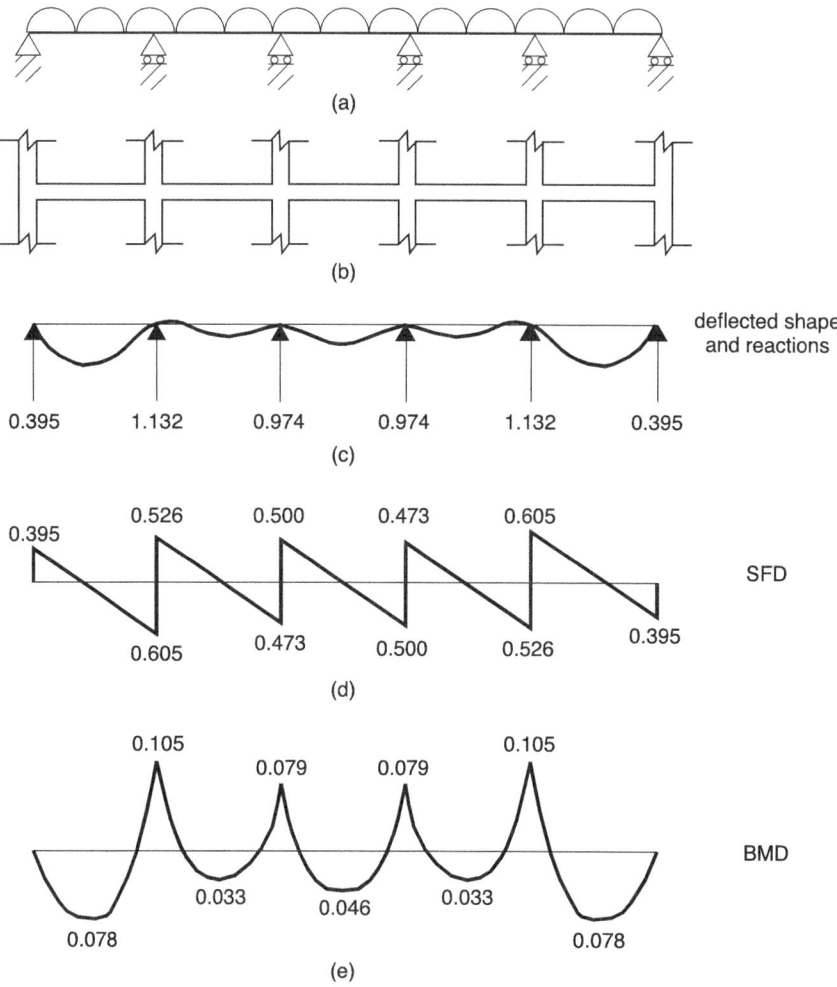

Figure 2.9   Five-span continuous beam, (a) deflections/reactions (×W), (b) shear force (×W) diagram, (c) bending moment (×WL) diagram

In Figure 2.9(a), the five-span continuous bridge shown is presumed to be of uniform section size; has equal spans, L; and carries a total load of W uniformly distributed on each span. From the reactions given in Figure 2.9(c), it may be seen that continuity alters the reaction coefficients somewhat from the central values of 1.0 and end values of 0.5 that would exist if the spans were individually simply supported. As an initial approximation, the beam can be thought of as comprising three central fixed-ended beams and two propped cantilevers at the ends. However, a comparison of Figure 2.9(e) with Figure 2.8 shows that the full propped cantilever support moment is not developed at the penultimate supports, since they are free to rotate, and will do so. Compared to the propped cantilever value, this leads to an increase in sagging moment in the end spans, and also creates a slight upward deflection just past the penultimate supports. The penultimate spans are also affected by the rotation of the penultimate supports such that the sagging moment is decreased relative to the fixed-end beam condition. It is not until the central span that the moments become close to those of the fixed-ended beam case.

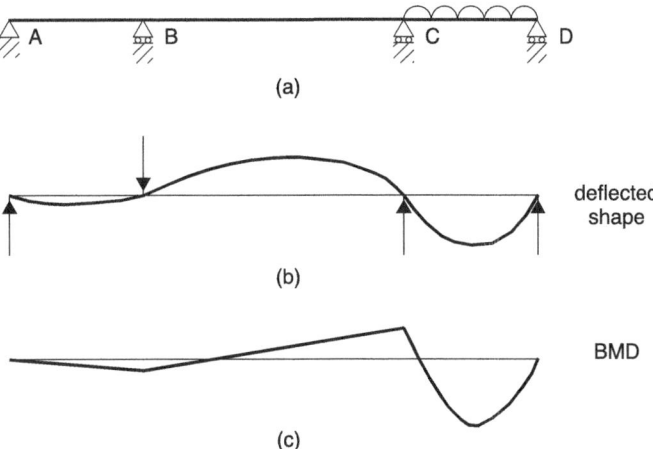

Figure 2.10   (a) Three-span beam example, (b) deflected shape, (c) bending moment diagram

Bending moment and deflection diagram sketching

Sketching bending moment and deflected shape diagrams for statically indeterminate beams is rather more difficult than for determinate systems. Spot values, for instance, cannot be readily calculated by hand and it may not even be possible to decide with complete confidence the correct directions of the reaction components. Nevertheless, use of the principles suggested for determinate beams generally allows diagrams to be produced that are adequate for the detection of gross errors in computer solutions. The tools that are generally of most help are: reference to standard cases (Figure 2.8); the deflection constraints; the likely direction of reaction components; the correspondence between bending moment and curvature.

Example 2.3   Bending moment and deflected shape diagrams

For the three-span beam shown in Figure 2.10(a), the loading and constraint conditions suggest that the deflected shape will take the approximate form shown in Figure 2.10(b). It may be noted that, if the zero displacement condition is to be maintained at B, the upward displacement of span BC will require a downward reaction at this support. The bending moment diagram will correspond to the curvature of the deflected shape and will be curved in the loaded span (CD), but linear elsewhere. In respect of likely magnitudes, span CD is somewhere between a simply supported beam and a propped cantilever, depending on the degree of rotational resistance provided by the rest of the beam (AC). The sagging and hogging moment coefficients in CD will therefore be bounded by $(0.125, 0)$ and $(0.070, 0.125)$, respectively. In view of the length and consequent relative flexibility of BC, the coefficients are likely to be closer to the simply supported than propped cantilever values, so that a probable bending moment diagram is as shown in Figure 2.10(c). This bending moment diagram may be compared with that for the suspended span, determinate version of the same beam (Figure W2.1.5), in which the articulation due to the hinges results in bending being sustained by the loaded span alone.

Worksheet 2.2 Bending moment and deflection diagrams

W2.2.1 State the degree of statical indeterminacy and indicate the directions of any force and moment reaction components at the supports of the beams shown in Figure W2.2.1. Sketch, also, deflected shapes and bending moment diagrams for the beams.

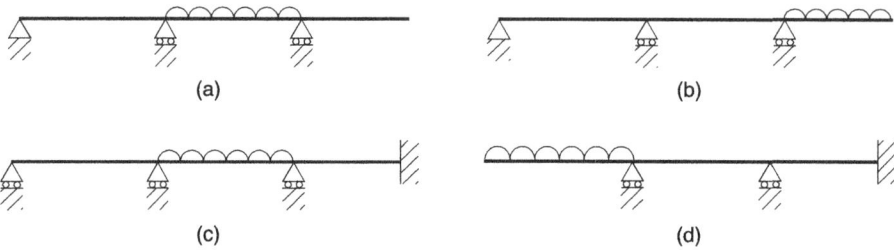

(a)

(b)

(c)

(d)

Figure W2.2.1   Worksheet 2.2.1

W2.2.2 For the beam shown in Figure W2.2.2(a), a computer-produced bending moment diagram is shown in Figure W2.2.2(b). What input errors were made?

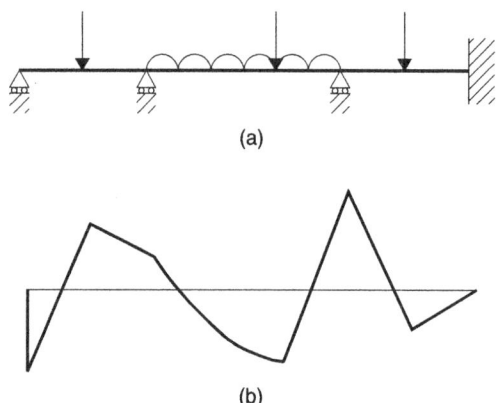

(a)

(b)

Figure W2.2.2   Worksheet 2.2.2

2.4   COMPUTER ANALYSIS

Global and local axes

For a horizontal beam, the global and local axes will be parallel, provided that care is taken to ensure that each member's axis (*u*-direction) is positive in the positive horizontal direction. It is therefore generally convenient to work in the global axes for the specification of loads, noting that the global vertical axis will normally be positive upwards, so that gravity loads, in particular, will usually need to be specified as negative. The positive sense of moments and rotations, which are referred to the global axes, and of bending moments, which are referred to local axes, need to be carefully established.

Member and material properties

Pure beam analysis only requires member second moments of area and material moduli of elasticity. Many computer packages will also require cross-sectional areas, since these will be required for the calculation of self-weight loadings. Some packages also allow for the incorporation of deflections due to shear, in which case 'shear areas' and shear moduli (G) values will also need to be specified. Simple shear deflection theory assumes that the shear stresses causing the deformation are constant over the cross-section. However, for all practical cross-sections, shear stresses vary. 'Shear areas', A_s, are therefore defined such that an approximating, constant shear stress can be obtained from F/A_s. For an I-section, the shear area may be taken to be equal to the area of the web, since the web resists essentially all the vertical shear force, and sustains an almost constant shear stress. For rectangular sections, theory (Megson, 1996) shows that $A_s = A/1.2$.

Restraints and mechanisms

Beam restraints are either fixed, pinned or roller supports. Pure beams should be restrained horizontally at one position only, so that no more than one pin or fixed support should strictly be employed. Any additional pins should be treated as rollers and, should there be a fixed support at both ends, then one should allow horizontal movement at one of the supports. In practice, however, the small-deflection assumption ensures that no horizontal movement whatever occurs in pure beams, so that the restraint of horizontal displacement is immaterial. As a general rule, however, it is good practice to model the restraints as closely as possible to the conditions that are expected in practice. Beam mechanisms usually result from the omission of horizontal restraint or the inclusion of surplus internal hinges.

Worksheet 2.3 Computer analysis of beams

W2.3.1 The multi-span bridge shown in Figure W2.3.1 is of constant section such that $A = 0.3\,\text{m}^2$ and $I = 0.02\,\text{m}^4$. Taking $E = 26\,\text{kN/mm}^2$, use a suitable computer package to produce deflected shape and bending moment diagrams caused by a uniformly distributed load of 70 kN/m applied across the complete structure. Use statics to determine the reactions at the supports and check against the computed values.

Re-analyse the bridge on the assumption that it is now made fully continuous by the removal of the three internal hinges. Relate the deflected shape to the relevant bending moment diagram for each design and compare the relative merits of the two designs.

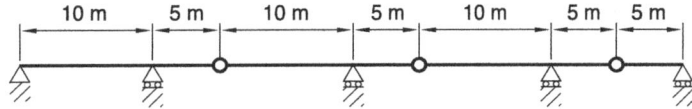

Figure W2.3.1   Worksheet 2.3.1

W2.3.2 The reinforced concrete beam bridge shown in Figure W2.3.2 has a rectangular, hollow cross-section with a constant breadth, but the depth varies from a minimum of 1 m to a maximum of 5 m, with a parabolic soffit profile. At the minimum depth, $A = 0.8\,\text{m}^2$ and $I = 0.1\,\text{m}^4$. The cross-sectional area, A, may be assumed to remain constant at other depths and the second moment of area, I, to vary as the square of the depth. Taking $E = 15\,\text{kN/mm}^2$, make use of symmetry to analyse one half of the bridge for the effects of a uniformly distributed load of 100 kN/m across the complete structure. A suitable computer package should be used and elements of length 2 m should be used over the variable section portions of the bridge.

Re-analyse the half bridge with a constant section being used of depth 2 m. Compare the respective shear force and bending moment diagrams for the two designs, and give reasons for any observed differences. If the reinforcing steel requirement for the bridge is assumed to be given by the ratio of bending moment to beam depth, compare the designs in terms of steel requirement at maximal positive and negative bending moment positions.

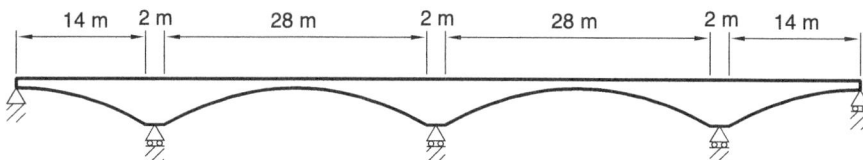

| 14 m | 2 m | 28 m | 2 m | 28 m | 2 m | 14 m |

Figure W2.3.2 Worksheet 2.3.2

2.5 INFLUENCE LINES

Sketching

Statically determinate beams

Influence lines for beams may be readily sketched by the application of Müller-Breslau's principle and sketches are often sufficient to determine the critical positions for application of live loads. To determine approximate influence lines for shear force and bending moment at point I on the suspended span structure shown in Figure 2.11(a), for example, application of the relevant unit displacements readily produces the required sketches (Figure 2.11(b), (c)). The sketches indicate that loading AC has no effect at I, since AC and I are isolated by the suspended span. Note that the influence line sign convention adopted results in downwards displacement on the lines corresponding to positive quantities for downwards loading. Thus, maximum positive shear force at I, due to uniformly distributed loading, is produced by loading CE and IF, and maximum negative shear force is achieved by loading EI. In the case of bending moment at I (Figure 2.11(c)), loading CE will result in the maximum negative moment and loading EF the maximum positive moment.

Statically indeterminate beams

If the suspended span of Figure 2.11(a) is now made continuous by removal of the hinged releases, the influence lines for point I may still be readily sketched but will

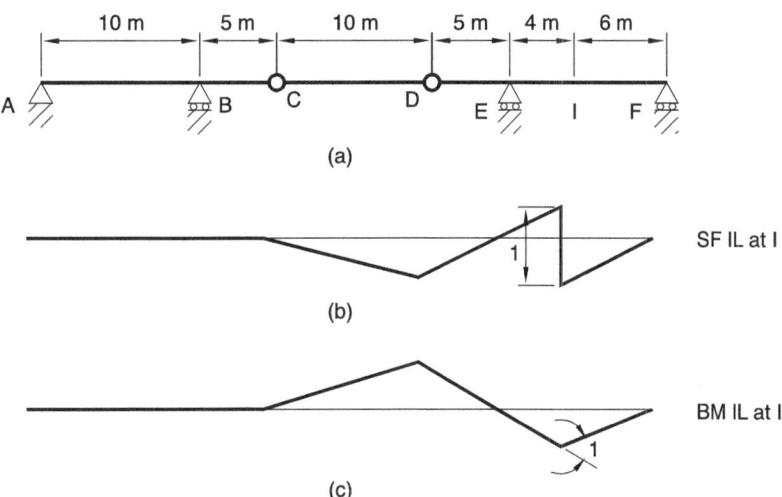

Figure 2.11   (a) Example beam, (b) IL for SF at I, (c) IL for BM at I

now be curved (Figure 2.12). Nevertheless, the same general conclusions may be drawn regarding placement of a distributed load for maximal effects, except that AC loading must now be appropriately included, since, albeit to a limited extent, it does produce actions at I.

Calculation of influence line values

Normally it is most convenient to use influence lines to determine the critical positioning of live loads and then to analyse appropriate load cases separately. However, it is possible to use the influence lines for analysis purposes and an outline of the relevant procedures is given in this section.

For statically determinate systems, it is possible to establish complete influence lines by geometric considerations once values have been found at a few critical locations, normally including the point under investigation. If the unit downwards load is placed at I in Figure 2.11(a), for example, then EF acts as a simply supported beam and the reaction at F, R_F, is readily shown to be 0.4. If the unit load is just to the **left** of I, therefore, the shear force at I will equal R_F and also be 0.4 in magnitude, which provides the influence line value at this point. Given the unit displacement at I, it follows that the influence line value just to the right of I will be 0.6. This conclusion could also have been arrived at by

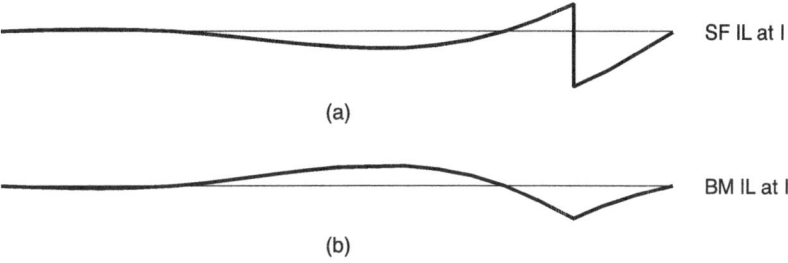

Figure 2.12   Continuous beam example: (a) IL for SF at I, (b) IL for BM at I

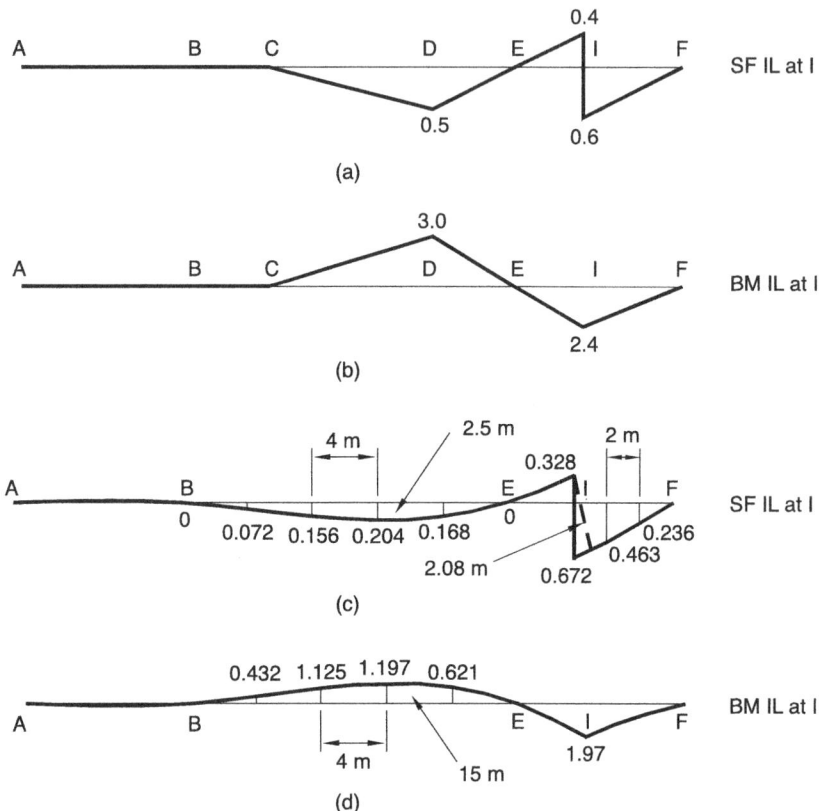

Figure 2.13   Determinate: (a) IL for SF at I, (b) IL for BM at I (×m); indeterminate: (c) IL for SF at I, (d) IL for BM at I (×m)

noting that the two sections of the influence line within span EF are parallel, since the unit lateral displacement at I is introduced without any relative rotation at I.

The remainder of the influence line (Figure 2.13(a)) may be obtained from the straight line property of determinate systems and the property that I is unaffected by loading AC. For the bending moment case, positioning the unit load at I again results in a value of 0.4 for R_F and hence a bending moment at I of $6 \times 0.4 = 2.4$ m. As before, this single value allows the remainder of the line to be determined (Figure 2.13(b)).

The determination of influence line values for statically indeterminate beams is considerably more arduous (Kennedy and Madugula, 1990), and it is normally more convenient to use a computer package, which is equally relevant to determinate or redundant structures. To obtain influence lines for shear force and bending moment it is necessary to apply nodal loads which would cause the appropriate unit displacement at the point under consideration. The resulting deflected shape will correctly represent the influence line at all nodal points, but will be incorrect over the member to which the end nodal forces are applied. It is therefore convenient to make this member a small proportion, say a tenth, of the span within which it occurs. Thus, to develop influence lines relevant to point I on the indeterminate version of the beam shown in Figure 2.11(a), a member, a–b, of a nominal metre length is inserted just to the right of point I. To produce the type of displacement shown in Figure 2.11(b), a 1 mm lateral displacement at I will be introduced by taking $w_a = 0.001$ m. Taking $E = 25 \, \text{kN/mm}^2$ and $I = 0.01 \, \text{m}^4$, the relevant stiffness equations give the nodal loads

$$\begin{Bmatrix} F_a \\ M_a \\ F_b \\ M_b \end{Bmatrix} = \begin{Bmatrix} 12EI/L^3 \\ -6EI/L^2 \\ -12EI/L^3 \\ -6EI/L^2 \end{Bmatrix} \{w_a\} = 25 \times 10^6 \times 0.01 \begin{Bmatrix} 12 \\ -6 \\ -12 \\ -6 \end{Bmatrix} \{0.001\}$$

$$= \begin{Bmatrix} 3 \\ -1.5 \\ -3 \\ -1.5 \end{Bmatrix} \times 10^3 \begin{matrix} \text{kN} \\ \text{kN m} \\ \text{kN} \\ \text{kN m} \end{matrix}$$

The resulting deflected shape (using mm output displacement units) will include the erroneous displacements in member a–b that are shown by the dashed line in Figure 2.13(c). These may be readily corrected by using the unit displacement property at I to provide the influence line shown in full in the figure. The creation of the bending moment influence line requires the application of the forces needed to create $\theta_a = 0.001$ rad, which are given by

$$\begin{Bmatrix} F_a \\ M_a \\ F_b \\ M_b \end{Bmatrix} = \begin{Bmatrix} -6EI/L^2 \\ 4EI/L \\ 6EI/L^2 \\ 2EI/L \end{Bmatrix} \{\theta_a\} = 25 \times 10^6 \times 0.01 \begin{Bmatrix} -6 \\ 4 \\ 6 \\ 2 \end{Bmatrix} \{0.001\}$$

$$= \begin{Bmatrix} -1.5 \\ 1 \\ 1.5 \\ 0.5 \end{Bmatrix} \times 10^3 \begin{matrix} \text{kN} \\ \text{kN m} \\ \text{kN} \\ \text{kN m} \end{matrix}$$

For this case, the resulting deflected shape (Figure 2.13(d)) is continuous at node I and does not therefore require correction to represent the bending moment influence line.

As an application of the use of the influence lines just derived, assume it is required to derive the maximal positive shear force and maximal negative bending moment at I for both the determinate and indeterminate cases due to a uniformly distributed downwards live load of 50 kN/m. The determinate values are readily calculated from Figure 2.13(a), (b), by placing the live load on CE/IF for maximal shear force and CE alone for maximal bending moment to give

$$\text{maximum positive shear force at I} = 50 \times [(\tfrac{1}{2} \times 15 \times 0.5) + (\tfrac{1}{2} \times 6 \times 0.6)]$$

$$= 277.5 \, \text{kN}$$

$$\text{maximum negative bending moment at I} = 50 \times (\tfrac{1}{2} \times 15 \times 3) = 1125 \, \text{kN m}$$

The curvature of the indeterminate solution presents an additional difficulty in determining areas under the influence line. This can be overcome either by using a CAD plot to measure the areas or by numerical integration using Simpson's Rule (Stroud, 1995). By either of these means, the area values shown in Figure 2.13(c), (d) may be obtained and it follows that

maximum positive shear force at $I = 50 \times (2.5 + 2.08) = 229 \, \text{kN}$

maximum negative bending moment at $I = 50 \times 15 = 750 \, \text{kN m}$

The stiffening effect of the removal of the internal hinges therefore reduces both the values investigated from $277.5 \, \text{kN}$ and $1125 \, \text{kN m}$ to $229 \, \text{kN}$ and $750 \, \text{kN m}$ respectively.

Worksheet 2.4 Influence line sketching and calculation

W2.4.1 Sketch shear force influence lines at positions A, B^- (immediately to the left of support B), B^+ (immediately to the right of support B) and C for the four span beam structure shown in Figure W2.4.1. Sketch bending moment influence lines for positions A, B and C. On the same diagrams, sketch the corresponding influence lines if the three internal hinges are removed so that the beam is continuous throughout.

Calculate the principal ordinates on the shear force and bending moment influence lines at position C for the hinged version of the structure. Hence determine the maximum positive shear force and maximum negative bending moment which can be produced at C by a uniformly distributed live load of $60 \, \text{kN/m}$, which can be of any length and can be applied to any portion(s) of the structure.

Use a computer package to obtain shear force and bending moment influence lines for position C for the continuous version of the structure. Determine the maximum positive shear force and negative bending moment at C for this design and compare with the values obtained for the hinged system.

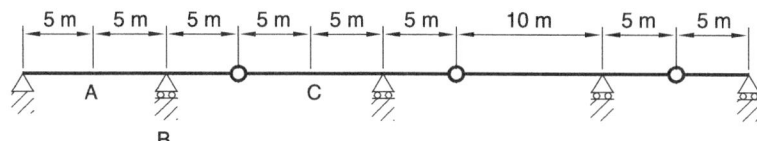

Figure W2.4.1   Worksheet 2.4.1

2.6   CASE STUDIES

C2.1.1 The concrete supporting beam for the concourse area of the Sydney Opera House (lower right in Figure C2.1.1(a)) was originally designed (Holgate, 1986) to be supported at two intermediate positions (Figure C2.1.1(b)), but architectural objections required the removal of these supports (Figure C2.1.1(c)). Assuming a uniform beam of the type shown in section E (Figure C2.1.1(e)), examine the structural implications of the decision to remove the supports by employing a computer package to analyse the idealisations of the concourse beam system shown in Figure C2.1.1(b), (c) for the effects of a uniformly distributed, vertical load of $20 \, \text{kN/m}$ of beam.

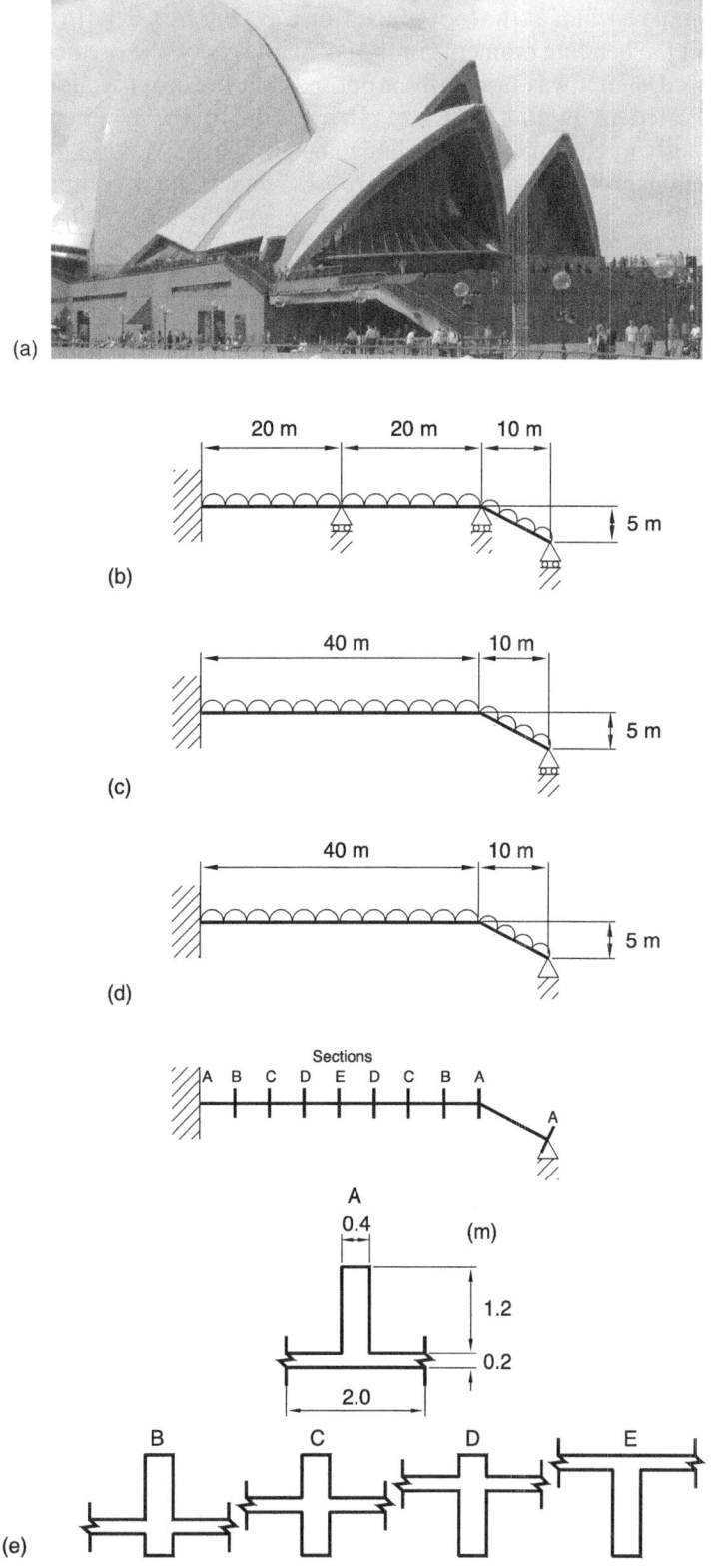

Figure C2.1.1   Case study 2.1.1

To ameliorate the effects of the removal of the supports, it was decided to introduce a horizontal restraint at the foot of the stairway (Figure C2.1.1(d)). Predict how this is likely to affect the bending moments in the beam and check the prediction by computer analysis. For this design it was originally proposed that the cross-section of the beam should vary as indicated in Figure C2.1.1(e). In relation to the bending moment diagram, what advantages might this 'variable flange height' have? Making use of the gross dimensions for section property evaluation purposes, undertake a computer analysis to check the effect of the variable section proposal on the deflected shape and bending moment distribution.

C2.1.2 The three span concrete bridge shown in Figure C2.1.2 is of uniform section such that its self-weight is 30 kN/m. The bridge is constructed in two halves, each being supported on formwork during construction. Calculate the dead-load bending moments when the left-hand half only has been constructed and its formwork removed. Use a computer package to determine the moments in the complete bridge due to the dead load of the right-hand half once it has been constructed and its formwork removed. Hence determine the resultant dead-load bending moment distribution in the full bridge.

If a hinged connection were provided between the two halves, what effect would this have on the dead-load bending moment distribution? What would the dead-load bending moment distributions be for the continuous and hinged designs if the bridge was fully constructed before the formwork was removed?

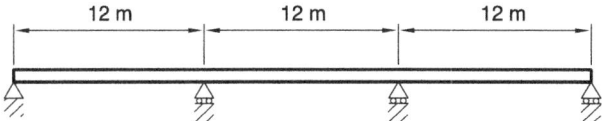

Figure C2.1.2   Case study 2.1.2

C2.1.3 The three-span beam concrete beam shown in Figure C2.1.3 is of uniform section and the rotational stiffness of the columns may be ignored. The loading regimes for the beam comprise a uniform dead load within the range 20–40 kN/m and a uniform imposed (live) load of 0–15 kN/m.

Employ influence line sketches to identify the intensities of dead and live load that should be applied to each of the spans to produce maximum positive and negative bending moments at in-span and support positions. Use a computer package to analyse the critical loading conditions and thus to determine the maximal bending moments. From these values sketch a bending moment envelope for the beam and, if the facility is available, use the package to verify the sketch.

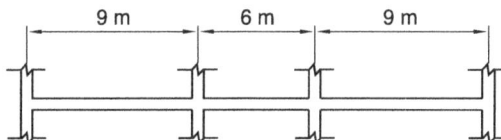

Figure C2.1.3   Case study 2.1.3

C2.1.4 The Britannia Tubular Steel Bridge (Figure C2.1.4(a)) crosses the Menai Straits to link Wales to the island of Anglesey. The bridge opened in 1851 and was innovative in both design and construction, being one of the first bridges to which elastic, continuous beam theory was applied in construction to achieve design savings.

(a)

(b)   (c)

Figure C2.1.4   Case study 2.1.4

The bridge comprised two parallel tubes and had four spans, the two central spans being 140 m and the two outer spans 70 m. The spans were initially erected (Figure C2.1.4(b)) as individual, simply supported sections. Subsequently, continuity was introduced by joining the spans at the intermediate piers. During the course of this operation pre-stressing was introduced, with the aim of reducing the maximum dead-load moment in the longer, central spans. A simplified version of the scheme adopted (Charlton, 1982) is:

Stage 1   An end span was lifted at its bank support so that the dead-load rotations of the span and its neighbouring central span were eliminated and the two spans were then joined.

Stage 2   The remote end of the second central span was lifted so that the dead-load rotation at its near end and the now reduced dead-load end-rotation of the two connected spans were eliminated. The sections were then joined.

Stage 3   The remote bank end of the second end span was lifted until its near end dead-load rotation and the reduced dead-load end-rotation of the connected section were eliminated. A final connection was then made.

Taking $I_{end} = 9\,m^4$ and $I_{centre} = 14\,m^4$ for the cross-sections (Figure C2.1.4(c)) of the end and centre spans, respectively, and assuming a constant dead load of 100 kN/m, use a computer package to determine the bending moment distribution at the start and end of each of the stages of construction described. Compare the bending moment diagram at completion with the original diagram for the individual simply supported spans before the pre-stressing operation was undertaken. Use this comparison to assess the benefits that were achieved by the pre-stressing.

2.7   REFERENCES

Charlton, T. M. (1982) *A History of the Theory of Structures in the Nineteenth Century*. Cambridge: Cambridge University Press – includes an account of the development of elastic continuous beam theory and its application to the Britannia Bridge.

Holgate, A. (1986) *The Art in Structural Design*. Oxford: Clarendon Press – describes the design of the Sydney Opera House, including the concourse area.

Kennedy, J. B. and Madugula, M. K. S. (1990) *Elastic Analysis of Structures*. New York: Harper and Row – Chapter 14 of this reference deals with influence lines in general, including lines for continuous beams.

Megson, T. H. G. (1996) *Structural and Stress Analysis*. London: Arnold – Chapter 6 of this reference (pp. 363–366) discusses beam deflections due to shear.

Stroud, K. A. (1995) *Engineering Mathematics*. Basingstoke: Macmillan – covers approximate (numerical) integration, including the application of Simpson's Rule.

2.8   FURTHER READING

Owens, G. A. and Knowles, P. R. (eds) (1992) *Steel Designers' Manual*. 5th edn. Oxford: Blackwell Scientific Publications – appendices provide bending moment, shear force and deflection results for a range of beam types and influence lines for uniform continuous beams of up to four equal spans.

3 Plane trusses

3 Plane trusses

3.1 INTRODUCTION

Plane trusses are sometimes known as triangulated or pin-jointed frameworks and all of these names are helpful in understanding their structural behaviour. Plane truss implies a two-dimensional arrangement of members which are capable of withstanding applied loads by axial (tension/compression) effects only, with bending moments and shear forces being absent. Triangulated arrangements of members certainly promote this type of structural response. If bending effects are to be eliminated, pin-joints are necessary, since the rotation of a rigid joint would produce bending moments in the connected members. Also, loads must be applied to the joints, since loading applied to a member would cause it to act as a simply supported beam and hence incur bending. Members should also be arranged so that their centroidal axes intersect at joints, since eccentricity induces bending.

The pin-joints and joint loading assumptions are rarely fulfilled in practice. In the nineteenth century, attempts were often made to provide pin-joints to match the theoretical assumption, but bolted or welded connections are now generally preferred, although pinned connections are occasionally made (Figure 3.1). Welded (and to a lesser extent bolted) connections will introduce joint rigidity, and hence bending in the members. Bending effects due to joint rigidity are usually ignored because the (*secondary*) stresses, which result, are normally small in comparison with the *primary* stresses, which arise from the axial loads. The lower bound theorem of plasticity also ensures that adding stiffness to a structure cannot reduce its collapse load. Neglecting joint rigidity therefore provides additional safety. Loads are certainly most effectively supported if applied to nodes. Occasionally this is not convenient, in which case the bending effects produced by the loads must be taken into account.

Pin-joints and joint loading will be assumed in the present chapter, so that plane truss internal force analysis will involve the determination of the axial loads in the members only. The ability to determine which members carry tension (*ties*), and which compression (*struts*), is important in view of the danger of strut buckling, which, in metal trusses, results in much more substantial members generally being required for the struts. Designers (Figure 3.1) sometimes emphasise the nature of the members by using slender rods for ties and much more substantial sections for struts.

Perhaps the most basic method for determining the type and magnitude of the axial forces in a statically determinate truss is to apply the conditions of joint

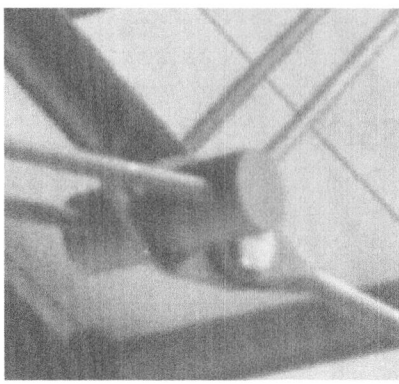

Figure 3.1   Pinned truss joint

equilibrium (Kennedy and Madugula, 1990). As part of an evaluation of the nature of the forces existing in the members of a truss, equilibrium conditions are particularly useful in establishing member force relationships at unloaded joints. Common situations at unloaded joints are shown in Table 3.1.

In assessing overall structural behaviour, it is convenient to divide trusses into two main categories, depending upon whether the top and bottom members are approximately parallel (Figure 3.2(a)), or whether they have significant inclination and are therefore *pitched* (Figure 3.2(c)). Parallel extreme members produce a truss that behaves in a very similar manner to a beam and many features appropriate to beams are therefore relevant. Pitched truss behaviour has more in common with that of arches and therefore provides concepts which can be utilised in that context. Some idea of the difference in behaviour may be obtained from the response of the two trusses of Figure 3.2(a), (c) to a central point load. As shown in Figure 3.2(b), (d), the entire perimeter and most of the internal members of the parallel-sided truss carry axial loads due to the application of the point load, but only the perimeter members are loaded in the case of the pitched truss. The nature of the forces in the internal members may be established from the rules of Table 3.1. For the parallel-sided truss, the left-hand end vertical is clearly a strut due to the support reaction effect. Joint equilibrium at the top left-hand joint then

*Table 3.1   Member forces at **unloaded** joints*

Case	Type	Example	Result	Proof
1.	Two members only		Both members unloaded	Resolution normal to each member
2.	Three members – two collinear		Force in non-collinear member is zero	Resolution normal to collinear members
3.	Four members – two collinear		Forces in non-collinear members are of opposite type	Resolution normal to collinear members

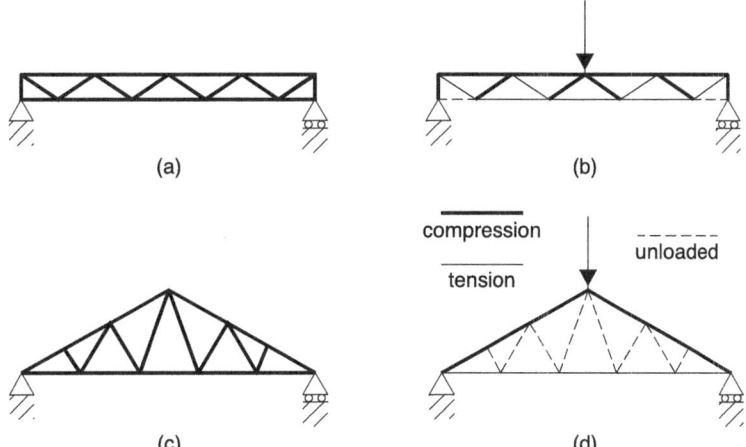

Figure 3.2   Parallel sided: (a) truss, (b) under point loading. Pitched: (c) truss, (d) under point loading

indicates that the first diagonal member must be in tension. The nature of the remaining internal members then follows by application of Case 3 of Table 3.1 to the next three joints. For the pitched truss, Case 2 shows that the left-hand internal member is unloaded and the use of Case 3 at the next three joints shows the remaining internal members to be similarly unaffected by the load. In passing, it should be mentioned that the fact that particular members are unloaded does **not** imply that they do not provide a vital service. Removal of the two unloaded members in Figure 3.2(b), for example, would create a mechanism. Removal of the internal members of the pitched truss would leave a viable structure, but would almost certainly lead to the buckling of the long inclined struts.

3.2   PARALLEL SIDED TRUSSES (LATTICE GIRDERS)

Horizontal systems

If arranged horizontally, as a bridge or building beam, for example, the top and bottom parallel members are normally referred to as the top and bottom *chords* (or *booms*), while the internal members are the *bracing* (*lattice* or *web*) members. The Warren truss shown in Figure 3.3(a) uses an isosceles triangle pattern for the bracing members. Another common arrangement is the *Pratt* (or *N-*) truss (Figure 3.3(b)), in which a right-angled triangle pattern of bracing members is employed, so dividing the truss up into *panels*, the joints of which are sometimes referred to as *panel points*. A rather less common form is the *Howe* truss (Figure 3.3(c)), which reverses the orientation of the diagonal bracing members used in the N-truss.

The trusses of Figure 3.3 are primarily intended to resist vertical, gravity loads and, as previously mentioned, act in a beam-like fashion. If only three reaction components are provided, as with the simple supports of Figure 3.3, then the

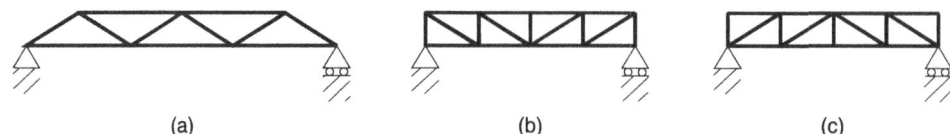

(a) (b) (c)

Figure 3.3   Bracing systems: (a) Warren, (b) Pratt (N-), (c) Howe

forms shown in the figure are all statically determinate, as may be verified by the application of determinacy principles. As with beams, the introduction of further supports will introduce redundancies. Whether redundant or not, the internal forces may be readily obtained from computer analysis and the nature of the forces and their relative magnitudes can be verified by considering the beam-type action of the truss. To establish an analogy between the action of the truss and a beam, the internal forces shown in Figure 3.4 can be considered. From a comparison of Figure 3.4(a) and (b), it follows, by moments about the intersection of T and S, that is p in Figure 3.4(b), that

$$M = C \times d \quad \text{or} \quad C = \frac{M}{d} \tag{3.1}$$

and, resolving vertically in Figure 3.4(b)

$$F = S \sin \theta \quad \text{or} \quad S = \frac{F}{\sin \theta} \tag{3.2}$$

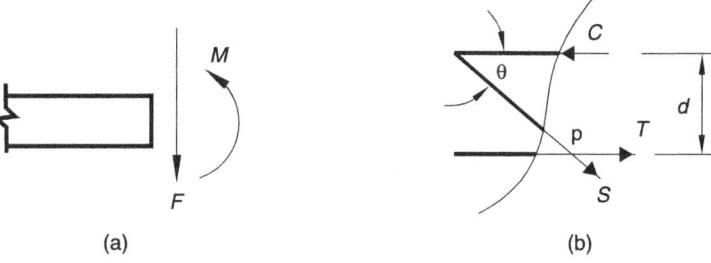

(a) (b)

Figure 3.4   (a) Beam section, (b) truss section

The forces in the parallel-sided (*boom*) members are therefore proportional to the bending moment (and inversely proportional to the truss depth), while the diagonal bracing forces are proportional to the shear force (and inversely proportional to the angle the brace makes with the horizontal). It follows that the variations in magnitude of the boom and diagonal brace forces may be obtained from bending moment and shear force diagrams respectively. The nature of the forces induced will depend on the type of bending and shear, as shown in Figure 3.5 for positive bending and shear actions.

Example 3.1   Truss internal force evaluation

Deflected shape and bending moment diagrams for the truss of Figure 3.6(a) are given in Figure 3.6(b), (c). From the deflected shape and bending moment

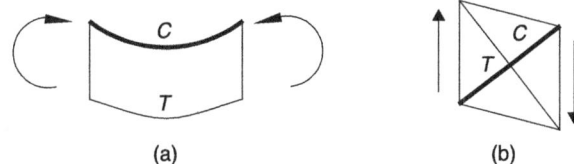

Figure 3.5   Nature of (a) boom, (b) bracing forces

diagrams, the positive bending in span A–B will produce boom forces of the type shown in Figure 3.5(a). Within span B–C, negative bending is experienced and the nature of the boom axial loads therefore reverses, before reverting to the positive system in span C–D (Figure 3.6(e)).

From the reaction directions indicated in Figure 3.6(b), a likely shear force diagram is as shown in Figure 3.6(d). The positive shear in A–B will result in alternating compression and tension in the bracing members, following the pattern of Figure 3.5(b). These will then reverse in B–C due to the negative shearing action, while the shear in C–D is such as to produce tension in each diagonal brace (Figure 3.6(e)).

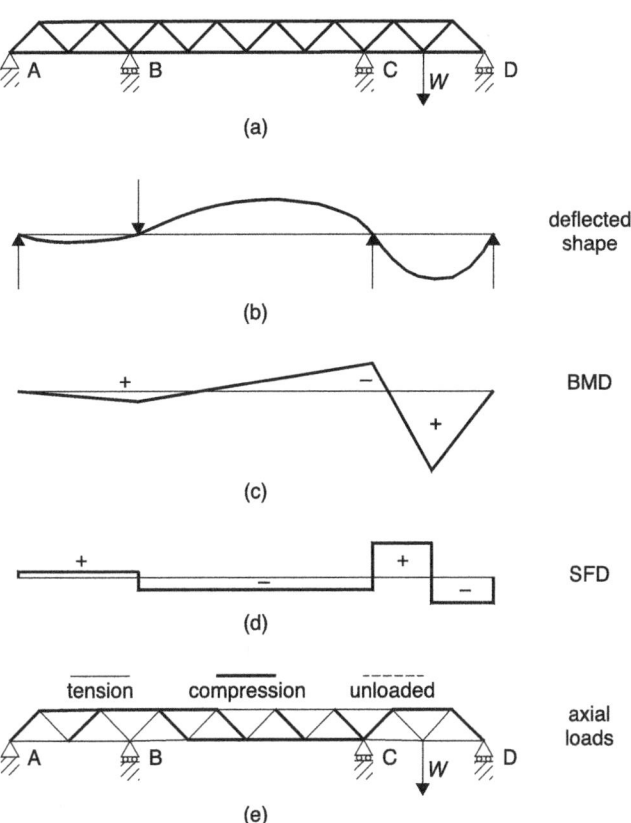

Figure 3.6   (a) Example truss, (b) deflected shape, (c) BM diagram, (d) SF diagram, (e) axial loads

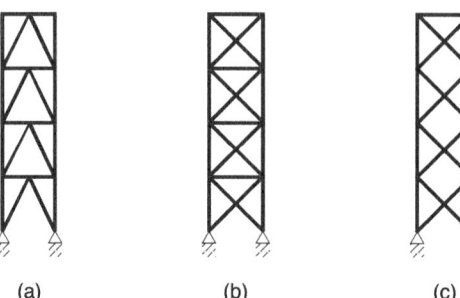

(a)   (b)   (c)

Figure 3.7   Vertical bracing: (a) K-braced, (b) cross(counter)-braced, (c) open cross(X)-braced

Vertical systems

Vertical trusses are used to resist lateral forces in buildings and also find widespread application in tower construction. The bracing arrangements shown in Figure 3.3 for horizontal trusses are also used for vertical ones, but the further K- and cross-braced patterns of Figure 3.7 are also used, both for vertical and, occasionally, horizontal trusses. Vertical trusses are often subjected to gravity as well as lateral loads, but the axial load distribution due to vertical loading is normally readily assessed since this will be carried purely by compression in the vertical members.

The nature of the forces induced in the cross-braced truss shown in Figure 3.8 may be established by separating the vertical and lateral loading effects as indicated. The lateral loading causes the tower to act in a cantilever mode, producing the indicated tensions and compressions in the verticals. The shear is of constant sign and is such as to produce the tension/compression system indicated in Figure 3.8(c). The nature of the forces in the horizontal braces is somewhat less easy to determine without calculation, but the nature of the lateral loads suggest these will be struts.

Cross-braced towers are statically indeterminate and, by comparison with the determinate singly-braced trusses of Figure 3.3(b), (c), the degree of indeterminacy

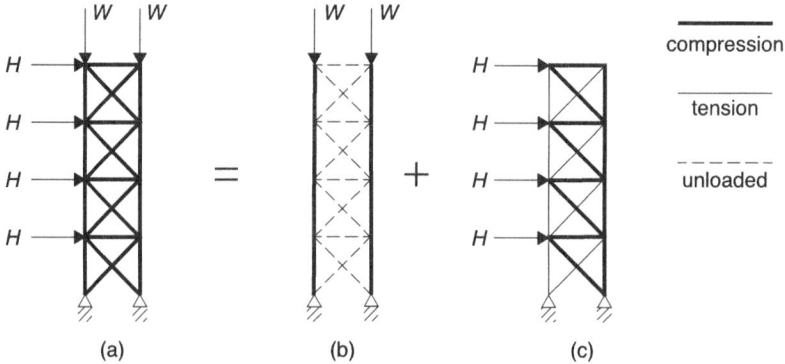

(a)   (b)   (c)

Figure 3.8   Forces in cross-braced tower

is clearly equal to the number of cross-braces introduced. Prior to the advent of computers, it was therefore usual to simplify the analysis of cross-braced trusses by ignoring the diagonal compression braces. This is also necessary if the bracing design is to be based on the provision of rod members, since they have very low buckling capacity and are therefore essentially ineffective in compression. If the diagonal braces are to be designed to withstand compression, an approximate evaluation of the forces in the braces may be obtained by assuming that the shear in a panel is divided equally between the braces. This concept may also be applied, with rather less accuracy, to the singly-redundant, open, cross-braced form of Figure 3.7(c).

Example 3.2 Tower under lateral loading

Large tower structures generally have splayed verticals to improve their stability. This also produces a progressive increase in lever arm, and thereby to reduce the maximum forces induced (equation (3.1)). Cross bracing in the higher panels is often replaced by K-bracing lower down to reduce the length of the compression bracing. The simplified example tower shown in Figure 3.9(a) demonstrates both of these features, but omits the *secondary bracing* (Figure 3.10) that would normally be used to further reduce the unsupported lengths of compression members.

For this example, the nature of the axial forces produced by lateral loading will be established, and a quick estimate of the maximum leg and bracing loads in the upper and lower sections will be obtained for checking purposes. The cantilever action of the tower produces tension and compression in the left-hand and right-hand legs, respectively. The shear is resisted by the diagonal braces and the concept of Figure 3.5(b) may be used to establish the nature of the forces induced. Property 2 of Table 3.1 may then be applied to determine the natures of the forces in the horizontal bracing of the lower section, those in the upper

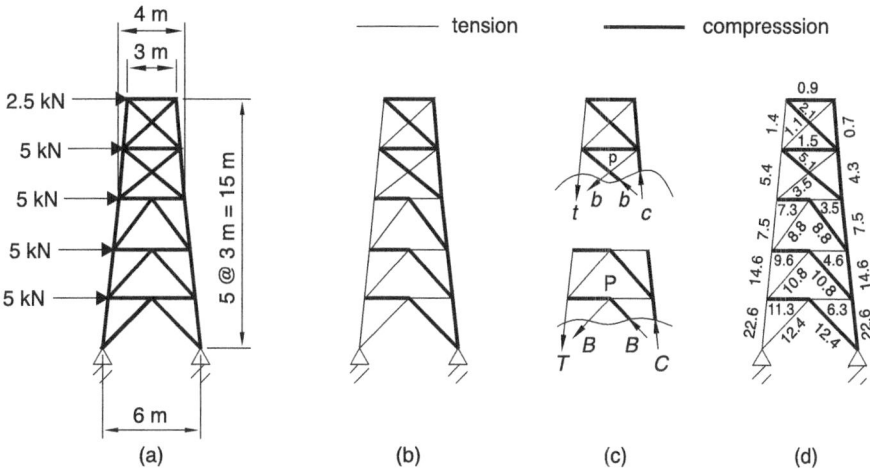

Figure 3.9 *(a) Example tower, (b) axial load types, (c) sections, (d) computed loads*

Figure 3.10   Transmission tower bracing

section being deduced from the direction of the loading. The resulting overall force distribution is as shown in Figure 3.9(b).

To estimate the maximum loads induced in the upper part of the tower, the top section of Figure 3.9(c) can be used. First, by moments about p, neglecting leg inclination, gives

$$\left(c \times \frac{3.75}{2} \right) + \left(t \times \frac{3.75}{2} \right) = (2.5 \times 4.5) + (5 \times 1.5) = 18.75\,\text{kN m}$$

Taking the bracing loads as being equal in magnitude, it follows that the leg loads are similarly equal, so that

$$c = t = \frac{18.75}{3.75} = 5\,\text{kN}$$

Resolving horizontally for the section, again neglecting leg inclination, results in

$$2 \times b \times \cos\left(\tan^{-1}\left(\frac{3}{3.75} \right) \right) = 2.5 + 5 = 7.5 \text{ whence } b = 4.8\,\text{kN}$$

Similarly, using the lower section of Figure 3.9(c)

$$\left(C \times \frac{5.33}{2} \right) + \left(T \times \frac{5.33}{2} \right) = (2.5 \times 12) + (5 \times (9 + 6 + 3))$$

so that $C = T = 22.5\,\text{kN}$

and

$$2 \times B \times \cos\left(\tan^{-1}\left(\frac{3}{3}\right)\right) = 2.5 + 5 + 5 + 5 + 5 \text{ so that } B = 15.9\,\text{kN}$$

These estimated maximal forces are sufficiently accurate to show that no gross error has been made in generating the computer results of Figure 3.9(d).

Worksheet 3.1 Parallel-sided trusses

W3.1.1 Use a colour code to produce diagrams illustrating the nature of the axial loads induced in the trusses of Figure 3.3 under the action of equal, downwards point loads acting at each of the lower panel points. Comment on the utility of any unloaded members.

If it is wished to minimise the length of compression bracing members, how do the trusses of Figure 3.3 compare in this respect? Why would a modification to the N-truss of the form, shown in Figure W3.1.1(a), be undesirable?

Under the same loading, what advantages might be gained by introduction of vertical braces into the Warren arrangement, as shown in (a) Figure W3.1.1(b) and (b) Figure W3.1.1(c).

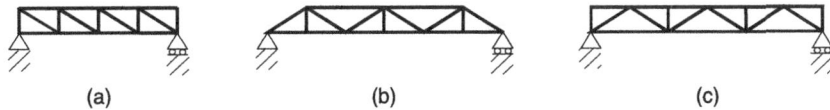

(a) (b) (c)

Figure W3.1.1   Truss modifications: (a) N-, (b) Warren, (c) Warren

W3.1.2 Prove that the trusses shown in Figure W3.1.2 are all statically determinate. Produce suitable sketches to indicate the **nature** of the forces induced in the members of the trusses by the loadings shown. The magnitudes of the forces are **not** required.

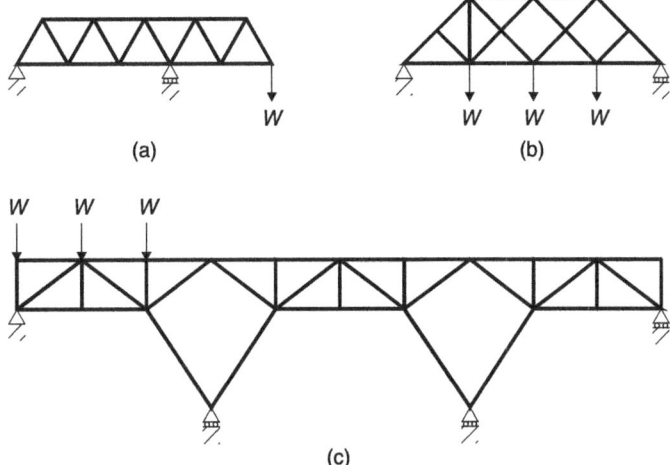

Figure W3.1.2   Worksheet 3.1.2

W3.1.3 Determine the degree of statical indeterminacy for the four towers shown in Figure W3.1.3. Establish the nature of the forces in the members of the towers of

Figure W3.1.3(a), (b) and comment on any differences. How are the force distributions in the towers of Figure W3.1.3(c), (d) likely to compare with those of Figure W3.1.3(a), (b)?

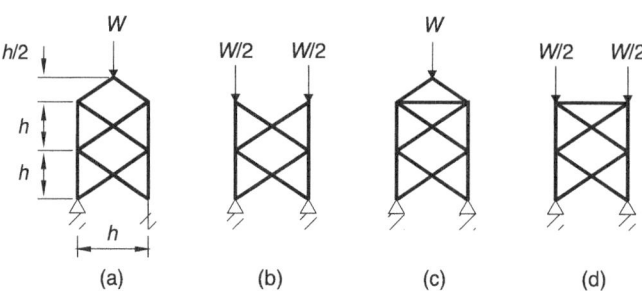

Figure W3.1.3   Worksheet 3.1.3

W3.1.4 For the tower structure of Figure W3.1.4(a) use the sections shown in Figure W3.1.4(b), (c) to determine the forces F_1, F_2, F_5, F_6. Hence, determine F_3 and F_4. How would the forces in the legs and bracing be expected to vary in the lower panels of the tower? How can a beam analogy be used to confirm the predictions?

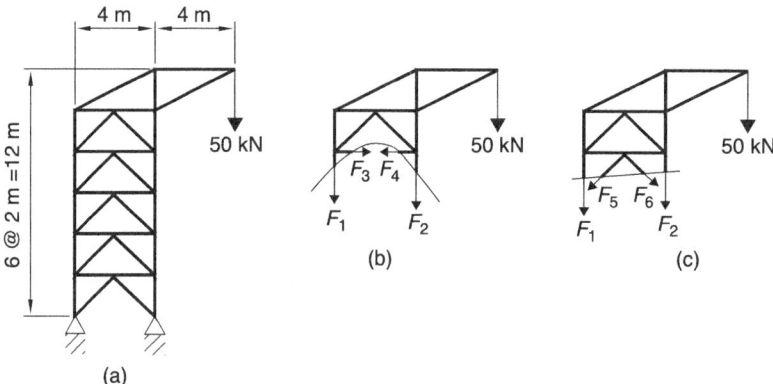

Figure W3.1.4   Worksheet 3.1.4

3.3   NON-PARALLEL-SIDED TRUSSES

Pitched systems

Pitched trusses of the form shown in Figure 3.2(c) are commonly used for roofing purposes. The internal bracing shown in Figure 3.2(c) is of the Warren form, but the Pratt or Howe types (Figure 3.3(b), (c)) are also used. For larger spans, a *fink* arrangement (Figure 3.11(a)) is often employed. This comprises two separately braced sub-systems, which are linked together to form a *compound* truss. Whatever bracing arrangement is provided, a vertical load at the apex (Figure 3.2(d)) will be supported by the main pitched members (*rafters*) in compression and the bracing will remain unloaded. Such a system represents the most direct transmission of the load to the supports. If, as is usual, the supports are of a simply supported nature (Figure 3.2(d)) then the rafter compressions will tend

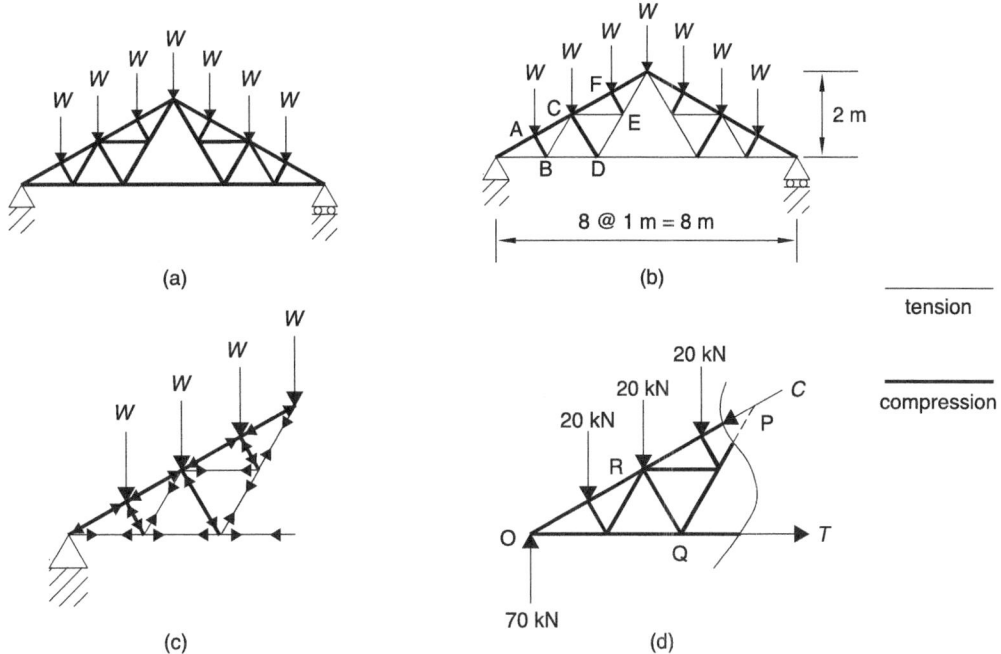

tension

compression

Figure 3.11   (a) Fink truss, (b) example truss, (c) internal forces, (d) section

to spread the supports, and this tendency is contained by the lower member (*main tie*) becoming tensioned. Under this loading, therefore, the function of the bracing is to provide restraint, so that the buckling capacity of the rafters is increased.

Internal bracing systems do, however, also provide additional nodal points on the rafter to which longitudinal beams (*purlins*) can be attached without causing bending of the rafter. Under this form of loading (Figure 3.11(a)) the bracing members certainly sustain load and the nature of the forces induced may be determined by use of the concepts of Table 3.1.

Example 3.3   Forces in fink truss

As an example, the natures of the forces and the magnitudes of the maximum and minimum compression in the rafter and the maximum and minimum tension in the main tie will be established for the fink truss of Figure 3.11(b), taking $W = 20$ kN. As with apex loading, the rafters and main tie will be in compression and tension, respectively. The nature of the forces in the bracing may be established by noting that, from joint equilibrium at A, AB must be a strut. Principle 3 of Table 3.1 then shows that BC is a tie. Similar arguments show that FE is a strut and EC a tie. Joint equilibrium at C then indicates that CD is a strut and Principle 3 at D finally indicates that DE is a tie. The complete force distribution is as shown in Figure 3.11(c), and the typical alternating strut/tie pattern of the bracing forces should be noted.

The compression in the rafters will increase from the apex due to the application of the lower loads and the maximal compression will therefore occur at the supports. From symmetry, each support will provide a vertical reaction of $7 \times W/2 = 7 \times 20/2 = 70\,\text{kN}$.

By resolving vertically at a support

$$\text{maximal rafter compression} = \frac{70}{\sin\left(\tan^{-1}\left(\frac{2}{4}\right)\right)} = \frac{70}{\sin 26.6°} = 157\,\text{kN}$$

The minimum rafter compression may be obtained by use of the section shown in Figure 3.11(d), in which

$$OR = 2/\cos 26.6° = 2.24\,\text{m}; \quad RQ = OR\tan 26.6° = 1.12\,\text{m};$$

$$OQ = OR/\cos 26.6° = 2.5\,\text{m}$$

Taking moments about Q for the section gives

$$C \times RQ = (70 \times 2.5) + (20 \times 0.5) - (20 \times (0.5 + 1.5)) \text{ so that } C = 130\,\text{kN}$$

The rafter compression therefore varies from 130 kN at the top to 157 kN at the bottom.

It is perhaps not so clear as to whether the maximal main tie tension occurs in the support or central section. However, a consideration of the directions of the bracing forces (Figure 3.11(c)) shows that the tension will decrease towards the centre.

Hence, resolving horizontally at a support gives

$$\text{maximum main tie tension} = 157 \times \cos 26.6° = 140\,\text{kN}$$

The tension in the central portion of the tie may again be obtained from the section shown in Figure 3.11(d). By moments about P

$$\text{central main tie tension} = \frac{(70 \times 4) - (20 \times (3 + 2 + 1))}{2} = 80\,\text{kN}$$

The main tie tension therefore ranges from 140 kN to 80 kN.

Bowed systems

Figure 3.12(b) shows a truss in which the compression chord has been bowed to a parabolic profile to produce a form sometimes used for roofing and bridge applications. The figure also compares the performance of such a system with that of a pitched form (of similar span and rise) under both concentrated apex loading and the same total load when uniformly distributed horizontally. The striking feature of the comparison is that, as seen previously, the bracing is unloaded for the pitched system under apex loading, but it is the bowed system bracing that is unloaded under distributed load.

The questions therefore arise as to why the systems respond in different ways to the different loading systems, and whether further loading types will be associated

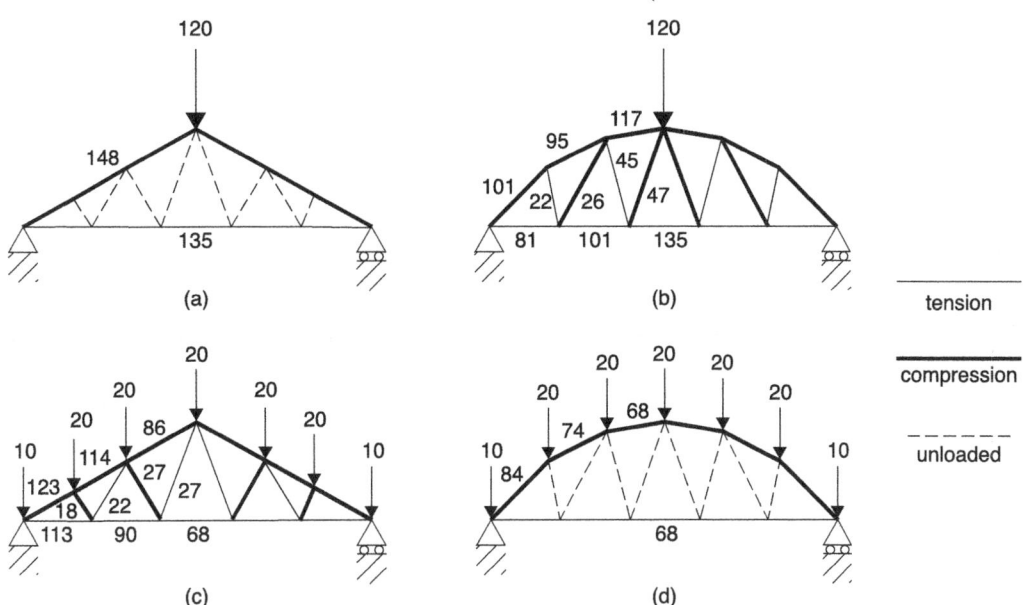

Figure 3.12   Forces (kN) in pitched and bowed systems: (a), (b) point loading, (c), (d) distributed loading

with truss geometries that result in unloaded bracing systems? To investigate these matters, the typical section shown in Figure 3.13 is considered. Taking moments about P gives

sum of moments of external loads and reactions to left of P = $BM_\mathrm{P} = T \times h$

$$(3.3)$$

In the absence of bracing forces, the main tie tension, T, will be constant (Figure 3.12(d)), and equation (3.3) therefore shows that the profile required for zero bracing forces has the same shape as the bending moment diagram for the loading system – hence the triangular outline for the point load and parabolic one for the distributed load.

It may also be noted from Figure 3.12 that the bowed system reduces the loads in the compression chord, even under the point loading. Appropriate sections show that the force in the centre of the main tie must be the same for both the pitched and bowed formations under a given loading, but, when the main tie

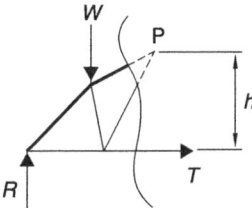

Figure 3.13   Bowed truss section

force varies, it reduces to the supports for the bowed truss but increases for the pitched one.

If headroom considerations allow, the bowed truss can be inverted to produce a rather more efficient form in which the compression chord is now straight and is more conveniently restrained. This is achieved by the deck, for instance, in the case of a bridge.

Worksheet 3.2   Non-parallel-sided trusses

W3.2.1   Use a colour code to produce diagrams illustrating the nature of the axial loads induced in the trusses of Figure W3.2.1 under the action of both apex point loading (Figure W3.2.1(a)) and distributed loading (Figure W3.2.1(b)). Compare the distributions with the results for a Warren-braced pitched truss and with the results obtained for the parallel-sided trusses of Figure 3.3 obtained in Worksheet 3.1.

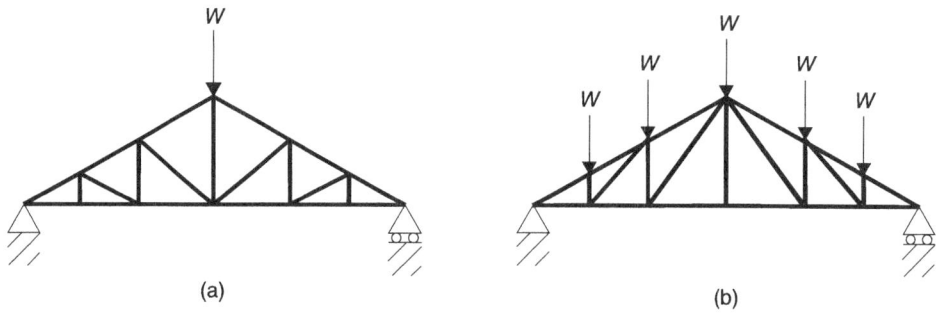

Figure W3.2.1   Worksheet 3.2.1

W3.2.2   For the N-braced truss shown in Figure W3.2.2 determine the height, h, which results in the bracing system being unloaded. Determine the maximum tension and compression in the truss (a) under the loading shown and (b) if the loads of 60 kN and 30 kN are replaced by equal downward loads of 45 kN.

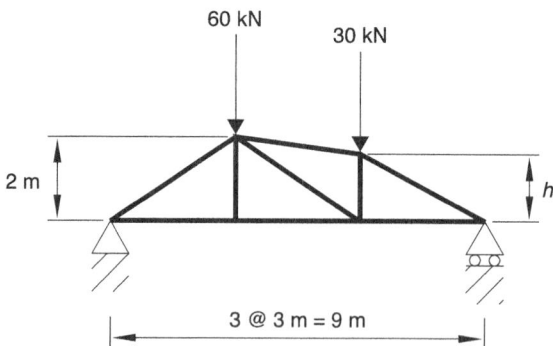

Figure W3.2.2   Worksheet 3.2.2

W3.2.3   Calculate the force in the brace AB of the truss shown in Figure W3.2.3 if the end height, h, is 1 m. Hence, produce a colour-coded diagram to illustrate the nature of the forces in all the members of the truss. For (a) 1 m $< h \leq$ 2 m and (b) 0 $< h <$ 1 m, state the **nature** of force in AB and hence sketch diagrams illustrating the nature of the forces induced in the truss members in these cases.

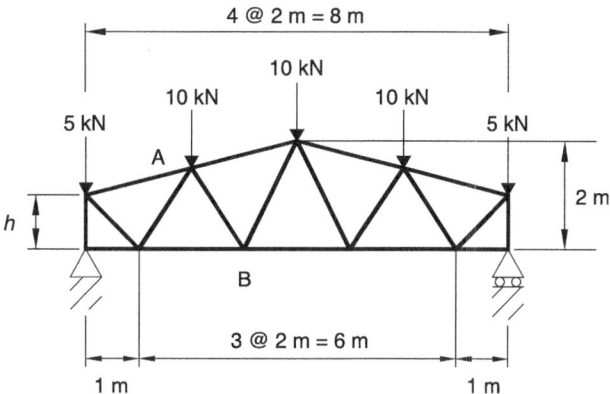

Figure W3.2.3   Worksheet 3.2.3

W3.2.4 Secondary bracing may be introduced into a bowed Pratt (N-) bridge truss in either of the ways shown in Figure W3.2.4 to produce a *Petit* truss. The function of the secondary bracing is mainly to reduce the distance between panel points on the truss and hence reduce the size of the cross-girders needed to support the bridge floor and transfer load to the side trusses.

In Figure W3.2.4, the upper nodes on the upper compression chord have been set out parabolically and the trusses carry equal vertical loads at each of the lower panel points. State why the members indicated as dashed carry no load in this situation. Hence, sketch diagrams showing the nature of the forces in all the members of each truss.

In a central panel of each truss, how will the forces in the compression boom compare for the two bracing arrangements and, in the same panel, how will the main tie forces compare?

Which bracing arrangement is likely to be the more economical?

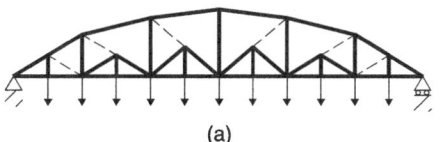
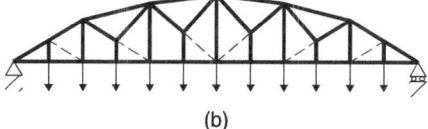

(a)   (b)

Figure W3.2.4   Worksheet 3.2.4

3.4   COMPUTER ANALYSIS

Modelling of pin-joints

It is usual to analyse trusses as being pin-jointed (Steel Construction Institute, 1995), and therefore to ignore secondary (bending) effects. This requires that nodal loads only are applied, so that local member bending is not induced. Computer packages that provide a customised sub-system for the analysis of plane trusses are usually the most convenient. Should such a sub-system not be available, then a plane (rigid) frame package may be used, but it will be necessary to model the pin-joints in some way. This may be possible automatically, if such a facility is provided, or pins may be inserted at the ends of members, provided it is recalled that one member at each joint must remain unpinned. This latter approach is tedious and error-prone, so that a better tactic may be to remove bending stiffness from the analysis. Pin-jointed trusses rely solely on axial stiffness and plane truss programs will therefore generally only require cross-sectional areas to be provided as member properties. Plane frames rely on bending stiffness as well as axial stiffness, and hence require second moments of area, I, values to be specified. Bending effects may therefore be effectively eliminated by using reduced I values. Dividing true values by a thousand will usually be appropriate.

Global/local axes and member properties

Global and local axes rarely cause significant problems with truss analysis. The truss geometry and nodal loading will be specified relative to the chosen global axes. Local axes will normally be used for the specification of the internal force output, and the appropriate convention must be understood if the output does not otherwise distinguish between compression and tension members. Alternatively, and usually preferably, reliance should be placed on graphical output, which typically employs a simple convention for distinguishing tension from compression members.

The only essential member properties for truss analysis are cross-sectional areas and modulus of elasticity values. Sometimes packages incorporate a facility whereby members may be designated to be operative under tension only. If 'tension only' members cannot be specified directly, then the relevant members may be omitted from the analysis, if it is clear from inspection that they are inoperative in all load cases. Alternatively, preliminary analyses may be undertaken to establish those load cases in which the members are subject to compression.

Restraints, symmetry and mechanisms

Unlike pure beams, it is vital that horizontal restraint conditions be appropriately modelled for truss analysis. Bridge bearings are commonly designed to be either of a pin or a roller variety and can therefore be directly modelled. Roof trusses are normally either column or wall supported and the actual degree of restraint provided is much less obvious. For truss analysis purposes, however, horizontal restraint should be provided at one, arbitrary position only. For a single span truss, this means that a simple support system (Figure 3.2, for example) is used. Pinning such a truss at either side will produce major changes to its behaviour, and practical supports cannot normally provide this degree of lateral restraint.

If symmetry is invoked to reduce the size of the analysis, then displacement normal to the plane of symmetry should be restrained. It is therefore necessary to apply lateral restraint on the line of symmetry, as shown in Figure 3.14(a), (b). Should the plane of symmetry cut a truss member (Figure 3.14(c), (d)), then a fictitious support member must be provided that will remain unloaded in the analysis. If this is not done, the cut member will be a local mechanism and may produce spurious results.

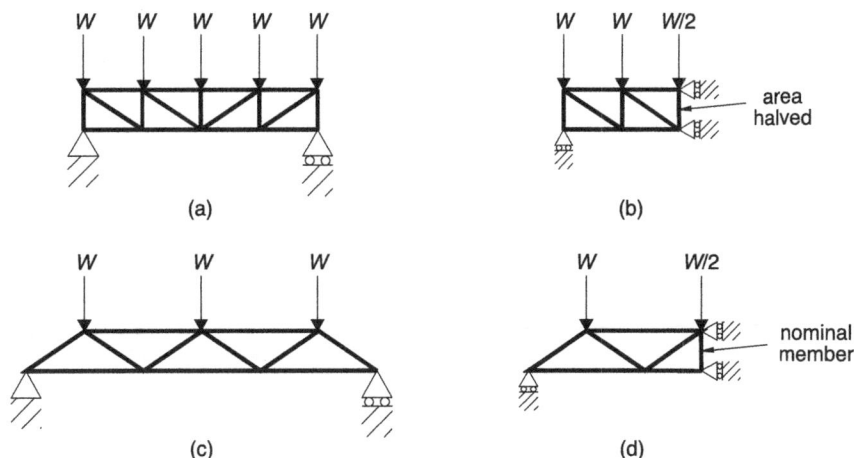

Figure 3.14 N-truss: (a) truss, (b) symmetry model; Warren truss: (c) truss, (d) symmetry model

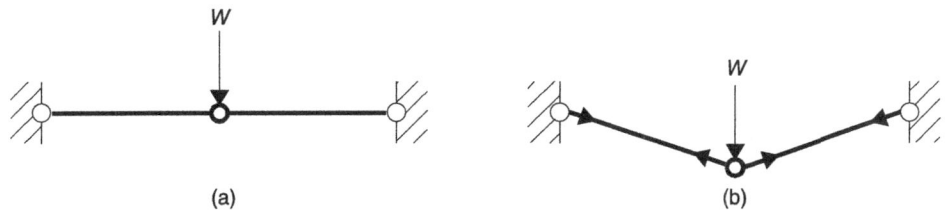

Figure 3.15 Two-bar truss: (a) small displacements, (b) large displacements

Mechanisms are unlikely to be created if standard bracing forms are employed, but are quite likely if unusual truss configurations are explored. Mechanisms for which the degree of statical indeterminacy, q, is calculated to be equal to or greater than zero are particularly difficult to predict in advance. Some of these may belong to a sub-category of structures that are mechanisms under the assumptions of small deflection theory, but stable at large deflections. The simplest such example is the two-bar truss of Figure 3.15(a), for which it is impossible to satisfy vertical equilibrium at the central node in the unloaded (small displacement) geometry, but the load can be sustained by rod tensions in the loaded (large displacement) geometry (Figure 3.15(b)). To avoid mechanism problems, unexpectedly large displacements or warnings regarding the failure of an equilibrium check should always be investigated thoroughly.

Worksheet 3.3 Computer analysis of plane trusses

W3.3.1 The fan-braced roof truss of Figure W3.3.1(a) is subjected to vertical, dead loading (*DL*) such that $W = 5\,kN$. The imposed loading (*LL*) produces $W = 7\,kN$. Uplift wind loading (*WL*) results in normal loading of the form shown in Figure W3.3.1(b), such that $N = 6\,kN$. Using symmetry, analyse one-half of the truss using a plane truss computer package. Consider combined loading cases of maximum downward loading ($DOWN = (1.4 \times DL) + (1.6 \times LL)$) and maximum uplift ($UP = DL + (1.4 \times WL)$).

State, with reasons, how the member cross-sectional areas chosen will affect the results of the analysis in respect of reactions, displacements and axial loads. Tabulate the axial load in the truss members for both loading cases and hence determine which of the two loading cases will be critical for the design of each member.

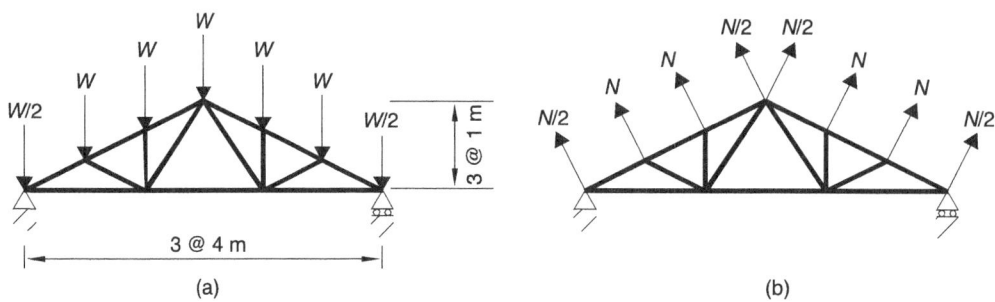

Figure W3.3.1   Worksheet 3.3.1

W3.3.2 Use a plane truss computer package to analyse the truss shown in Figure W3.3.2 by (a) taking $h = 2\,\text{m}$ and (b) taking $h = 4\,\text{m}$. Discuss the reasons for the different results obtained.

Show that the degree of statical determinacy, q, of the truss is equal to zero. A hand test for the stability of a structure possessing $q = 0$ is that, for an unstable structure, it will be possible to find a self-stressing system such that a set of internal forces and reactions are in equilibrium without external loading. Assuming unit compression in member AB, show that it is possible to derive a self-stressing system for the truss of Figure W3.3.2 if $h = 2\,\text{m}$, but not if $h = 4\,\text{m}$.

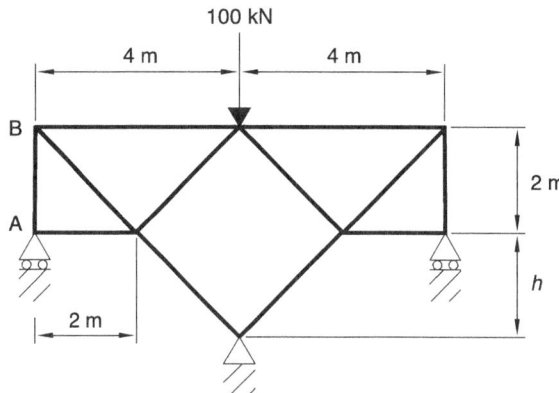

Figure W3.3.2   Worksheet 3.3.2

W3.3.3 The cantilever framework shown in Figure W3.3.3(a) is subjected to a dead load of 3 kN/m (on the slope) and an imposed rafter load of 5 kN/m (on plan). Determine equivalent nodal loads for the combined dead and imposed effects.

Assuming rigid joints throughout, use a **plane frame** computer package to analyse the truss. Take $E = 206\,\text{kN/mm}^2$; $A = 1000\,\text{mm}^2$ and $I = 1 \times 10^6\,\text{mm}^4$ for truss members; and $A = 4000\,\text{mm}^2$ and $I = 50 \times 10^6\,\text{mm}^4$ for the columns. Make a partial check on the validity of the results by a hand determination of the reactions at the column supports and by using a suitable section to determine the axial load in member AB.

In practice, the load is transmitted to the truss through a purlin system such that loads are applied, as shown in Figure W3.3.3(b). Calculate the values of P and Q. To determine the bending effects of the non-nodal application of the loads P, re-analyse the truss. Why do the bending moments from the two analyses differ much more than the axial loads? Compare the maximum bending moment results for the left-hand rafter with those obtained by use of Figure W3.3.3(c), (d), assuming a pin support at the lower end and fully fixed supports at the other nodes of the rafter. Give possible reasons for any discrepancies.

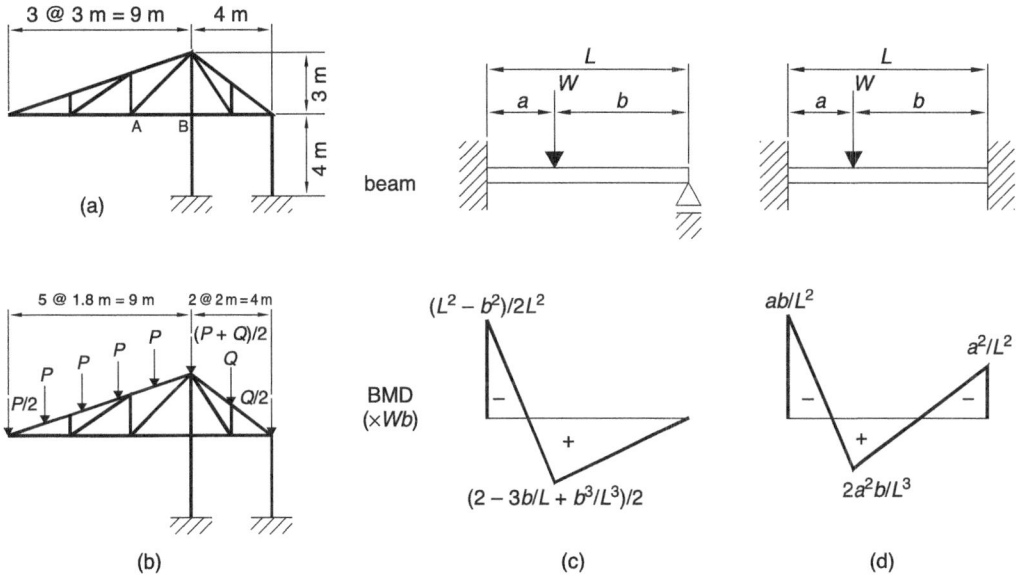

Figure W3.3.3   Worksheet 3.3.3

W3.3.4 A computer plane truss analysis of the tower structure, shown in Figure W3.3.4(a), produced the axial load/reaction distribution, shown in Figure W3.3.4(b). Identify the four errors made in the input specification for the analysis.

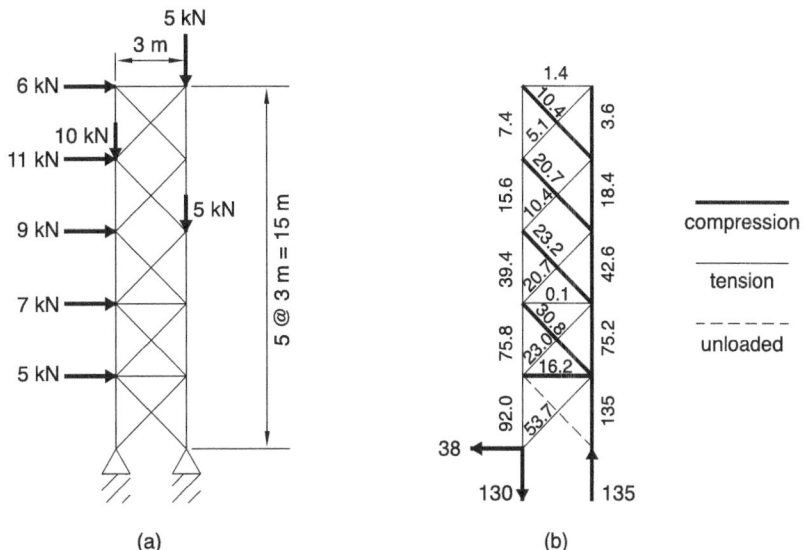

Figure W3.3.4   Worksheet 3.3.4

3.5   INFLUENCE LINES

Sketching

Influence lines for complete trusses are not so readily sketched as for beams. However, the approach is the same, namely that the required diagram may be produced by the application of Müller-Breslau's principle. For a truss member, it is, of course, axial load that is the internal force of interest and the unit displacement must therefore be applied in the direction of the axis of the member.

Example 3.4   Truss influence line diagrams

For the example truss of Figure 3.16(a), Müller-Breslau's principle will be used to construct influence line diagrams for the axial loads in members CE, DF and CF. The application of the principle to a truss member may be visualised as being equivalent to cutting the member and applying a unit displacement across the resulting gap in the direction of the member. Thus, if CE is cut, the section ACFDB of the truss (Figure 3.16(b)) is unaffected. This represents the null effect on CE of loading this region, since any load will be transmitted directly to the supports. Introduction of the unit displacement along CE will cause the section EGIHF of the truss to rotate about F until the unit displacement is produced. It should be noted that, for 'small' displacements, E moves perpendicularly to the original line FE and does **not** displace vertically during the rotation. By Müller-Breslau's principle, Figure 3.16(b) is the required influence line diagram for CE.

With the sign convention that positive values correspond to positive work done by the unit load, the diagram will work for loading in any direction. For vertical loading, the angle of rotation of EGIHF is $\tan^{-1}(1/3)$, so the vertical displacement of G or H is one, and is downwards. Presuming the loading is also downwards, it follows that a positive (tensile) unit load is produced in CE for unit vertical load at G or H. Similarly, unit downward load at I will produce a tension

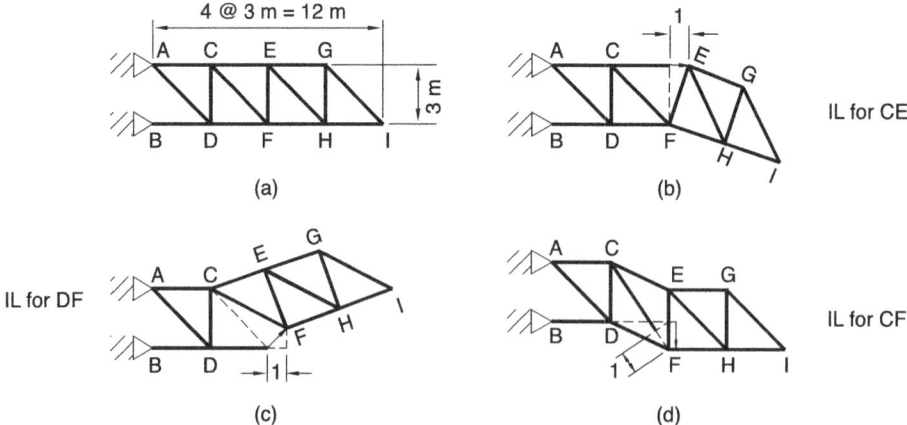

Figure 3.16   (a) Example truss, (b) IL for CE, (c) IL for DF, (d) IL for CF

of two in CE. If horizontal loading to the right is considered, then, for small displacements, H and I do not displace horizontally in Figure 3.16(b), so no force is induced in CE. However, both E and G have unit displacements to the right, so unit loading at these points will produce unit tensions in CE.

The influence diagram for DF is similarly obtained by cutting the member and introducing a unit displacement in its axial direction (Figure 3.16(c)). In this case portion ACDB is unaffected, while portion CEGIHF rotates about C until F has moved unit distance in the direction of DF (horizontally). Node F will move perpendicularly to the line that connects it to the centre of rotation (CF). Given the 45° orientation of CF, this means F will move vertically by a unit distance, as well as horizontally. The angle of rotation of CEGIHF is again $\tan^{-1}(1/3)$, so that unit downwards load applied to either E or F will produce unit **compression** in DF, due to the differing directions of the load and displacement. Similarly, unit downwards loading at G or H produces a compression of two in DF and loading I gives a compression of three. It is, of course, horizontal loading in the lower boom which produces load in DF, as evidenced by the null horizontal displacements of E and G and the unit horizontal displacements of F, H and I.

Cutting CF results in a different distortion of the truss in which panel CEFD shears and the influence diagram is produced when nodes E and F have descended, such that unit extension has been achieved in CF. The truss geometry requires that E and F must descend by $\sqrt{2}$ to extend CF by unity (Figure 3.16(d)). For small displacements, E and F do not move horizontally during the distortion. In this case loading ACDB leaves CF unaffected. Vertical downwards unit loading on EGIHF will produce a tension of $\sqrt{2}$ in CF, representing the constant shear produced in this panel by loading to its right. Horizontal loading will leave CF unloaded since EGIHF does not displace horizontally at all.

Influence lines by calculation

Influence line diagrams such as Figure 3.16(b), (c), (d) have the advantage of generality and relate to any direction of the unit load applied to any node of the truss. In many cases, however, the direction and points of application of the load may be more restricted, a common situation being vertical loading applied to one or other of the booms of the truss. In such a case, the influence diagram reduces to a set of inter-connected lines for a statically determinate truss and the construction of the diagram by calculation is relatively straightforward.

Example 3.5 Truss influence lines by calculation

As an example, the influence lines for the forces in members DF, EF and EG of the Warren truss shown in Figure 3.17(a) will be established by calculation, it being assumed that loading is applied vertically to the lower boom. The influence line is readily derived from an appropriate section, it being most convenient to use whichever section does not include the unit load for a given value of the load

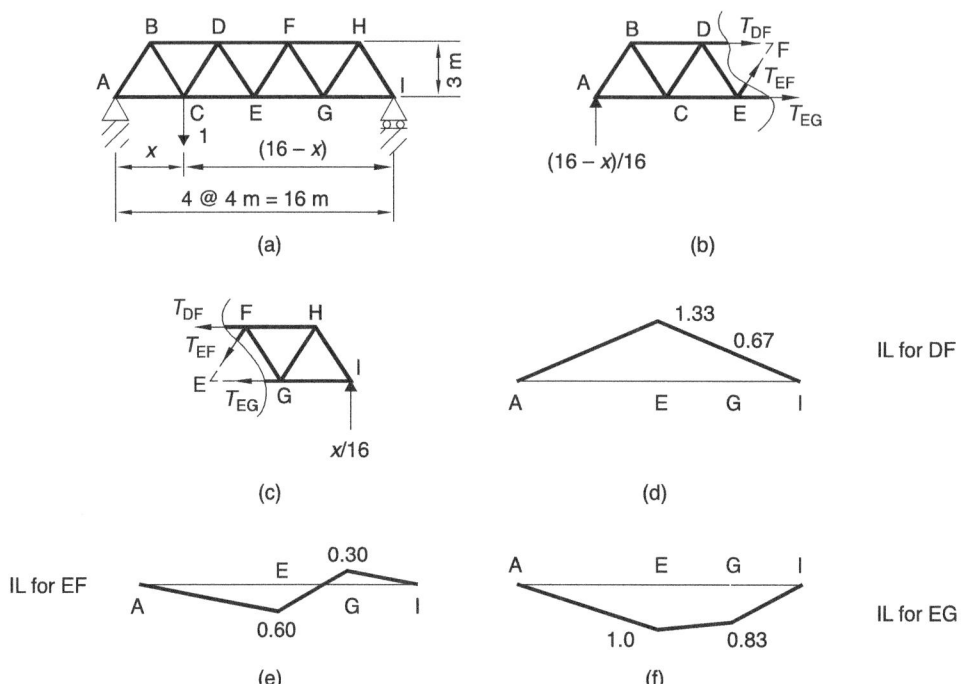

Figure 3.17   (a) Example truss, (b) left-hand section, (c) right-hand section, (d) IL for DF, (e) IL for EF, (f) IL for EG

position, x. Thus, for $x \leq 8$ m, the section of Figure 3.17(c) is used and gives:

By moments about E

$$(3 \times T_{DF}) + \left(8 \times \frac{x}{16}\right) = 0 \quad \text{hence} \quad T_{DF} = -\frac{x}{6}$$

By resolving vertically

$$T_{EF} \times \cos\left(\tan^{-1}\left(\frac{2}{3}\right)\right) = \frac{x}{16} \quad \text{hence} \quad T_{EF} = 0.075x$$

By moments about F

$$3 \times T_{EG} = 6 \times \frac{x}{16} \quad \text{hence} \quad T_{EG} = \frac{x}{8}$$

For $x \geq 12$ m the section of Figure 3.17(b) is relevant and gives:

By moments about E

$$(3 \times T_{DF}) + \left(8 \times \frac{(16-x)}{16}\right) = 0 \quad \text{hence} \quad T_{DF} = -\frac{(16-x)}{6}$$

By resolving vertically

$$T_{EF} \times \cos\left(\tan^{-1}\left(\frac{2}{3}\right)\right) + \frac{(16-x)}{16} = 0 \quad \text{hence} \quad T_{EF} = -0.075(16-x)$$

By moments about F

$$3 \times T_{EG} = 10 \times \frac{(16 - x)}{16} \quad \text{hence} \quad T_{EG} = 10 \times \frac{(16 - x)}{48}$$

The AE portions of the influence lines shown in Figure 3.17(d), (e), (f) are therefore obtained from the first set of three equations above and the GI portions come from the second three equations. The linear nature of influence lines for statically determinate structures may then be invoked to connect these portions over the sectioned length EG. In drawing the influence lines, it must be recalled that the sign convention is that if the value is offset in the direction of the load, then positive value is indicated. With the assumed downwards direction of the unit load, this means that tension (positive) values are plotted **below** AI. The general shape of the influence lines may be checked by a consideration of the vertical displacement that would be produced along AI due to cutting the three members in turn and applying unit displacements across the cuts.

Example 3.6 Computer-derived influence lines for redundant truss

Influence line sketches and calculation of spot values are most useful for validating the results of computer-generated influence diagrams and lines. The use of an automated approach is often convenient for statically determinate trusses and essential to cope with the additional complexities introduced by redundancy. As an illustration, the truss of Example 3.4 (Figure 3.16) will be reconsidered with the addition of counter-bracing (Figure 3.18(a)), so that three redundancies are introduced. Following the general approach for computer generation of influence lines, it is necessary to apply loads to the member of interest that are such that, if applied to the member alone, a unit extension would be developed. Since the axial load in a pin-jointed truss member is constant, the loads need not be applied at a 'point', but can more conveniently be applied at the member's end nodes.

It will be assumed that all the members of the truss are similar and are such that $E = 200 \, \text{kN/mm}^2 = 200 \times 10^6 \, \text{kN/m}^2$; $A = 1000 \, \text{mm}^2 = 1 \times 10^{-3} \, \text{m}^2$. The same

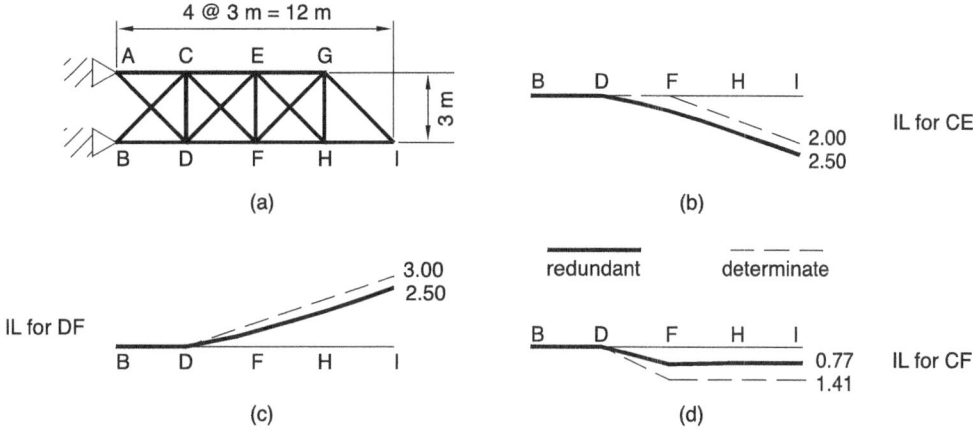

Figure 3.18   (a) Indeterminate truss, (b) IL for CE, (c) IL for DF, (d) IL for CF

three members will be considered as in Example 3.4, namely CE, DF and CF. The force required to produce a unit (1 mm) extension of the members may be obtained by taking $u_a = 0$ and $u_b = 1 \times 10^{-3}$ m. Thus, in the case of CF, for example

$$
\begin{Bmatrix} N_C \\ N_F \end{Bmatrix} = \frac{EA}{L} \begin{bmatrix} 1 & -1 \\ -1 & 1 \end{bmatrix} \begin{Bmatrix} u_C \\ u_F \end{Bmatrix}
$$

$$
= \frac{200 \times 10^6 \times 10^{-3}}{3 \times \sqrt{2}} \begin{bmatrix} 1 & -1 \\ -1 & 1 \end{bmatrix} \begin{Bmatrix} 0 \\ 10^{-3} \end{Bmatrix} = \begin{Bmatrix} -47.14 \\ 47.14 \end{Bmatrix} \text{kN}
$$

The nodal loads to be applied at C and F may then be applied by resolution or, equivalently and more formally, by noting that the inclination angle of member CF, $\alpha = -45°$. Hence

$$
\begin{Bmatrix} f_{xC} \\ f_{zC} \\ f_{xF} \\ f_{zF} \end{Bmatrix} = \begin{bmatrix} \cos\alpha & 0 \\ \sin\alpha & 0 \\ 0 & \cos\alpha \\ 0 & \sin\alpha \end{bmatrix} \begin{Bmatrix} -47.14 \\ 47.14 \end{Bmatrix} = \begin{Bmatrix} -33.33 \\ 33.33 \\ 33.33 \\ -33.33 \end{Bmatrix} \text{kN}
$$

Similarly, for both CE and DF ($\alpha = 0$), it may be shown that

$$
\begin{Bmatrix} f_{xa} \\ f_{za} \\ f_{xb} \\ f_{zb} \end{Bmatrix} = \begin{bmatrix} \cos\alpha & 0 \\ \sin\alpha & 0 \\ 0 & \cos\alpha \\ 0 & \sin\alpha \end{bmatrix} \begin{Bmatrix} -66.67 \\ 66.67 \end{Bmatrix} = \begin{Bmatrix} -66.67 \\ 0 \\ 66.67 \\ 0 \end{Bmatrix} \text{kN}
$$

Applying the above nodal loads as three separate loading cases produces the influence diagrams as the deflected shapes of the truss. In Figure 3.18(b), (c), (d), influence lines (actually curves in the case of indeterminate structures) for the three members are shown for the case of vertical loading applied to the lower boom. The lines are obtained from the vertical displacement values (in mm) at nodes B–I, produced by the relevant analyses. The previous results from the determinate truss are also shown in the figure and it may be noted that the introduction of the cross-bracing approximately halves the force in the diagonal brace CF, since vertical shear is now shared between the two counter-braces. The improved load dispersion achieved by the counter-bracing results in the bending action due to vertical load being equally shared between CE and DF, as evidenced by their similar influence lines.

Worksheet 3.4 *Influence line sketching and calculation*

W3.4.1 Sketch influence diagrams for the members X, Y, Z of the plane trusses shown in Figure W3.4.1. Numerical values are not required. Hence, state if the members would be in compression, tension or unloaded under the action of (a) a downward vertical load at point A, and (b) a horizontal load at point A acting to the right.

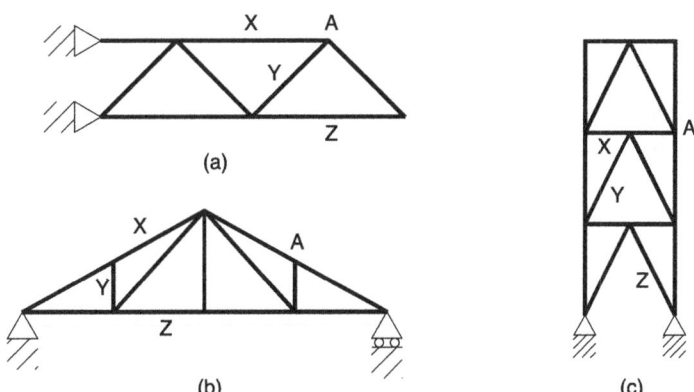

Figure W3.4.1   Worksheet 3.4.1

W3.4.2 Construct influence lines for members X, Y and Z of the truss shown in Figure W3.4.2(a) under bottom boom vertical loading. State principal values on the diagrams. Hence, calculate the maximum tension and compression that can occur in the members due to the load train shown in Figure W3.4.2(b) if either load may lead.

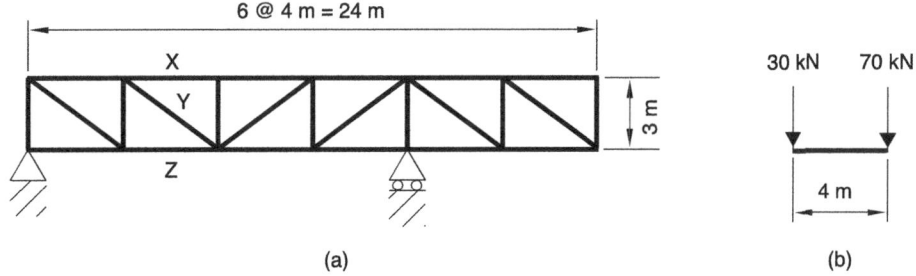

Figure W3.4.2   Worksheet 3.4.2

W3.4.3 The members of the bridge truss shown in Figure W3.4.3 have $E = 200 \, \text{kN/mm}^2$ and cross-sectional areas: $A_c = 50 \times 10^3 \, \text{mm}^2$ (chords); $A_v = 25 \times 10^3 \, \text{mm}^2$ (verticals); $A_d = 10 \times 10^3 \, \text{mm}^2$ (diagonals). Use a suitable computer package to run analyses of the truss from which influence lines may be obtained for the diagonal bracing members x, y and z under bottom boom vertical loading.

The loading is applied to the lower chord panel points. The dead load consists of loads of 150 kN at each panel point. The live load comprises 75 kN at panel points (any number of which may be loaded) plus a single load of 50 kN, which may be applied to any panel point. If the diagonal bracing members are to be designed to resist tension only, would any of the panels containing members x, y and z need to be cross-braced? Are any other panels likely to need cross-bracing?

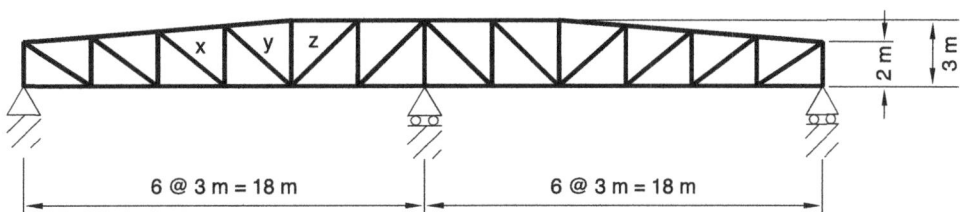

Figure W3.4.3   Worksheet 3.4.3

3.6   CASE STUDY

C3.1.1   State the necessary conditions for a structure to act as a pure truss, in which the members are subject to only axial loading with no bending effects. Discuss the extent to which the nineteenth century roof truss shown in Figure C3.1.1 fulfils the conditions. Construct a diagram showing the nature of the loads carried by the truss members under the loading shown in Figure C3.1.2. Why have different cross-sectional shapes been chosen for the various members of the truss?

Figure C3.1.1   Nineteenth-century roof truss

Use a computer package to analyse the truss under the action of the loading shown in Figure C3.1.2. The simply supported conditions indicated in Figure C3.1.2 may not be achieved in practice since the truss is built into masonry supporting walls. Repeat the analysis, therefore, assuming that pinned supports exist at either side.

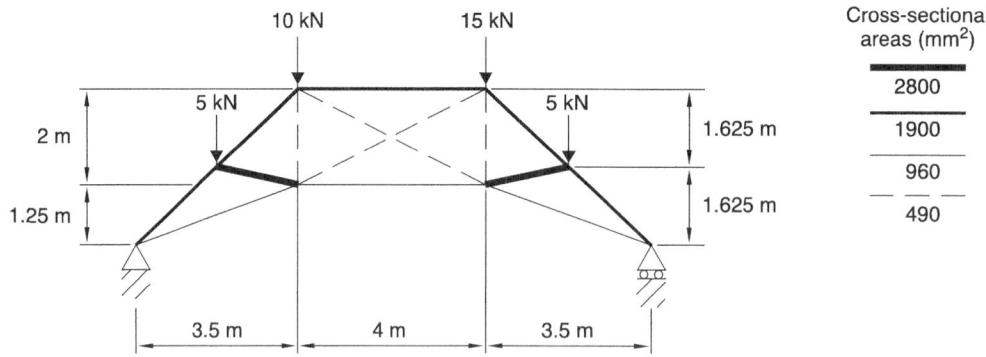

Figure C3.1.2   Case study 3.1.1

Compare the results of the two analyses, explaining significant differences. Can member loads be predicted with any great accuracy for this structure if it is presumed that the loads and dimensions given in Figure C3.1.2 are exact?

3.7   REFERENCES

Kennedy, J. B. and Madugula, M. K. S. (1990) *Elastic Analysis of Structures*. New York: Harper and Row – Chapter 4 of this reference covers truss analysis. The stability and determinacy of trusses are also treated in detail in its Chapter 3.

Steel Construction Institute (1995) *Modelling of Steel Structures for Computer Analysis*. Ascot: Steel Construction Institute – general information on computer analyses and detailed advice on modelling a variety of practical details.

4 Arches

4 Arches

4.1 INTRODUCTION

The arch is perhaps best known from historical masonry structures. The fit between the arch and masonry arises due to the match between a structural form that, when suitably designed, sustains load principally through compressive forces and a material that has significant compressive strength but very limited tensile (or, consequently, bending) resistance. The arch may be regarded as a bowed truss from which the bracing has been removed and the top chord made continuous. If the main tie is retained, then the arch is said to be *tied* (Figure 4.1(a)). Alternatively, should suitable foundations be available, horizontal restraint can be provided by the supports (Figure 4.1(b)). In either case, it is the provision of the horizontal restraining force that is all-important to the development of arch action. A curved shape alone is **not** sufficient. The structure of Figure 4.1(c), for example, will act primarily as a beam and resist applied loads mainly by bending and shear. In fact, by statics, the bending moments (*BM*) shown on the right-hand half of Figure 4.1(c) will be exactly the same as those in the counterpart horizontal beam, although the diagram has become curved due to a convention being followed that diagrams are plotted perpendicularly to the line of the structure.

The importance of horizontal support was brought home with particular force to Charles Gordon in 1863 when he ordered that one of the arches of the 53-span Baodai Bridge in China be removed in order that his ship could pass. He subsequently reported that '*I am sorry to say that twenty-six of the arches fell in yesterday like a pack of cards, killing two men, ten others escaped by running as the arches fell one after another as fast as a man could run*' (Knapp, 1993).

If complete horizontal restraint is provided (Figure 4.1(b)), the horizontal component of reaction developed at the supports will resist the bending effects but produce an increased compressive axial load (*AL*) in the arch, as indicated on the left-hand sides of the figures (Figure 4.1(b), (c)). A similar effect is produced by the introduction of the tie (Figure 4.1(a)), but to a somewhat reduced degree since the flexibility of the tie will prevent full horizontal restraint being achieved.

Pin-jointed trusses can resist vertical loads purely by compression if the shape of the bending moment diagram matches the shape of the structure. The shape required for any given set of loads can also be obtained by considering the form a cable would take up under the particular loading system. This follows because

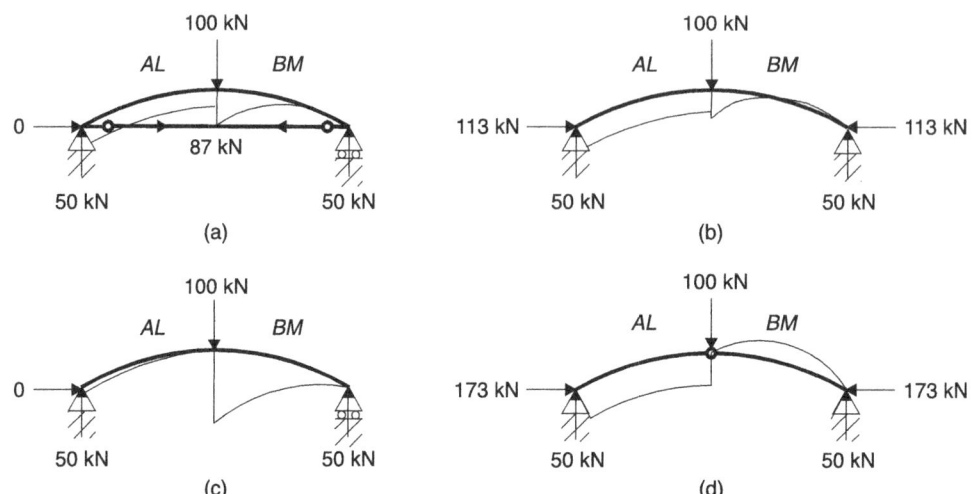

Figure 4.1   (a) Tied, (b) pinned arch, (c) curved beam, (d) three-pinned arch

a pure cable is unable to resist bending to any significant degree and must therefore take up a geometry that allows it to resist loads by tension alone. Reversing the shape the cable would take up will give an arch that is capable of resisting the loads under pure compression. Such a shape is sometimes known as a *funicular* arch. Since loads are more efficiently carried by direct forces than bending moments, it is advantageous to use an arch shape that is funicular for the majority of the loading. A bridge may be subjected to loading which is, in the main, uniformly distributed horizontally, for example. A parabolic arch shape would therefore be indicated. However, an isolated heavy vehicle will produce point axle loads, corresponding to a non-parabolic bending moment diagram. Non-funicular loading results in the arch needing transverse resistance. This is provided by the bracing forces in a bowed truss or by transverse shear forces (and corresponding bending moments) in the case of an arch.

4.2   LINE OF THRUST

In many cases, the form of the bending moment distribution in an arch can be conveniently obtained by use of a *line of thrust* concept. The line of thrust is the locus of the points of application of the resultant load acting on the arch, considering loads to only one side. Presuming it is possible to locate such a line, the bending moment diagram under vertical loading readily follows. Considering the typical line of thrust shown in Figure 4.2, then, for horizontal equilibrium under only vertical loading, the resultant thrust, *P*, must have a constant horizontal component equal to the horizontal reaction component, *H*. Thus, using the positive bending moment convention shown in the figure

$$BM_X = -H \times d \tag{4.1}$$

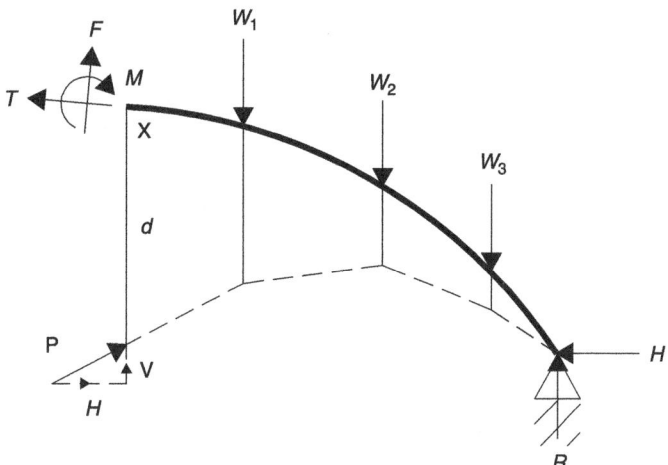

Figure 4.2 Line of thrust/bending moment relationship

Since H is constant, it follows that the bending moment distribution is provided by the vertical distances between the arch and the line of thrust, but that negative (hogging) moments will be plotted underneath the arch, rather than above, as required by the tension-side plotting convention adopted here. The offset distances therefore need to be reversed to match the convention. It also follows that a funicular arch is one for which the arch shape is defined by the line of thrust, so that there are no offsets and hence no bending moments.

Lines of thrust are particularly easy to establish if the arch is statically determinate, which generally implies a three-pinned arch of the form shown in Figure 4.1(d). The statical determinacy of arches is normally readily established from the support arrangements. The simple supports of Figure 4.1(c), for example, provide a determinate structure. Adding a further reaction component (Figure 4.1(b)) or tie force (Figure 4.1(a)) makes a two-pinned or tied arch, respectively, both of which are singly redundant. Introducing a pin release into the two-pinned arch eliminates the redundancy and makes the three-pinned arch determinate (Figure 4.1(d)).

To obtain the line of thrust for the three pinned-truss of Figure 4.1(d), it is simply necessary to note that the line must pass through the central pin and the supports, since the zero bending moments at these points imply zero offsets for the thrust line. These conditions are sufficient to define the line in this case (Figure 4.3(a)). Reversing the offsets and plotting normally to the arch then provides the bending moment diagram (Figure 4.1(d)), from which it may be observed that the central load causes the arch segments to bulge outwards and be subjected to hogging moments. In contrast, the tied arch sags under the central load (Figure 4.1(a)). In line of thrust terms, this represents a much steeper line of thrust (Figure 4.3(b)) due to the lower horizontal restraint force (87 kN instead of 173 kN). Finally, the two-pinned arch (Figure 4.3(c)) has an intermediate slope of thrust line that results in a central, sagging region with hogging moments towards the supports.

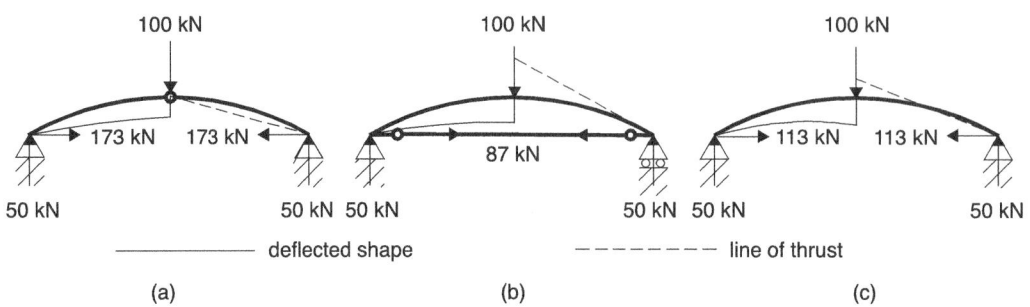

Figure 4.3   Lines of thrust for (a) three-pinned, (b) tied, (c) two-pinned arches

Determining reaction components manually, and hence establishing a precise position for a line of thrust, is involved for the redundant tied and two-pinned forms. In these cases, sketching lines of thrust is probably most valuable for validating computer analyses. For the determinate three-pinned arch, however, statics allows lines of thrust to be relatively straightforwardly derived, either graphically (Prakash Rao, 1997) or by calculation.

In sketching lines of thrust it is helpful to keep the following properties in mind:

1. *On unloaded sections of arches the line of thrust is linear.*
2. *A point load causes a 'kink' in a line of thrust due to the change in direction of the resultant force.*
3. *A uniformly distributed load causes a curved (parabolic) variation in a line of thrust due to the continuous change in the direction of the line of action of the resultant force.*
4. *The line of thrust must pass through any points of zero bending moment such as pinned supports or internal hinges.*

Example 4.1   Thrust and bending moment diagrams for three-pinned arch

The three-pinned arch shown in Figure 4.4(a) comprises two circular arcs which have a common horizontal tangent at the internal hinge. Sketches of the line of thrust and bending moment diagrams are required. First, it should be noted that application of a downward load on one arc of a three-pinned arch will cause the loaded arc to sag and the unloaded arc to bulge outwards and therefore to hog. In the present case the load on the right-hand arc (75 kN total) is substantially greater than the load on the left-hand arc (25 kN). The left-hand arc is therefore likely to hog and the right-hand arc to sag. In this case, the line of thrust will commence underneath the arch at the left-hand support, so as to represent a hogging moment, as shown to an exaggerated vertical scale in Figure 4.4(b). The point load will alter the slope of the line, which must then pass through the internal pin position. The uniformly distributed load will then cause a parabolic line of thrust, which must eventually return to the right-hand pin support. The vertical ordinates between the line of thrust and the arch are proportional to the bending moments. When plotted conventionally on the tension side and normally to the arch (Figure 4.4(c)), it may be noted that the shape of the bending moment diagram

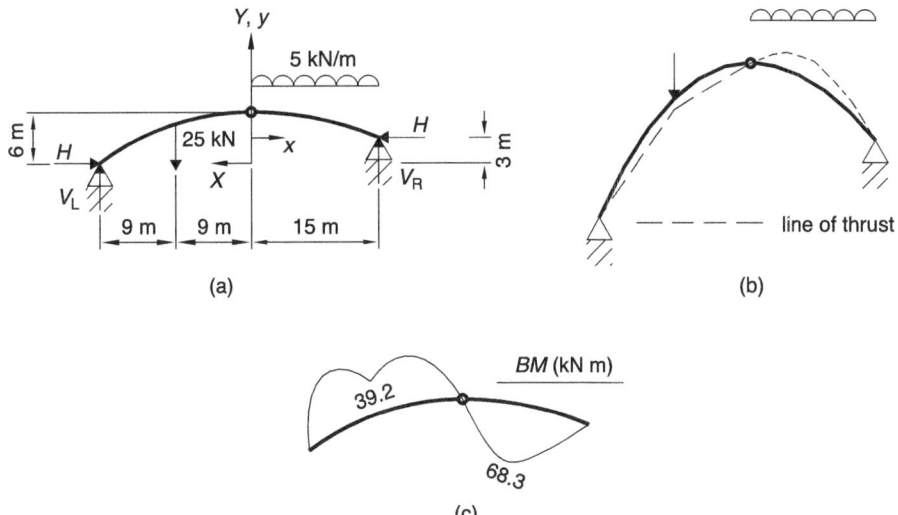

Figure 4.4   (a) Three-pinned arch, (b) line of thrust, (c) bending moment diagram

inevitably changes. To verify the form of the diagram, particularly the left-hand arc hogging/right-hand arc sagging presumption, a spot bending moment in each arch segment will be calculated. To do this, it is first necessary to determine the reaction components. Hence, by moments about the left-hand support

$$33V_R + 3H = (25 \times 9) + ((5 \times 15) \times 25.5) \tag{4.2}$$

Using the zero bending moment condition at the internal pin

$$15V_R - 3H = (5 \times 15) \times 7.5 \tag{4.3}$$

Solving equations (4.2) and (4.3) simultaneously by adding gives

$$V_R = 56.25\,\text{kN} \quad \text{and} \quad H = 93.75\,\text{kN}$$

The component V_L could now be obtained by vertical equilibrium, but, since all subsequent calculations depend vitally on the reaction components, it is better to retain vertical equilibrium as a check. So, taking moments about the right-hand support

$$33V_L - 3H = (25 \times 24) + ((5 \times 15) \times 7.5)$$

Using the known value for H gives $V_L = 43.75\,\text{kN}$ and the vertical equilibrium check is

$$56.25 + 43.75 = 25 + 75 = 100$$

The equations of the two circular arcs related to the axes X, Y and x, y (Figure 4.4(a)) are (in m units)

$$X^2 + (Y + 24)^2 = 30^2 \quad \text{and} \quad x^2 + (y + 36)^2 = 39^2$$

Bending moments will be calculated under the point load ($X = 9\,\mathrm{m}$) and at $x = 7.5\,\mathrm{m}$. When $X = 9\,\mathrm{m}$, $Y = 4.618\,\mathrm{m}$, so that, from Figure 4.4(a)

$$BM_{X=9} = (V_\mathrm{L} \times 9) - (H \times 4.618)$$
$$= (43.75 \times 9) - (93.75 \times 4.618) = -39.2\,\mathrm{kN\,m} \qquad (4.4)$$

Also, when $x = 7.5\,\mathrm{m}$, $y = 2.272\,\mathrm{m}$, so that

$$BM_{x=7.5} = (V_\mathrm{R} \times 7.5) - (H \times 2.272) - ((5 \times 7.5) \times 7.5/2)$$
$$= (56.25 \times 7.5) - (93.75 \times 2.272) - ((5 \times 7.5) \times 7.5/2)$$
$$= 68.3\,\mathrm{kN\,m} \qquad (4.5)$$

Equations (4.4) and (4.5) verify the nature of the bending moments presumed in the two arcs of the arch, and can also be used to validate a computer derivation of the full bending moment diagram (Figure 4.4(c)).

Worksheet 4.1   Lines of thrust and bending moment diagrams for arches

W4.1.1   The arches shown in Figure W4.1.1(a), (b), (c) have the same parabolic profile such that $24y = x^2$. The 100 kN and 25 kN loads are applied to the arches at $x = -6\,\mathrm{m}$ and $x = 6\,\mathrm{m}$, respectively. Sketch probable line of thrust diagrams, and hence bending moment diagrams, for the arches. Numerical values are not required.

Validate the sketches for the arch of Figure W4.1.1(c) by calculating the values of bending moment under the point loads.

In Figure W4.1.1(d), the point loads are in the same horizontal positions as before. Determine the heights, H and h, if the bending moments at the points of application of the loads are to be zero. Why is the resulting linear segment arch of a *funicular* form and why is its outline the line of thrust diagram for Figure W4.1.1(c)?

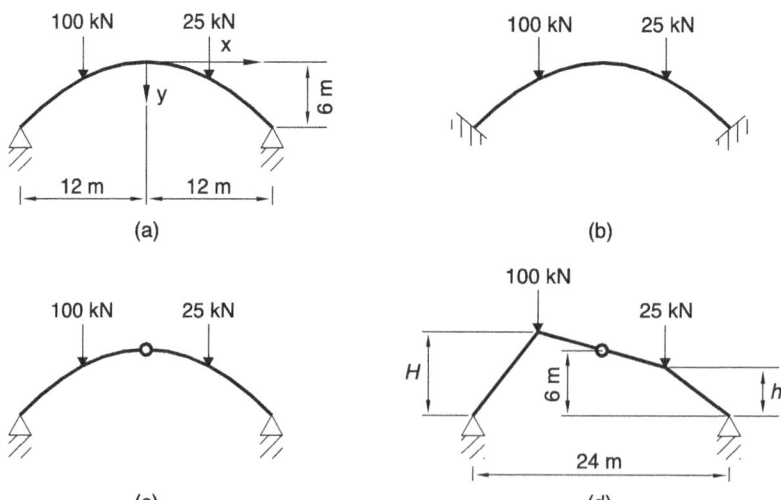

Figure W4.1.1   Worksheet 4.1.1

W4.1.2 The three-pinned arch shown in Figure W4.1.2(a) comprises a parabolic segment and a linear segment. Sketch probable line of thrust diagrams, and hence bending moment diagrams, when the arch is subjected to loads (i) as shown in Figure W4.1.2(a) and (ii) as shown in Figure W4.1.2(b). Numerical values are not required.

If the parabolic segment is tied by a rod, which is rigid in tension, but incapable of resisting compression, as shown in Figure W4.1.2(b), how would the lines of thrust and bending moment diagrams for the two loading cases alter?

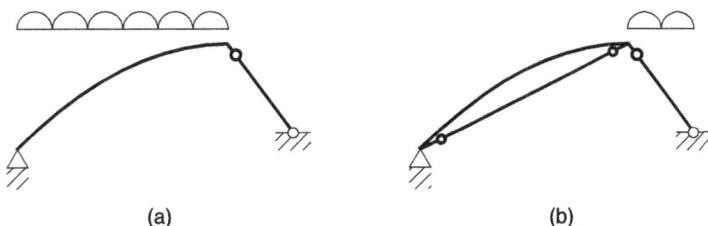

(a)   (b)

Figure W4.1.2   Worksheet 4.1.2

4.3   COMPUTER ANALYSIS

Stiffness method applied to plane frames

The combination of axial (direct) force and bending effects present in arch action means that a plane frame formulation of the stiffness method is required. The formulation is essentially a combination of the truss (providing the axial effects) and the pure beam (providing the bending effects) formulations. The general principles of the stiffness method also apply, so that it is only necessary to redefine appropriately the member stiffness matrix related to the local axes, $\mathbf{K}_m$, and the transformation matrix, $\mathbf{T}$.

For a plane frame there are three displacement components at each node (u, w, θ) and three stress resultants at each end of the member (N, F, M) (Figure 4.5(a)). The stiffness matrix in local axes, which connects these sets of quantities, may be obtained by combining the relevant truss and beam stiffness

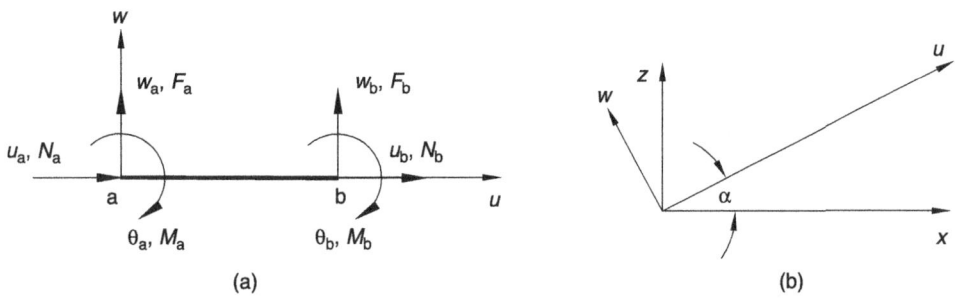

(a)   (b)

Figure 4.5   (a) Plane frame member in local axes, (b) local and global axes

matrices to give

$$
p = \begin{Bmatrix} N_a \\ F_a \\ M_a \\ N_b \\ F_b \\ M_b \end{Bmatrix} = E \begin{bmatrix} A/L & 0 & 0 & -A/L & 0 & 0 \\ 0 & 12I/L^3 & -6I/L^2 & 0 & -12I/L^3 & -6I/L^2 \\ 0 & -6I/L^2 & 4I/L & 0 & 6I/L^2 & 2I/L \\ -A/L & 0 & 0 & A/L & 0 & 0 \\ 0 & -12I/L^3 & 6I/L^2 & 0 & 12I/L^3 & 6I/L^2 \\ 0 & -6I/L^2 & 2I/L & 0 & 6I/L^2 & 4I/L \end{bmatrix}
$$

$$
\times \begin{Bmatrix} u_a \\ w_a \\ \theta_a \\ u_b \\ w_b \\ \theta_b \end{Bmatrix} = \mathbf{K}_m e \tag{4.6}
$$

It may also be shown (Kennedy and Madugula, 1990) that the transformation matrix, which relates quantities in local axes to their equivalents in the global axes, is given (Figure 4.5(b)) by

$$
\mathbf{T} = \begin{bmatrix} \mathbf{T}_0 & \mathbf{0} \\ \mathbf{0} & \mathbf{T}_0 \end{bmatrix} \quad \text{where} \quad \mathbf{T}_0 = \begin{bmatrix} \cos\alpha & \sin\alpha & 0 \\ -\sin\alpha & \cos\alpha & 0 \\ 0 & 0 & 1 \end{bmatrix} \tag{4.7}
$$

Global/local axes and member properties

Even more care than with beams is required if the sign conventions adopted by computer packages for member forces are to be correctly interpreted. The forces are normally expressed in relation to the local member axes and the relevant conventions must be consulted, particularly in relation to tabulated output. Shear force conventions typically give most problems and results are usually easiest to interpret if it is arranged that the member u axis is directed in the positive x direction (or positive z direction in the case of a vertical member).

From equation (4.6), the member properties needed will be the modulus of elasticity (E), cross-sectional area (A) and second moment of area (I). Many arch members are, of course, curved and will need to be represented by a number of straight sub-members, if the package does not provide for curved members. A typical recommendation (Steel Construction Institute, 1995) is that one straight member should be provided for every $15°$ change in direction. This may be taken to be a minimum and it is generally advisable, and no real inconvenience, to exceed the recommendation.

Member loading

Since the local member axes are no longer parallel to the global axes, as is the case for beams, it becomes important to distinguish between member loads, which are more appropriately specified in respect of the local axes, and those loads for which global axes are preferable. Generally the z-axis is arranged vertically, so that gravity loads are conveniently specified in the global (negative z) direction. However, pressure loads (notably wind) act normally to members and are often most easily specified in the local w-axis (Figure 4.5), unless the member happens to be parallel to a global axis.

Most computer packages will allow member loading to be specified in either global or local directions, but different conventions exist for the treatment of the intensity of distributed loads. Some packages require the intensity of distributed loads to be specified per unit length of member, but in other cases per unit length normal to the direction of loading is used. Care is therefore needed to establish the system adopted by any particular package and load intensities may need to be suitably converted before input. If the package adopts a load per unit length of member, w_m, convention, for example, and the load is specified per unit length horizontally, w_h, then, for the small element of an inclined member shown in Figure 4.6, equating the load applied gives

$$w_h \times \mathrm{d}x = w_m \times \mathrm{d}L \quad \text{or} \quad w_m = w_h \times \frac{\mathrm{d}x}{\mathrm{d}L} = w_h \times \cos \alpha \qquad (4.8)$$

Restraints, releases, symmetry and mechanisms

Arch supports are normally fully-fixed, (x, z, θ), or pins, (x, z). Three-pinned arches will need an internal member pin (moment release) to be incorporated and packages will usually incorporate facilities for this. Such a release should not be used as a support, in conjunction with a pin or roller, since these restraints incorporate moment release anyway.

To take advantage of symmetry requires that displacement be restrained in a direction that is normal to the line of symmetry, but in the plane of the arch. Furthermore, rotation should be restrained about an axis normal to the line of

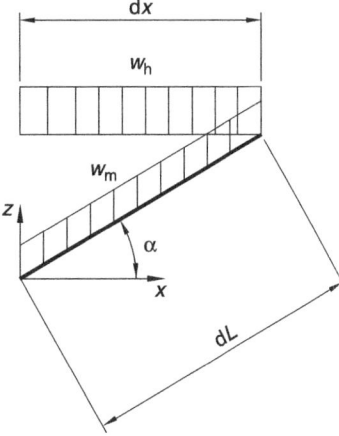

Figure 4.6   Distributed member load specification

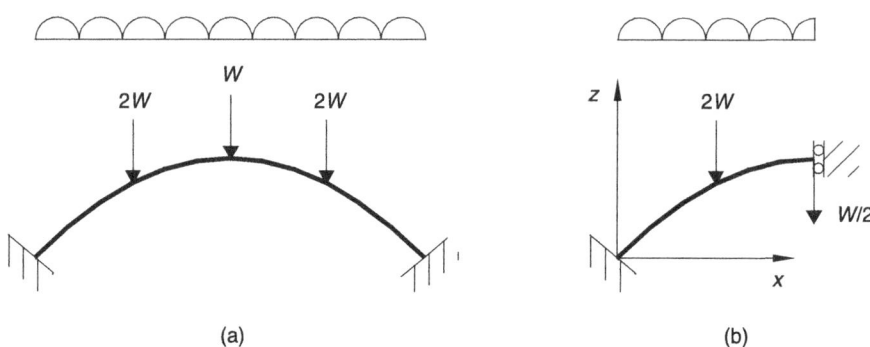

Figure 4.7   (a) Symmetrical arch, (b) half-arch model

symmetry and perpendicular to the plane of the arch. Forces applied in the direction of the line of symmetry will need to be halved. For the z-axis line of symmetry of the arch shown in Figure 4.7(a), it will therefore be necessary to introduce (x, θ) restraints at the symmetry position and to reduce the point load at this position by a half (Figure 4.7(b)).

Mechanisms are relatively unusual in arch analysis, but, if suspected, inadequate support specification or over-provision of internal releases is the most probable cause.

4.4   INFLUENCE LINES

Although it is possible to obtain influence lines for arches by graphics (Prakash Rao, 1997) or by manual calculation (Kennedy and Madugula, 1990), it is generally more convenient to employ computer analysis. Influence lines may be generated for any of the three relevant internal forces by use of Müller-Breslau's principle in conjunction with the appropriate unit displacements for axial load, shear force and bending moment influence lines, respectively. The end forces due to the application of the unit displacements may be obtained from equation (4.6), the displacement being applied to, say, end a of a nominal member. The nominal member needs to be short due to the errors involved over the length of the member and also so that arch curvature, if present, is reasonably modelled.

Example 4.2 Influence line derivation by computer analysis

As an example, the two-pinned, uniform, circular arch shown in Figure 4.8(a) will be considered. The arch has equation (in m units): $x^2 + (z + 12)^2 = 20^2$; and the following properties will be assumed: $A = 10 \times 10^{-3}\,\text{m}^2$; $I = 200 \times 10^{-6}\,\text{m}^4$; $E = 206\,\text{kN/mm}^2$. The influence line for bending moment at the point on the arch (I) defined by $x = 8\,\text{m}$ will be obtained and used to determine the portion of the arch to which a gravity loading that is uniformly distributed horizontally should be applied to produce the maximum positive and negative bending moments at I.

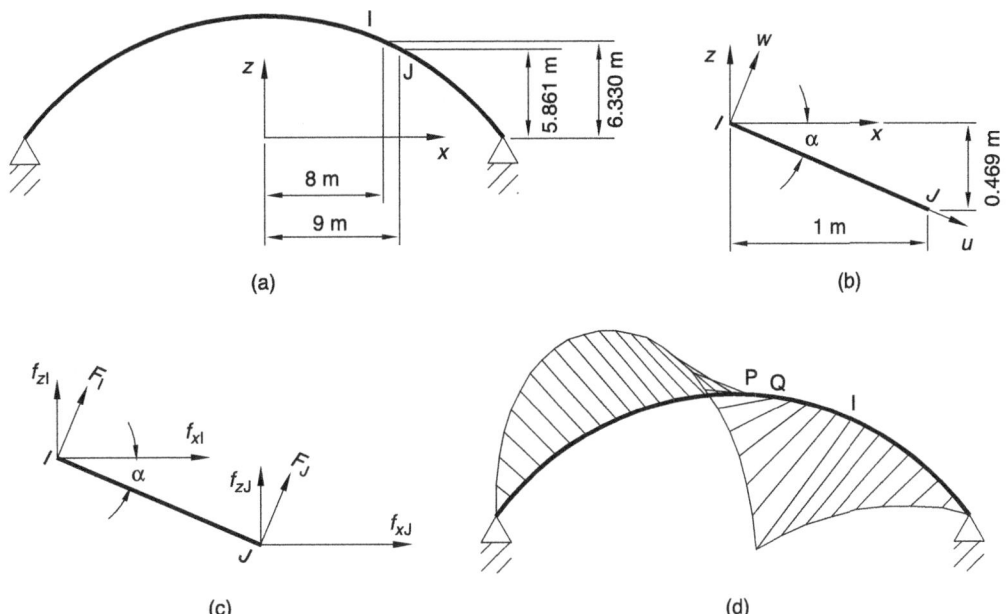

Figure 4.8 *(a) Example arch for influence line, (b) member I–J, (c) force resolution, (d) influence line for* BM$_I$

For the purposes of computer analysis, the arch is divided into 32 linear members, each of which has a unit projection on the x-axis. The member to which forces will be applied, such as to cause a unit rotation at I, will be taken as the member to the immediate right of I. If mm output units are presumed, then the bending moment influence line will be obtained from the displacements due to the introduction of a rotation, $\theta_a = 0.001$ rad, to the member. From Figure 4.8(b) the length of the member, L, is 1.105 m, and the end forces arising from the application of this rotation may be obtained from equation (4.6) as

$$\begin{Bmatrix} N_I \\ F_I \\ M_I \\ N_J \\ F_J \\ M_J \end{Bmatrix} = E \begin{Bmatrix} 0 \\ -6I/L^2 \\ 4I/L \\ 0 \\ 6I/L^2 \\ 2I/L \end{Bmatrix} \{\theta_a\} = 206 \times 10^6 \times 200 \times 10^{-6} \begin{Bmatrix} 0 \\ -6/L^2 \\ 4/L \\ 0 \\ 6/L^2 \\ 2/L \end{Bmatrix} \{0.001\}$$

$$= \begin{Bmatrix} 0 \\ -202.5 \\ 149.1 \\ 0 \\ 202.5 \\ 74.6 \end{Bmatrix} \begin{matrix} kN \\ kN \\ kN\,m \\ kN \\ kN \\ kN\,m \end{matrix}$$

These forces are related to the local axes for member I–J and, if they are to be applied as nodal loads to I and J, will need to be converted to the global axes. This may be done either by employing the transformation of equation (4.7), using $\alpha = \tan^{-1}(0.469) = 25.13°$, or by resolution of the forces (Figure 4.8(c)) as

$$f_{xI} = F_I \cos \alpha = -202.5 \cos 25.13° = -86.0 \, \text{kN}$$

$$f_{zI} = F_I \sin \alpha = -202.5 \sin 25.13° = -183.3 \, \text{kN}$$

In a similar fashion, f_{xJ} and f_{zJ} may be determined and the full set of nodal loads to be applied is then given by

$$\begin{Bmatrix} f_{xI} \\ f_{zI} \\ m_{yI} \\ f_{xJ} \\ f_{zJ} \\ m_{yJ} \end{Bmatrix} = \begin{Bmatrix} -86.0 \\ -183.3 \\ 149.1 \\ 86.0 \\ 183.3 \\ 74.6 \end{Bmatrix} \begin{matrix} \text{kN} \\ \text{kN} \\ \text{kN m} \\ \text{kN} \\ \text{kN} \\ \text{kN m} \end{matrix}$$

The deflected shape arising from a computer analysis using these nodal loads is shown in Figure 4.8(d). When interpreting such a diagram, it is important to remember that, for vertical loads, it is the vertical displacements that provide the influence line values. As indicated, a displacement diagram of the form shown in Figure 4.8(d) will incorporate horizontal as well as vertical displacement components. To produce the maximum positive bending moment at I, therefore, the loading should be applied from Q (the point at which a downwards displacement component is first observed) to the right-hand support. Similarly, to produce maximum negative bending moment at I the loading should be applied from the left-hand support to P (the point at which upwards displacement is last observed).

Worksheet 4.2 Computer analysis of arches

W4.2.1 Making use of symmetry, analyse one half of the uniform arch shown in Figure W4.2.1(a), employing 12 approximating linear members having equal projections of 1 m on the x-axis. The arch has a circular profile of equation (in m units): $x^2 + (z + 9)^2 = 15^2$. Repeat the analysis for the support arrangements shown in Figure W4.2.1(b), (c), (d). For the arch take: $A = \times 10^{-3} \, \text{m}^2$ and $I = 2 \times 10^{-3} \, \text{m}^4$. For the tie take: $A = 120 \times 10^{-6} \, \text{m}^2$. Use $E = 206 \, \text{kN/mm}^2$ for both the arch and tie members.

Compare and contrast the axial load and bending moment distributions for the analyses undertaken, giving reasons for significant differences. For the different cases, sketch lines of thrust that are compatible with the bending moment diagrams obtained. Validate the analyses of Figure W4.2.1(c), (d) by calculating the bending moment under an off-centre 100 kN load. State, with reasons, the differences made to the analyses if the I value for the arch is taken to be $I = 20 \times 10^{-3}\,\mathrm{m}^4$, all other properties remaining as before.

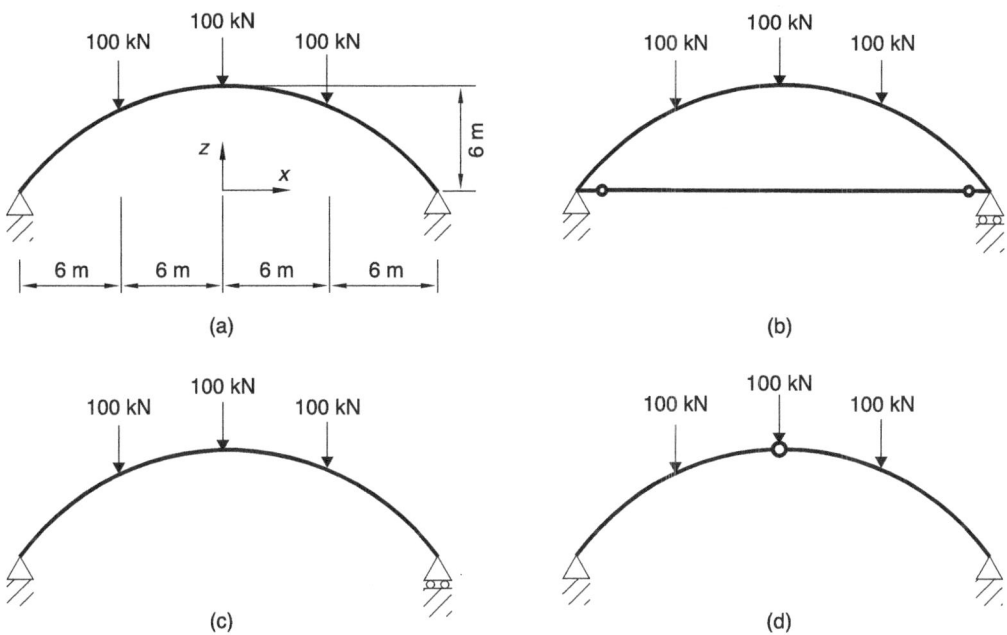

Figure W4.2.1   Worksheet 4.2.1

W4.2.2 The three-pinned, uniform, parabolic arch shown in Figure W4.2.2 has equation (in m units): $40z = x^2$; and the following properties can be assumed: $A = 200 \times 10^{-3}\,\mathrm{m}^2$; $I = 20 \times 10^{-3}\,\mathrm{m}^4$; $E = 25\,\mathrm{kN/mm}^2$. Employ a computer analysis based on 20 approximating linear members, each having a projection of 2 m on the x-axis to obtain the influence line for bending moment at the point on the arch (I) defined by $x = 10\,\mathrm{m}$. Use the influence line to determine the positions of vertically downward loads comprising a point load of 120 kN and a uniformly distributed load (any length) of 20 kN/(horizontal)m, which will produce the maximum positive and negative bending moments at I. Analyse the arch for the loads in the critical positions to obtain the maximal moments at I.

Check the computer analysis by manually determining the bending moments at I for the critical positions of the loads.

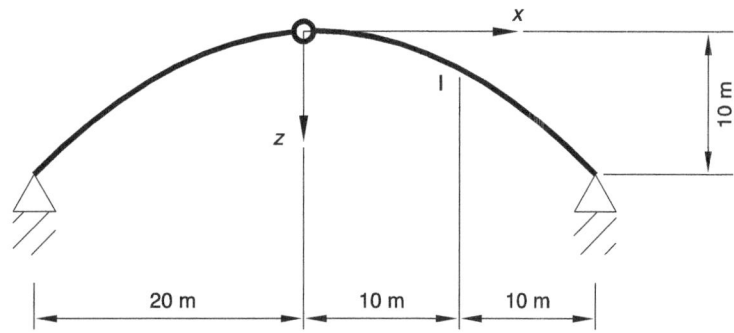

Figure W4.2.2   Worksheet 4.2.2

4.5   CASE STUDIES

Figure C4.1.1   Broadgate House

C4.1.1   Consult the paper 'Broadgate Exchange House: structural systems' by H. Iyengar *et al.* (*The Structural Engineer*, Vol. 71, No. 9, May 1993, pp. 149–159). From data provided in the paper, and other sources as necessary, make reasonable estimates of the dimension of an exterior arch, the dead and live service (unfactored) gravity loads acting on the arch, the sizes of the arch members and the nature of its supports and joints (see Figure C4.1.1).

Use a plane frame computer package to analyse the displacements and forces in the arch on the assumptions:

 (i) that neither intermediate nor diagonal ties are provided
 (ii) that intermediate ties are provided but diagonal ties are not
 (iii) that diagonal ties are provided but intermediate ties are not
 (iv) that both intermediate and diagonal ties are provided.

For each of the four analyses, the following three load cases should be considered:

 (i) dead and live loads acting on each of the column/tie
 (ii) dead load plus live load acting on the central four column/ties only
 (iii) dead load plus live load acting on the seven column/ties at the left-hand side only.

Compare the various analyses in respect of significant displacements and forces. State, with reasons, which of the four designs analysed would be the structurally preferable solution.

Sketch designs based on the use of a 10-storey cross-braced truss and a ten-storey catenary support system. Why is the load flow in these designs less 'direct' than in the arch solution?

Why is it generally considered desirable to minimise bending moments in arches? Why was a piece-wise linear parabolic shape chosen for the Broadgate Exchange arch? In what circumstances would it be appropriate to choose:

(i) a parabolic curve
(ii) a catenary curve
(iii) a triangle with a central apex for the shape of an arch?

C4.1.2 Select a roof arch for analysis from the paper 'Waterloo International Rail Terminal trainshed roof structure' by A. J. Hunt *et al.* (*The Structural Engineer*, Vol. 72, No. 8, 19 April 1994, pp. 123–130). Estimate the dimensions of the chosen arch, the nodal loading corresponding to a 2.5 kN/m^2 (on plan) inclusive ultimate (factored) roof loading, the sizes of the arch members and the nature of the supports and joints (see Figure C4.1.2).

By idealising the major and minor trusses as arch segments, perform a 'three-pinned' arch hand analysis of the roof structure to produce axial load, shear force and bending moment diagrams from which the maximum forces in the members of the major and minor trusses should be estimated.

Employing a two-dimensional idealisation of the structure, use a plane truss computer package to analyse the displacements and forces in the roof members, assuming that all connections are pinned. Re-analyse the roof, using a plane frame package, assuming all member connections are rigid.

In respect of significant displacements and forces, compare the various analyses both to each other and to the expected structural behaviour of the roof. Is the form of the structure appropriate to the internal forces it has to resist?

Determine the degree of statical indeterminacy, q, for the pinned and rigidly connected structures. Is the computer package used affected by the value of q? Do either the displacements or the internal forces produced by either of the computer analyses depend on the values of the member sizes that are assumed? Would it matter if the roof were designed

Figure C4.1.2   Waterloo International Rail Terminal trainshed roof

on the basis of the pinned connection analysis, even though the joints are actually rigidly welded?

Why was a three-dimensional arch form adopted in practice? Would you expect results from a three-dimensional analysis to vary greatly from the two-dimensional solution?

4.6   REFERENCES

Kennedy, J. B. and Madugula, M. K. S. (1990) *Elastic Analysis of Structures.* New York: Harper and Row – Chapter 12 of this reference gives transformation matrix details in connection with the stiffness method and its Chapter 14 deals with influence lines, including those for arches.

Knapp, R. G. (1993) *Chinese Bridges.* Hong Kong: Oxford University Press – Chapter 2 of this reference includes a description of the Baodai and other prominent Chinese arch bridges.

Prakash Rao, D. S. (1997) *Graphical Methods in Structural Analysis.* Hyderabad: Universities Press (India) – Chapter 6 of this reference deals with the application of graphical methods to arch analysis, including the construction of pressure (lines of thrust) lines. Its Chapter 8 deals with influence lines, including those for arches.

Steel Construction Institute (1995) *Modelling of Steel Structures for Computer Analysis.* Ascot: Steel Construction Institute.

5 Plane frames

5 Plane frames

5.1 INTRODUCTION

Plane frames are two-dimensional structures loaded in their own plane such that the members are subjected to axial load, bending moments and shear forces. Arches are a particular form of plane frame, which, if appropriately designed, maximise axial load action at the expense of bending action. Single-bay plane frame equivalents of the three-pinned arch are shown in Figure 5.1(a), (b), together with the lines of thrust corresponding to a load that is uniformly distributed horizontally across the spans. The enhanced divergence of the frames from the line of thrust, as compared with a curved arch form, indicates that the frames will bend more. Plane frames are therefore commonly characterised by bending deformation, axial deformations normally being secondary.

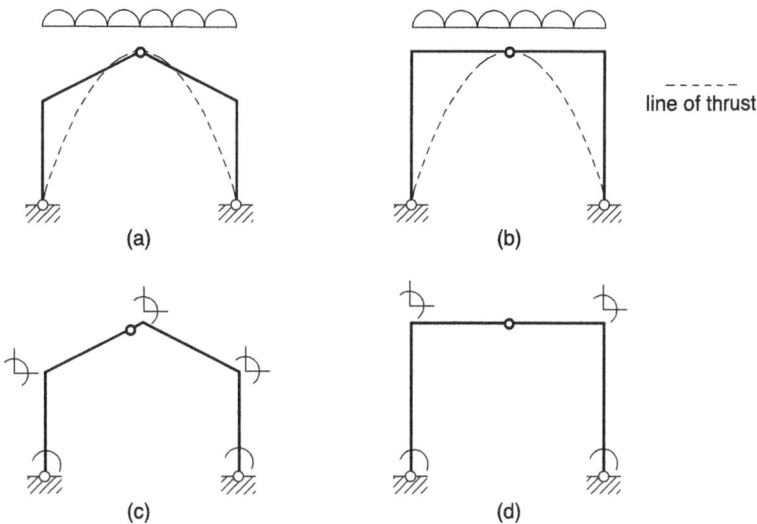

Figure 5.1   Three-pinned frames: (a), (c) pitched; (b), (d) flat

5.2 DETERMINACY OF PLANE FRAMES

Determinacy is not as vital an issue for plane frames as for trusses and beams, since there are only a few common determinate forms, such as inclined beams and the three-pinned arrangement shown in Figure 5.1. It is occasionally useful to establish the degree of indeterminacy of a frame, q, and this may be achieved in the usual fashion by finding the number of unknown stress resultants that may not be determined by statics. Each plane frame member will, by definition, sustain three stress resultants (axial load, bending moment and shear force). If there are h internal hinges at which the bending moment is known (zero value), then the number of unknown stress resultants for m members will be

$$\text{number of unknown stress resultants} = (3m - h) \qquad (5.1)$$

There will be three displacement degrees of freedom at each joint of the frame, corresponding to two translation components and a rotation. If there are d restrained displacement components, the total number of displacement degrees of freedom, and hence equations of joint equilibrium, at j joints will represent the degree of kinematic indeterminacy, k, and will be given by

$$\text{degree of kinematic indeterminacy } k = 3j - d \qquad (5.2)$$

$$\text{degree of statically indeterminacy (redundancy) } q = (3m - h) - k \qquad (5.3)$$

The determinacy of the three-pinned arrangement shown in Figure 5.1(c) may therefore be checked by noting that $j = 5$; $d = 4$; $m = 4$; and $h = 1$ giving $k = (3 \times 5) - 4 = 11$; $q = (3 \times 4) - 1 - 11 = 0$. It should be noted that joints are not taken at internal hinge positions, since these are internal to members. To ensure that the apex hinge (Figure 5.1(a)) is internal to a member, the structure has been remodelled as shown in Figure 5.1(c). In the case of the flat system (Figure 5.1(d)), the rafter may be directly modelled as a single member giving $k = (3 \times 4) - 4 = 8$; $q = (3 \times 3) - 1 - 8 = 0$. Alternatively, the degrees of kinematic indeterminacy may be obtained by inspection if the degrees of displacement freedom at the nodes are shown diagrammatically, as illustrated in Figure 5.1(c), (d). Summing the indicated degrees of freedom provides check results of $k = 11$ and $k = 8$ for the two frames.

Example 5.1 Degree of indeterminacy of single-storey frame

For the more general single storey frame shown in Figure 5.2(a), the kinematic indeterminacy may again be found by inspection if the degrees of freedom indicated on the figure are summed. Alternatively, $j = 6$; $d = 3 + 1 + 2 = 6$ so that $k = (3 \times 6) - 6 = 12$. There are five members so it follows that the statical indeterminacy $q = (3 \times 5) - 12 = 3$.

The degree of redundancy may also be determined by inspection if sufficient releases are introduced into the structure for it to become determinate. This is probably most easily achieved by initially making all the supports fully fixed and so increasing the redundancy by the number of extra restraints introduced. Thus, comparing the 'new' frame of Figure 5.2(b) with the 'old' frame of Figure 5.2(a):

$$q = q_{\text{new}} - 3 \qquad (5.4)$$

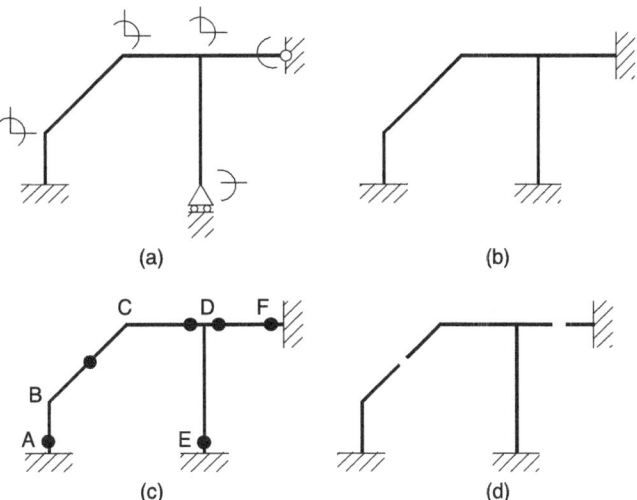

Figure 5.2   Single-storey frame indeterminacy example

The redundancy of the new frame may be found in one of two ways. Member EDF may be converted into a two-member truss by inserting three hinges (releases) and ABCD, then converted into a three-pinned frame by inserting a further three hinges (Figure 5.2(c)), giving a total of six introduced releases. The structure is now statically determinate, since both the two-member truss and the three-pinned frame are known to be determinate. It follows that $q_{new} = 6$ and, from equation (5.4), $q = q_{new} - 3 = 6 - 3 = 3$. Alternatively, complete 'cuts' may be made in sufficient members of the frame so that the frame is converted into a number of 'tree-type' cantilevers, as shown in Figure 5.2(d). At each cut three releases have been made since axial load, shear force and bending moment have all been set to zero. It again follows, therefore, that $q = q_{new} - 3 = (2 \times 3) - 3 = 3$.

Example 5.2   Degree of indeterminacy of multi-storey frame

The kinematic indeterminacy of the multi-storey frame shown in Figure 5.3(a) may be found to be 29, either by summing the displacement degrees of freedom given on the figure or as $k = (3 \times 13) - 10 = 29$, since $j = 13$ and $d = 10$. The degree of statical indeterminacy then follows as $q = (3 \times 15 - 2) - 29 = 14$, since $m = 15$ and $h = 2$. This result may be checked by creating a 'new', fully-rigid frame (Figure 5.3(b)), which involves the introduction of four extra redundancies, one each at the internal hinges and one at each of the pinned supports. Thus

$$q = q_{new} - 4 \tag{5.5}$$

To make the rigid version determinate by introducing hinge releases, the three storeys of the first bay are turned into three-pinned frames (Figure 5.3(c)) and the remaining bays are turned into two-member truss systems. This needs 18

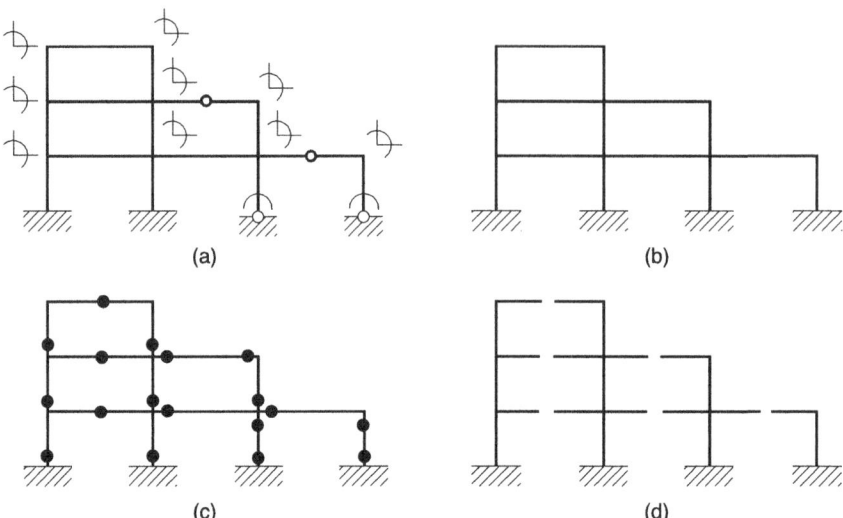

Figure 5.3   Multi-storey frame indeterminacy example

hinges in total, so that $q_{new} = 18$ and $q = 14$. Alternatively, the 'tree' system can be used (Figure 5.3(d)). This requires 6 cuts (18 releases) for determinacy, hence confirming the hinge solution.

Worksheet 5.1   Kinematic and statical determinacy of plane frames

W5.1.1   Calculate the degrees of kinematic and statical indeterminacy for the plane frames shown in Figure W5.1.1. State, also, whether the structures are statically determinate or indeterminate. If statically indeterminate, indicate a system of releases that will make the structure determinate and stable.

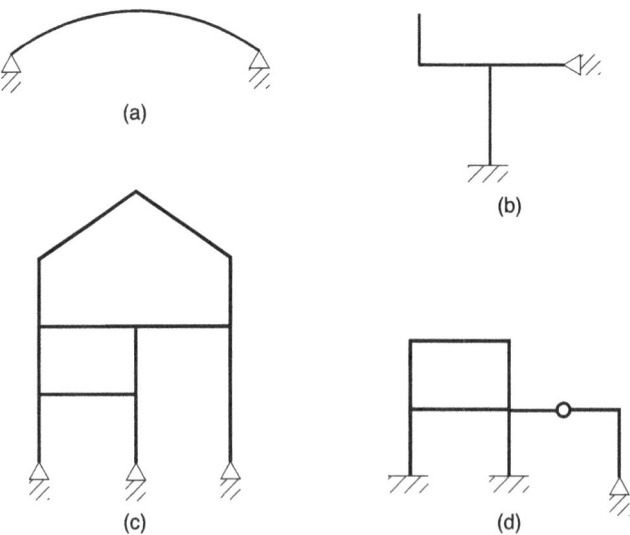

Figure W5.1.1   Worksheet 5.1.1

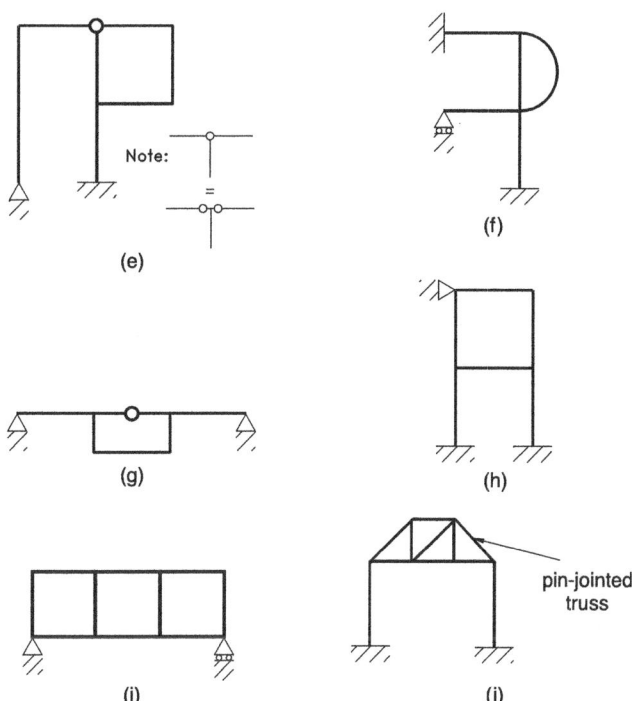

Figure W5.1.1 continued

5.3 STATICALLY DETERMINATE PLANE FRAMES

Axial load, shear force, bending moment and deflection diagrams

In constructing stress resultant and deflection diagrams for statically determinate plane frames, the general considerations used in conjunction with beam diagrams will still be relevant. The principal additional complexity arises from the varying orientation of the members, particularly sloping ones, and the effects of this are illustrated in the following examples.

Example 5.3 Stress resultant and deflected shape diagrams for a staircase

Stress resultant diagrams and a deflected shape diagram will be constructed for the staircase and landing shown in Figure 5.4(a) under self-weight loading, which is taken to be uniformly distributed and of intensity 6 kN/m. This type of structure is often categorised as an inclined beam, but it illustrates features that are common to more general plane frames. It may be noted that the loading, being gravitational, is purely vertical and therefore the horizontal component of reaction at the pinned base H is zero. The vertical reactions may be calculated in the usual manner by taking moments about each support in turn, thus

$$\text{total load on inclined stairs} = 6 \times \sqrt{6^2 + 4^2} = 43.3\,\text{kN}$$

$$\text{total load on horizontal landing} = 6 \times 2 = 12\,\text{kN}$$

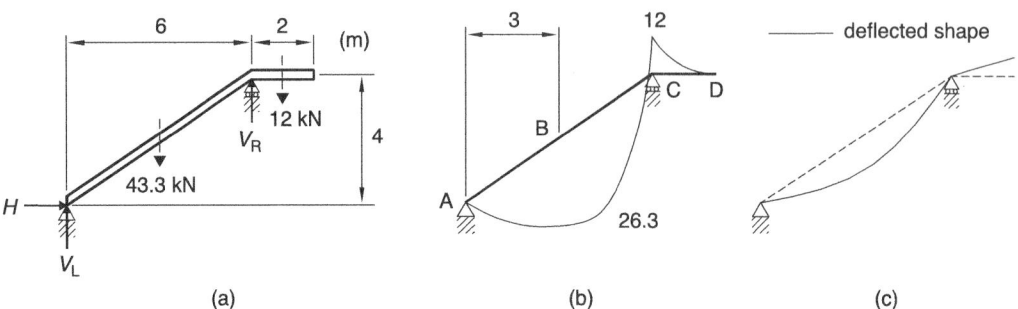

Figure 5.4   Staircase example: (a) arrangement, (b) BM diagram, (c) deflected shape

Hence, taking moments about the lower support

$$6V_R = (43.3 \times 3) + (12 \times 7) \text{ whence } V_R = 35.6 \, \text{kN}$$

Similarly, taking moments about the higher support

$$6V_L = (43.3 \times 3) - (12 \times 1) \text{ whence } V_L = 19.6 \, \text{kN}$$

The vertical equilibrium check then gives

$$43.3 + 12 = 55.3 \approx 35.6 + 19.6$$

In general, the structure can be expected to hog over the upper support and sag along the inclined stairs. To check this, the bending moments at B and C (Figure 5.4(b)) may be evaluated as

$$BM_B = ((19.6 \times 3) - (43.3/2)) \times 1.5 = 26.3 \, \text{kN m}$$

$$BM_C = -12 \times 1 = -12 \, \text{kN m}$$

This provides the bending moment diagram shown in Figure 5.4(b) and from this the deflected shape shown in Figure 5.4(c) may be sketched, making sure that sagging and hogging curvatures correspond to the relevant regions of the bending moment diagram. There could well be some uncertainty regarding whether the landing moves upwards or downwards and, in general, it is not possible to establish this type of effect without calculation. In the present case, however, the sagging of the stairs, due to the longer span, is certain to be more significant than the hogging of the landing. Thus, the rotation at the upper support caused by the sagging of the stairs will create the upward displacement of the landing.

   To sketch axial load and shear force diagrams it is essential to appreciate that these quantities act along and normal to the axis of any individual member. Members should therefore be considered individually and the forces acting on the frame resolved in the relevant axial and normal directions. If the stairs are treated first, the components of the applied loading and the reactions are as shown in Figure 5.5(a). It follows, either from the sign convention of Figure 1.26 or by inspection, that the reaction component along the axis at the lower support causes a compression of 10.9 kN. This initial compression is then progressively

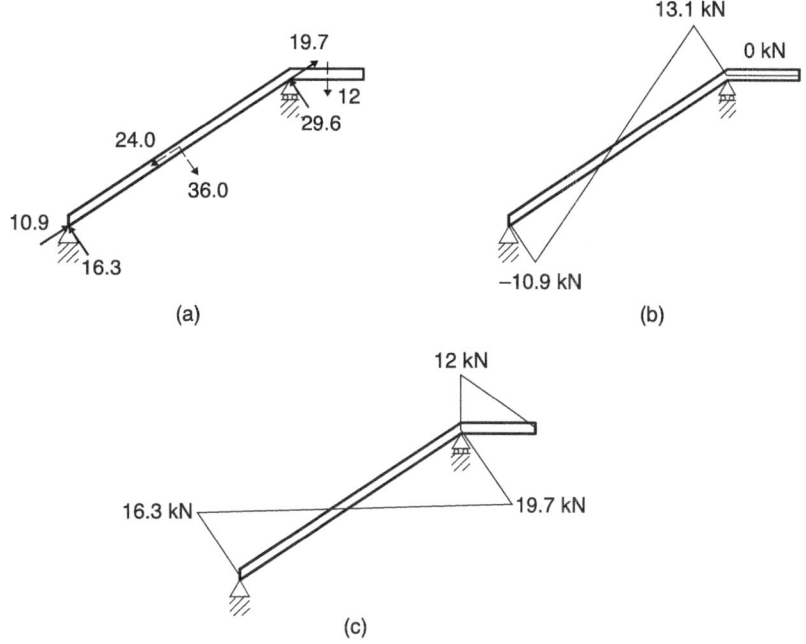

Figure 5.5   Staircase example: (a) resolved force components (kN), (b) axial load diagram (kN), (c) shear force diagram (kN)

reduced as progress is made up the stairs due to the axial component of the self-weight, which acts in the opposite direction and has a value of 24 kN. At the top of the stairs, therefore, the axial load has become tensile and of magnitude $24 - 10.9 = 13.1$ kN (Figure 5.5(b)). As with arches, these values are plotted normal to the direction of the stairs and such that negative values are below. The shear force diagram for the stairs (Figure 5.5(c)) may be readily obtained from the normal components of loading and reaction in exactly the manner that would be used for a horizontal beam. On reaching the landing, loading and reaction components are needed in the horizontal and vertical directions, since these are now along and normal to the element. The original layout diagram (Figure 5.4(a)) provides these load components. Since there is no horizontal loading, it follows, perhaps surprisingly, that there is no axial load in the landing. In respect of shear force, the cantilever form of the landing can be used to deduce the result shown in Figure 5.5(c).

From the stress resultant diagrams of Figure 5.5(b), (c), it should be noted that axial load and shear force are not, in general, continuous in value or sign at the joints of a plane frame. Indeed, the values vary very considerably at the upper support position in this case. This arises because force equilibrium requires that forces balance in a **specified** direction. The axial loads and shear forces in the stairs and landing relate to **different** directions and therefore cannot be expected to match where the members meet. On the other hand, bending moments, for a plane structure, are measured about an axis perpendicular to the plane and this is independent of the orientation of the members. Moment equilibrium will therefore require that the bending moment at a two-member joint such as the top of the stairs be continuous. Thus, the value of the moment will not change. It will also be of the same sign and will therefore appear on the same side of the frame (Figure 5.4(b)).

Worksheet 5.2 Stress resultant diagrams for sloping beams

W5.2.1 Sketch stress resultant (axial load, shear force and bending moment) diagrams for the sloping balcony beams shown in Figure W5.2.1, indicating principal values. Sketch, also, the expected displaced shape of the beams.

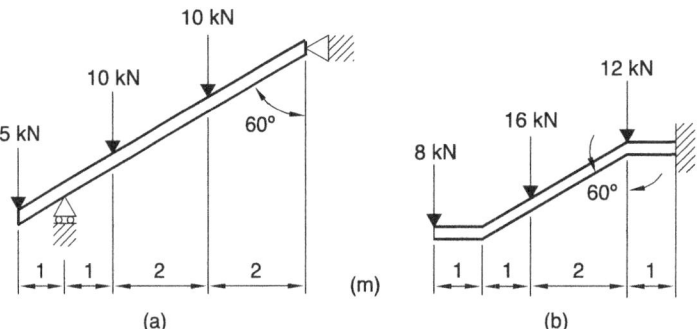

Figure W5.2.1   Worksheet 5.2.1

Example 5.4 Stress resultant and displaced shape diagrams for a three-pinned frame

As a slightly more complex example of a statically determinate plane frame, the stress resultant and displaced shape diagrams for the three-pinned frame shown in Figure 5.6(a) will be developed. The vertical reaction components for the frame are obtained, as usual, from moment equilibrium as

$$8V_R = ((4 \times 30) \times 6) + (2.5 \times 6) + (5 \times 3) \rightarrow V_R = 93.75\,\text{kN}$$

$$8V_L = ((4 \times 30) \times 2) - (2.5 \times 6) - (5 \times 3) \rightarrow V_L = 26.25\,\text{kN}$$

Clearly, the two components sum to 120 kN and satisfy vertical equilibrium with the total downwards load. The horizontal components of reaction may be obtained from the zero bending moment condition at the pin. This may be applied by considering moments to either side of the pin, but it is prudent to apply the condition to both sides and then to check for horizontal equilibrium. Proceeding in this way

$$6H_R - 4V_R + (120 \times 2) - (5 \times 3) = 0 \rightarrow H_R = 25.0\,\text{kN}$$

$$6H_L - 4V_L = 0 \rightarrow H_L = 17.5\,\text{kN}$$

The horizontal components of reaction are therefore in equilibrium with the resultant, horizontally applied load of 7.5 kN to the right, and may now be used to evaluate stress resultant diagrams with some confidence. For a frame, it is necessary to adopt a 'positive side' for each member to display the signs of axial loads and shear forces (bending moments will be placed on the tension side as before). Generally it is convenient to take the inside of the frame as the negative side and this is indicated by the dotted line in Figure 5.6(a). From the figure, the

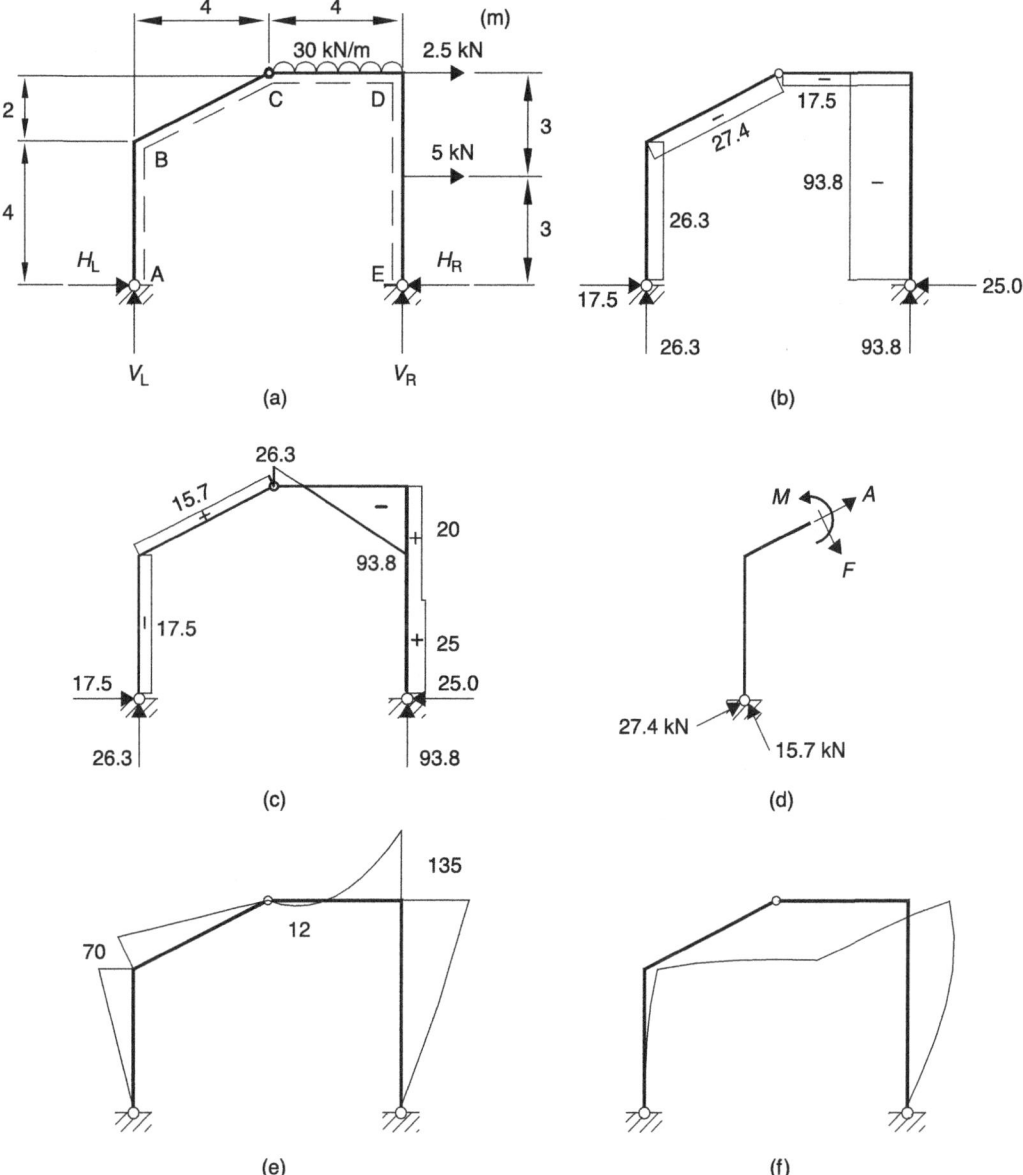

Figure 5.6   Example three-pinned frame: (a) general arrangement, (b) axial load diagram (kN), (c) shear force diagram (kN), (d) section along BC, (e) bending moment diagram (kNm), (f) deflected shape

axial load and shear force in **AB** are clearly given by V_L and H_L, respectively, and are shown in Figure 5.6(b), (c). For the sloping member **BC**, it is necessary to consider the section shown in Figure 5.6(d) and to resolve the left-hand reactions into components parallel and normal to **BC**. Since the slope of **BC** to the horizontal $\theta = \tan^{-1}(2/4) = 26.6°$, these components are

  parallel component $= 17.5 \cos \theta + 26.25 \sin \theta = 27.4\,\text{kN}$

  normal component $= -17.5 \sin \theta + 26.25 \cos \theta = 15.7\,\text{kN}$

From the section (Figure 5.6(d)) it follows that $A = -27.4\,\text{kN}$ and $F = 15.7\,\text{kN}$, and these values are shown on the appropriate stress resultant diagrams (Figure 5.6(b), (c)). The axial load diagram is completed by noting, from Figure 5.6(a), that the resultant load 'to the left' acting parallel to CD is H_L and that the resultant load 'to the right' acting parallel to DE is V_R. The shear force diagram may be completed by similarly noting that the resultant load normal to CD 'to the left' is V_L at C, and will then linearly decrease by 30 kN/m as CD is traversed. For DE, the resultant normal force to the left is $(H_L + 2.5)$ just past D and this is increased by 5 kN at the intermediate point load. The bending moment diagram is fixed in outline by establishing the values at C and D. Thus, using the same sign convention as for beams

bending moment at $\text{B} = -H_L \times 4 = -70\,\text{kN m}$

bending moment at $\text{D} = (-H_R \times 6) + (5 \times 3) = -135\,\text{kN m}$

The diagram will be linear along all the members apart from the uniformly loaded CD so that joining up the values at B and D with the known zero values at the hinges will complete the majority of the diagram. As with beams, the diagram will be on the tension side of the frame if positive values are plotted on the 'underneath' (dashes in Figure 5.6(a)) side. Along CD the shear force diagram indicates a maximum bending moment part way along. The position and magnitude of this moment may be determined as follows:

since the shear force decreases by 30 kN/m, position of maximum

moment from $\text{C} = 26.3/30 = 0.877\,\text{m}$

maximum bending moment

$\quad\quad$ = area under shear force diagram from zero moment position (C)

$\quad\quad = 0.5 \times 0.877 \times 26.3 = 12\,\text{kN m}$

The completed bending moment diagram therefore takes the form shown in Figure 5.6(e). The reasonableness of this bending moment distribution can be confirmed if it can be shown to correspond to an acceptable deflected shape. In this respect, it is first noted that the horizontal loads will ensure that the frame moves to the right overall. Axial deformations are normally negligible in frames, so that B and D will not be able to move vertically due to the axial stiffness of the column members AB and DE. Positions B and D therefore only move horizontally to the right. Since the bending moments are almost completely outside the frame it follows that each member is convex outwards, so that tension is created on the outside of the structure. Finally, using the rotation discontinuity that is possible at the internal hinge and the requirement that the original angles be preserved at the rigid joints B, D, the displacement diagram may be sketched as shown in Figure 5.6(f).

Worksheet 5.3 Stress resultant diagrams for statically determinate plane frames

W5.3.1 Calculate the horizontal and vertical components of reaction at the supports of the three-pinned frame shown in Figure W5.3.1(a). Hence sketch axial load, shear force and bending moment diagrams for the frame, indicating principal values. Why would it be possible to model the frame as shown in Figure W5.3.1(b)?

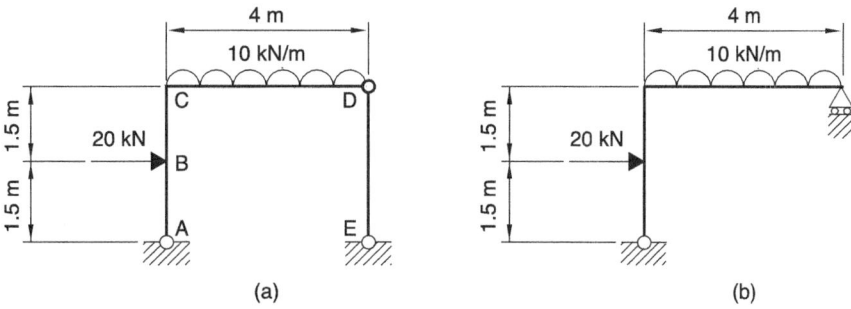

Figure W5.3.1   Worksheet 5.3.1

W5.3.2 The statically determinate frame shown in Figure W5.3.2 is pinned at A, C and D. Determine the reaction components at A and D and hence sketch axial load, shear force and bending moment diagrams for the frame, indicating principal values.

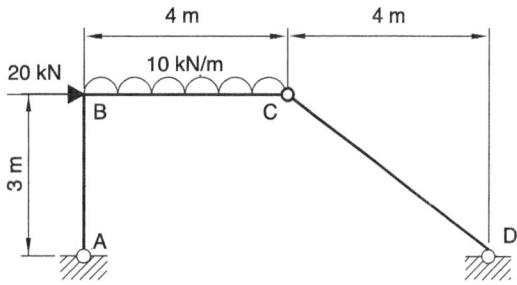

Figure W5.3.2   Worksheet 5.3.2

W5.3.3 Sketch axial load, shear force and bending moment diagrams for the frame shown in Figure W5.3.3, indicating principal values. Why does member CDE bend, whereas member CD in Figure W5.3.2 does not?

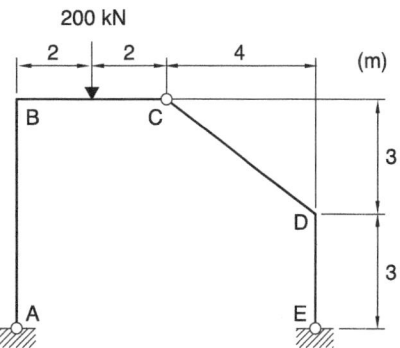

Figure W5.3.3   Worksheet 5.3.3

W5.3.4 Sketch diagrams showing the probable directions of the reactions and the displaced shapes of the statically determinate frames shown in Figure W5.3.4. Hence, or otherwise, show that the bending moment diagrams suggested in the figure are incorrect and sketch corrected figures. Relate the diagram for Figure W5.3.4(e) to the bending moment diagram for a simply supported beam of the same span. Sketch, also, probable axial load and shear force diagrams for the frames (adapted from Brohn, 1992 – originally based on a Chinese textbook).

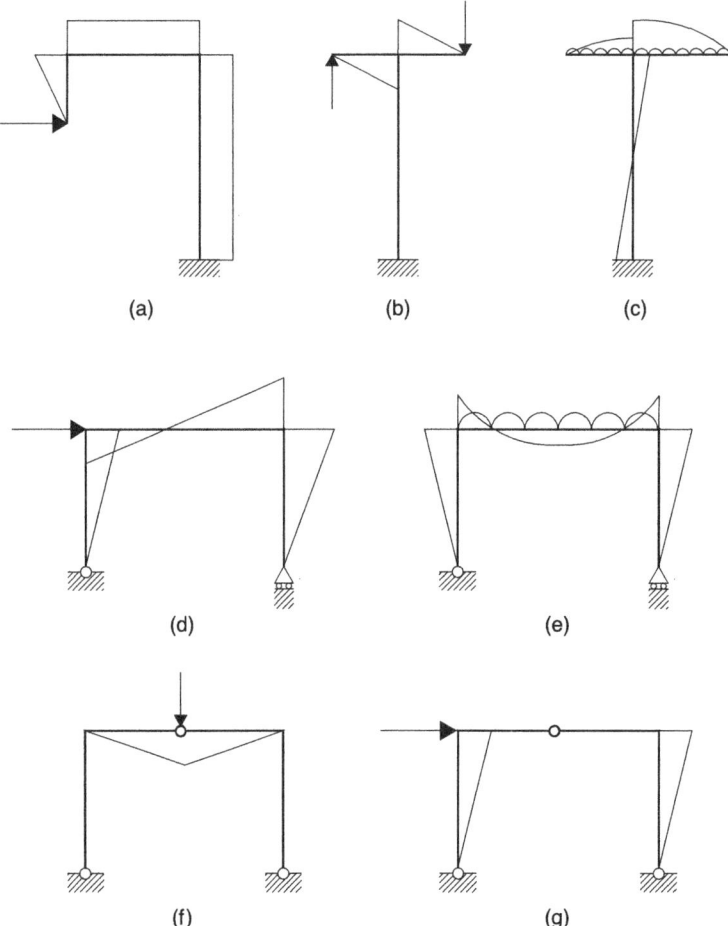

Figure W5.3.4   Worksheet 5.3.4

5.4   STATICALLY INDETERMINATE PLANE FRAMES

The quantitative analysis of statically indeterminate plane frames is laborious by hand calculation, since, by definition, they cannot be solved by equilibrium considerations alone. Computer solutions are therefore generally used, but it remains essential to be able to validate the results of such analyses, at least qualitatively. Attention will be given primarily to bending moments, since shear forces and axial loads are generally somewhat simpler to check. Many of the precepts outlined in respect of beams, arches and statically determinate plane frames are relevant to statically indeterminate frames, but it is also helpful to be able to recognise bending moment 'patterns', which occur when standard frame geometries are loaded in standard ways. The most commonly occurring situations are therefore considered next.

Single storey–single bay portal frames

Vertical loading

For a three-pinned portal frame, the line of thrust (Figure 5.7(a)) must pass through the pins to produce zero moment at these points. For the horizontal beam member, reversal of the line of thrust will produce the bending moment diagram on the tension side of the beam and the divergence of the thrust line from the frame indicates substantial bending effects at the eaves joint (Figure 5.7(d)). The bending moment diagrams for the columns may also be obtained by reversing the thrust line, although it should be noted that the bending moment in the vertical member will be given by the **vertical** component of the thrust multiplied by the **horizontal** offset, rather than the other way round, as is

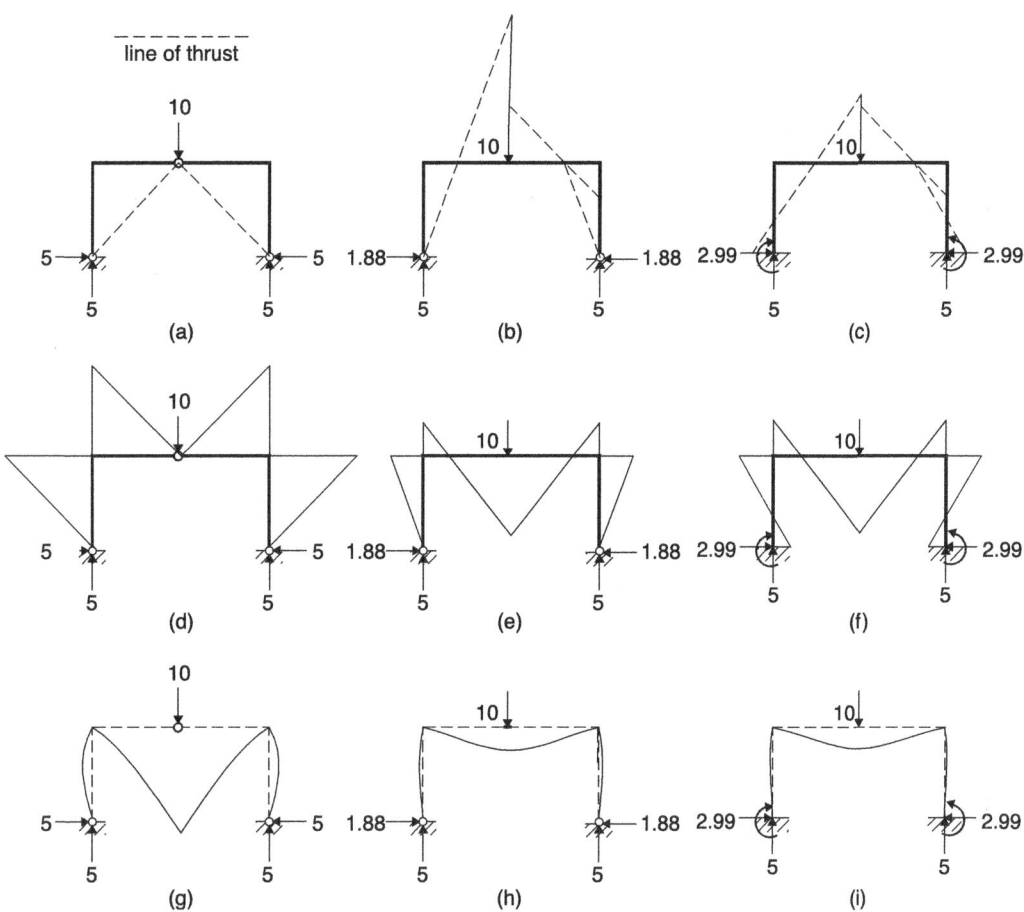

Figure 5.7 Single-storey/single bay portals: (a)–(c) thrust lines, (d)–(f) BM diagrams, (g)–(i) deflected shapes

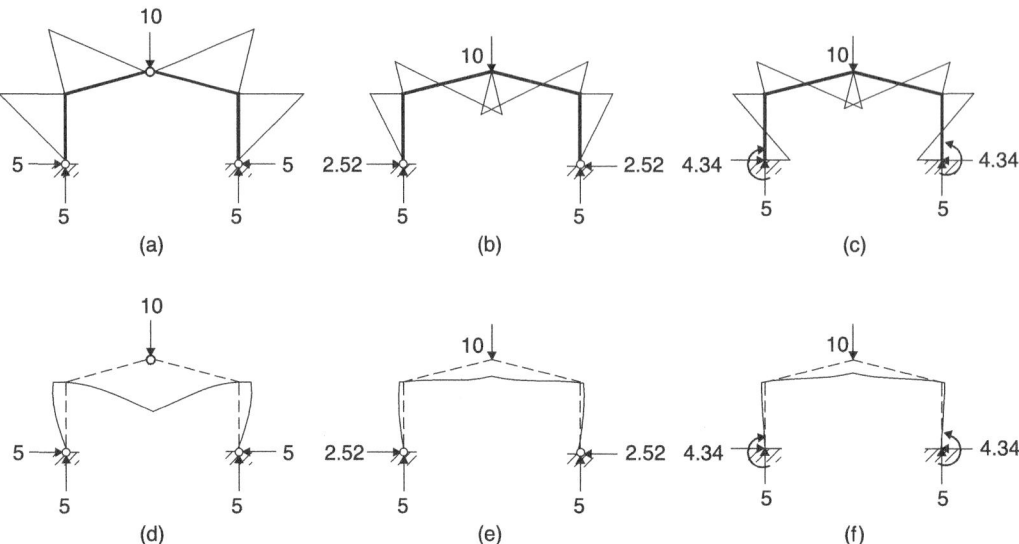

Figure 5.8   Gabled single-storey/single bay portal frames: (a)–(c) bending moment diagrams, (d)–(f) deflected shapes

the case in the beam member. For the example considered, for the equal height/ half span of the example frame, the horizontal and vertical components of the thrust are equal and the bending moment diagrams will be similarly scaled. However, if the central pin is removed (Figure 5.7(b)), the horizontal component of the thrust reduces considerably. The line of thrust will be as shown in the left-hand side of the figure, but if it is to be converted into a bending moment diagram, the vertical offsets on the horizontal beam must be scaled by the ratio of the thrust components (1.88/5). If this is done (right-hand side of the figure), and the new diagram reversed, then the bending moment diagram will result (Figure 5.7(e)). The diagram indicates substantially reduced eaves moments, but, of course, there is now a considerable bending moment at the centre of the beam. The same line of thrust argument may similarly be applied to the case of fixed feet (Figure 5.7(c)). Alternatively, the difference from the pinned base solution may be derived by considering the beam to be a member that is partially rotationally restrained (by the columns) at its ends. When the column feet are made fixed, this rotation restraint is enhanced, so that the eaves moment increases (in fact slightly) and the central sagging moment decreases correspondingly (Figure 5.7(f)).

Deflected shapes for the three frames are shown in Figure 5.7(g)–(i), which illustrate the relative flexibility of the three-pinned form and the general similarity of deflection magnitudes for the other two frames.

If, keeping the same overall height and span, the frame is gabled (Figure 5.8(a)), the frame will match the thrust line more closely and the eaves moment will be reduced for the three-pinned version. For the other two cases (Figure 5.8(b), (c)), it is the ridge moment that benefits, there being relatively little change in the eaves moments from the horizontal beam solutions. Comparison of the

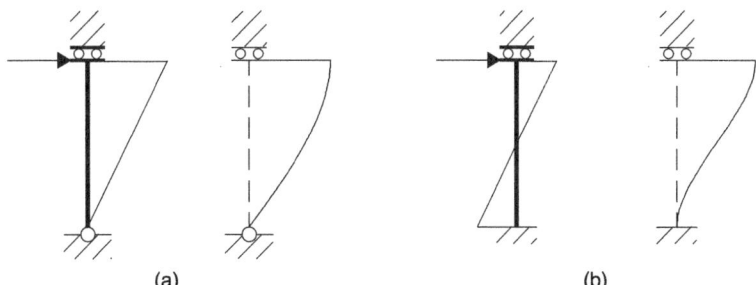

Figure 5.9   Single-storey column sway: (a) pinned base, (b) fixed base

deflected shapes (Figure 5.8(d)–(f)) with the corresponding previous results shows that gabling provides a more rigid arrangement, due to the reduced bending effects.

Horizontal loading
Sway loading is resisted primarily by the sway of a frame's columns (stanchions). If the beam member of a single-storey frame (Figure 5.7) is presumed extremely stiff then the columns will not rotate at the beam junctions, and their behaviour under sway loading will be as shown in Figure 5.9(a) for a pinned base and Figure 5.9(b) for a fixed base. In each case, the figure shows the deflected shape on the right and the corresponding bending moment diagram on the left. In practice, the beam will be flexible to some extent and there will be rotation (clockwise under the sway loading from the left shown) at the eaves positions. This will reduce the bending moment at the top of the columns and an equal bending moment will exist in the beam, giving rise to the bending moment and deflected shape diagrams shown in Figure 5.10 for frames of uniform section size.

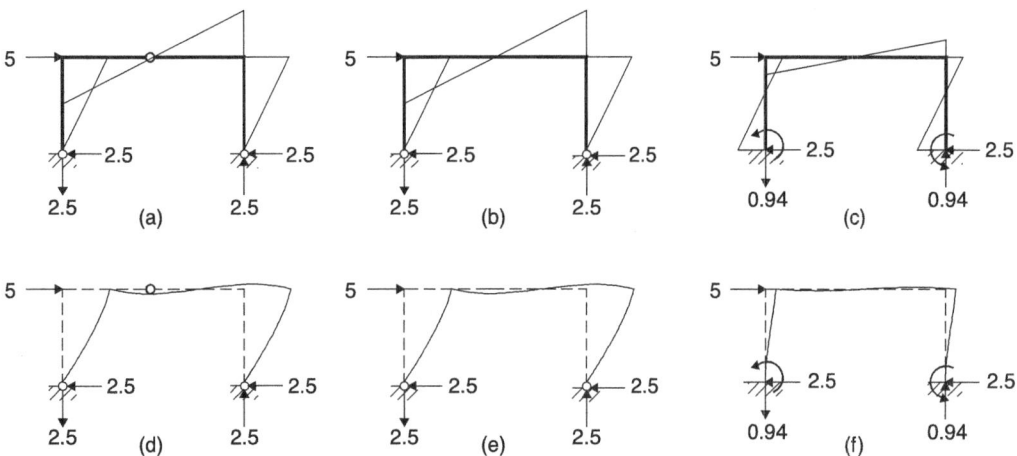

Figure 5.10   Single-storey/single bay portals under sway loading: (a)–(c) bending moment diagrams, (d)–(f) deflected shapes

One feature of the behaviour shown in Figure 5.10 is that the three-pinned and two-pinned frames behave identically under sway loading, since this produces zero bending moment at the centre of the beam in the two-pinned frame, and is where the third pin is situated in the three-pinned arrangement.

Single-storey multi-bay frames

Under vertical loading, the continuity provided by a multi-bay single-storey frame system results in interior columns being subjected to little or virtually no bending (Figure 5.11(a)), which means, in turn, that the interior beams act essentially in a fixed-ended manner. The two end columns do of course bend significantly due to the absence of continuity and the moments at the outer extremities of the end bay beams are therefore significantly reduced.

Under horizontal loading, multi-bay frames (Figure 5.11(b)) behave in a very similar fashion to single bay frames. The various bays resist the horizontal load approximately equally. In particular, the interior bays perform very similarly. The end bays, however, being rather more flexible, inevitably resist somewhat less of the sway than the internal ones.

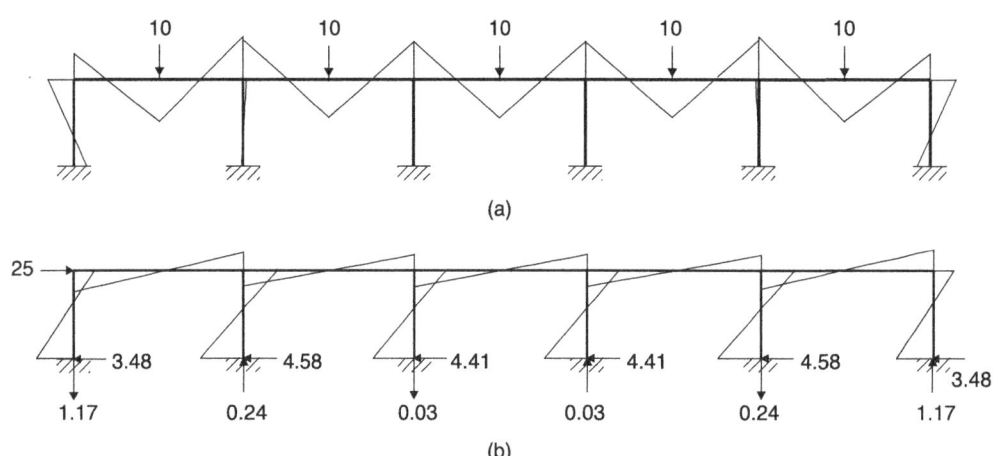

Figure 5.11 Single-storey/multi-bay portal frame: (a) BM diagram under vertical loading, (b) BM diagram under horizontal loading

Multi-storey single bay portals

Multi-storey, single bay portals (Figure 5.12) behave in very much the same manner as the corresponding single-storey frame under both vertical and horizontal loading. Multi-storey, multi-bay frames also act in a similar fashion to single-storey, multi-bay frames. These two forms do not therefore need to be considered separately in detail. In respect of horizontal loading, however, it may be noted from Figure 5.12(b) that the increasing shear towards the base results in a proportionate increase in column bending moments.

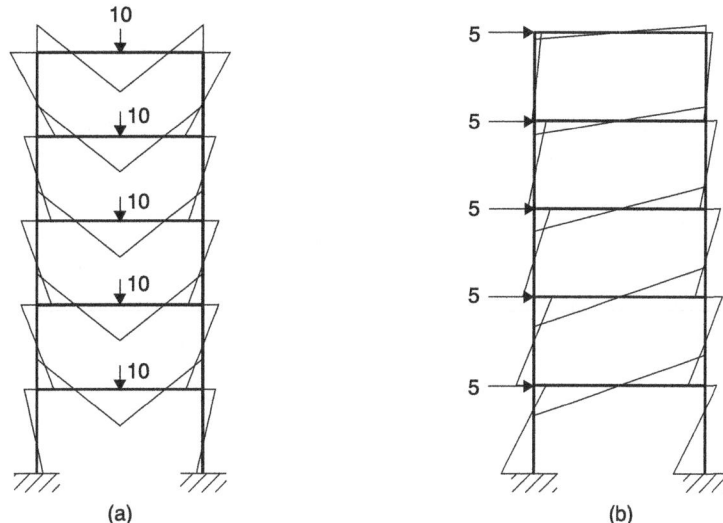

Figure 5.12 Multi-storey/single bay portal frame: (a) BM diagram under vertical loading, (b) BM diagram under horizontal loading

Vierendeel girders

The multi-storey frame behaviour under lateral loading (Figure 5.12(b)) may be related to the behaviour of a moment resisting (Vierendeel) frame (Figure 5.13), which may be obtained from the multi-storey frame by placing simple supports at the right-hand lowest and highest nodes. A Vierendeel girder therefore resists lateral load in a similar way to the multi-storey frame, namely by each bay swaying (Figure 5.13(b)). The sways induce bending moments (Figure 5.13(a))

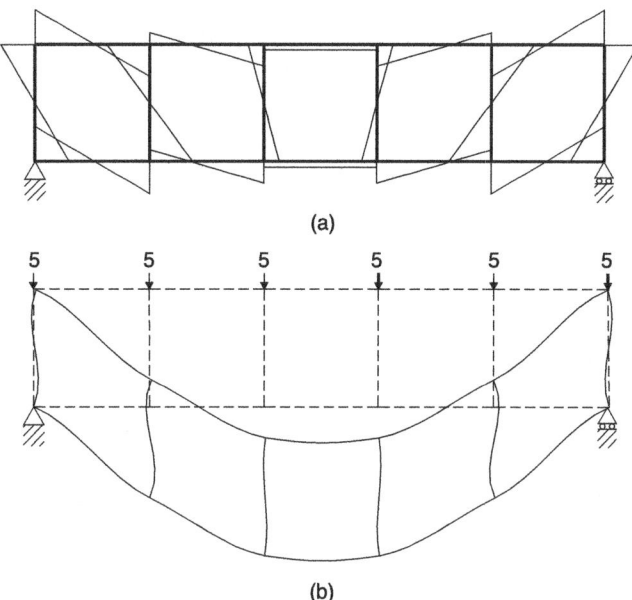

Figure 5.13 Vierendeel girder: (a) BM diagram, (b) deflected shape

such that the accompanying shear forces match the applied shear in the panel. For the uniform loading indicated, the shear is greatest in the end panels (hence the largest bending moments and sway deformations) and zero in the central panel (hence the absence of sway deformation and the constant bending moments in the booms, resulting in zero shear). Moment-resisting frames of this type are very flexible, as compared to equivalent truss structures in which diagonal bracing is provided in each panel, since, as seen, they rely on bending rather than axial deformation to resist lateral shear effects. The member stresses induced will also be greater. Nevertheless, this form of girder is sometimes employed for architectural reasons or for the lack of obstruction it offers to passing services through the structure.

Bending moment diagram sketching

In sketching bending moment diagrams for statically indeterminate frames, the precepts established for determinate frames will still apply and need to be used in conjunction with the typical forms of behaviour just demonstrated. The general form of the diagram may often be most readily developed from a predicted displaced shape. Assistance may also be gained from predicted reaction directions, although these are obviously less easy to establish than in determinate situations.

Example 5.5   Bending moment diagram sketching

Approximate bending moment and displaced shape diagrams for the frame shown in Figure 5.14(a) are most conveniently developed by initially assuming that lateral (sway) displacement is prevented. If this is the case, then the type of arguments used to develop the results shown in Figure 5.7 may be employed to predict the bending moment diagram shown in Figure 5.14(b). The non-symmetric nature of the load in the present example will clearly result in a sway movement of the frame, and initial intuition might suggest this is towards the loaded side, that is, to the right.

To check this, the nature of the column shears in the no-sway situation needs to be established. The end moments in the columns (found by considering the types of moment that will produce tension on the sides of the column shown in Figure 5.14(b)) are indicated in Figure 5.14(c). The directions of the shears in the columns (H_L and H_R) must then be such as to form a couple that will achieve equilibrium with the end moments. The shear directions that do this are also shown on the figure. From the relative magnitudes of the moments, it is clear that H_R will be greater than H_L in this case. Since no horizontal loading is present, H_L and H_R must be equal in magnitude under the actual sway conditions. It follows that the frame must sway **to the left** since this will increase H_L and reduce H_R. The nature of the bending moments resulting from a sway to the left may be deduced from the principles used to develop Figure 5.10 and are indicated in Figure 5.14(d). Combining these moments with the no-sway moments gives a resultant diagram of the form shown in Figure 5.14(e), corresponding to a deflected shape of the form shown in Figure 5.14(f).

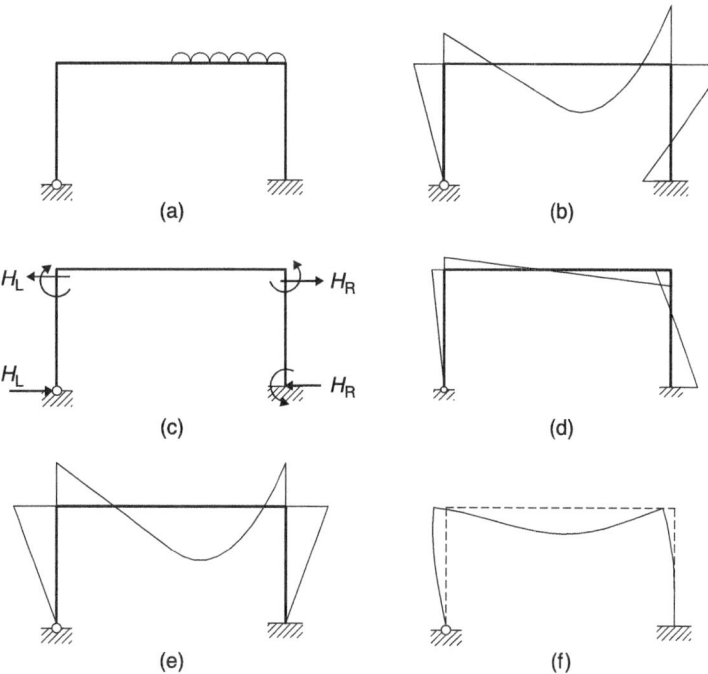

Figure 5.14 Example frame: (a) frame, (b) no-sway BM diagram, (c) column moments and shears, (d) sway BM diagram, (e) resultant BM diagram, (f) deflected shape

Worksheet 5.4 Stress resultant diagrams for statically indeterminate plane frames

W5.4.1 Sketch diagrams showing the probable directions of the reactions and the displaced shapes of the statically indeterminate frames shown in Figure W5.4.1. Hence, or otherwise, sketch probable bending moment diagrams for the frames. Uniform sections may be presumed for the frame members, unless indicated otherwise.

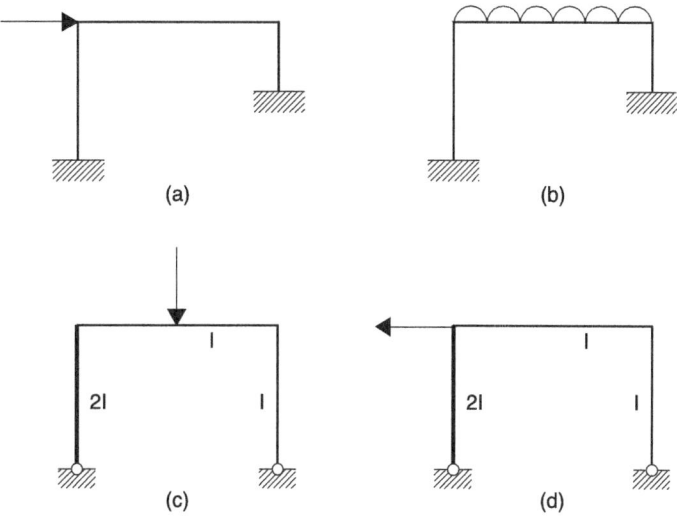

Figure W5.4.1   Worksheet 5.4.1

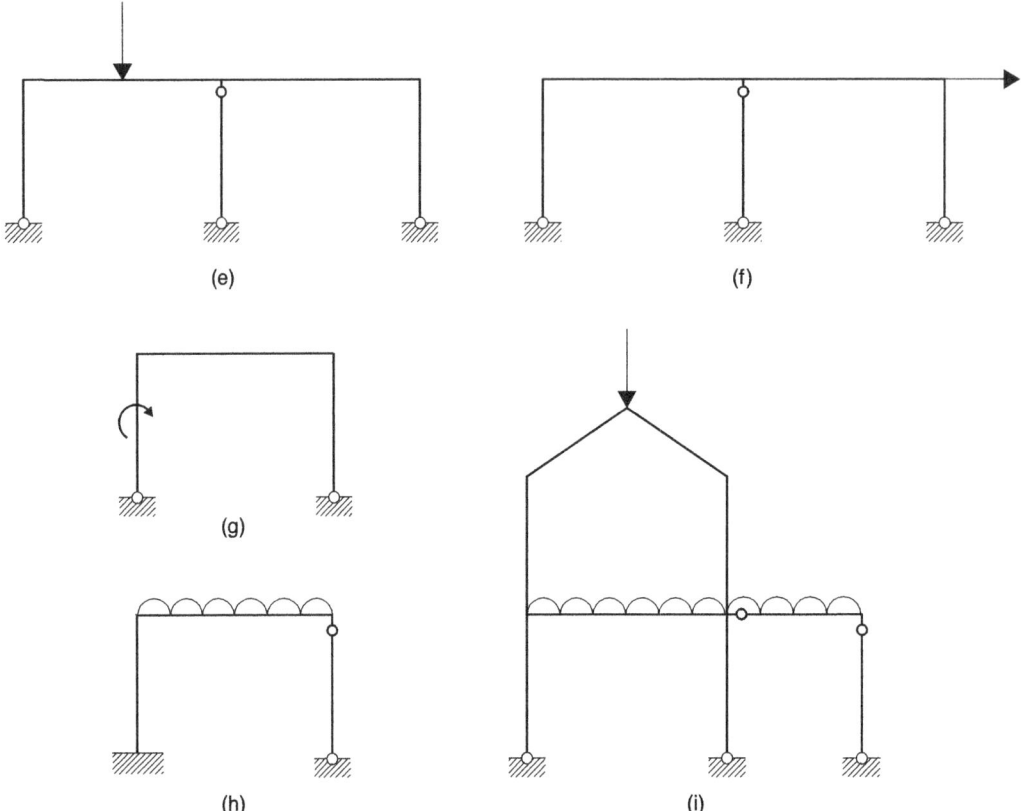

Figure W5.4.1 continued

5.5   COMPUTER ANALYSIS

The general provisions described in Section 4.2 apply equally to more general plane frames, and application of computer analysis to a plane frame structure is illustrated in the following example and worksheet.

Example 5.6   Computer analysis of existing frame for the effects of a new intermediate floor

The half portal frame shown in Figure 5.15(a) formed an assembly hall in a multi-storey building and is based on a description of the building's refurbishment (Vesey, 1987). As part of the refurbishment, a new floor was introduced at 4.4 m above the foot of the portal and resulted in the imposition of an additional load of 41 kN on the portal leg. The existing loads on the portal consisted of column loads from the upper floors of 202 kN from the edge columns and 838 kN from a central column. The edge columns were eccentric to the corner node (6) of the computer model, which produced a point moment of 100 kN m. Since only half of the frame is to be analysed, the central column load is halved. The uniformly distributed floor loading was represented by an equivalent

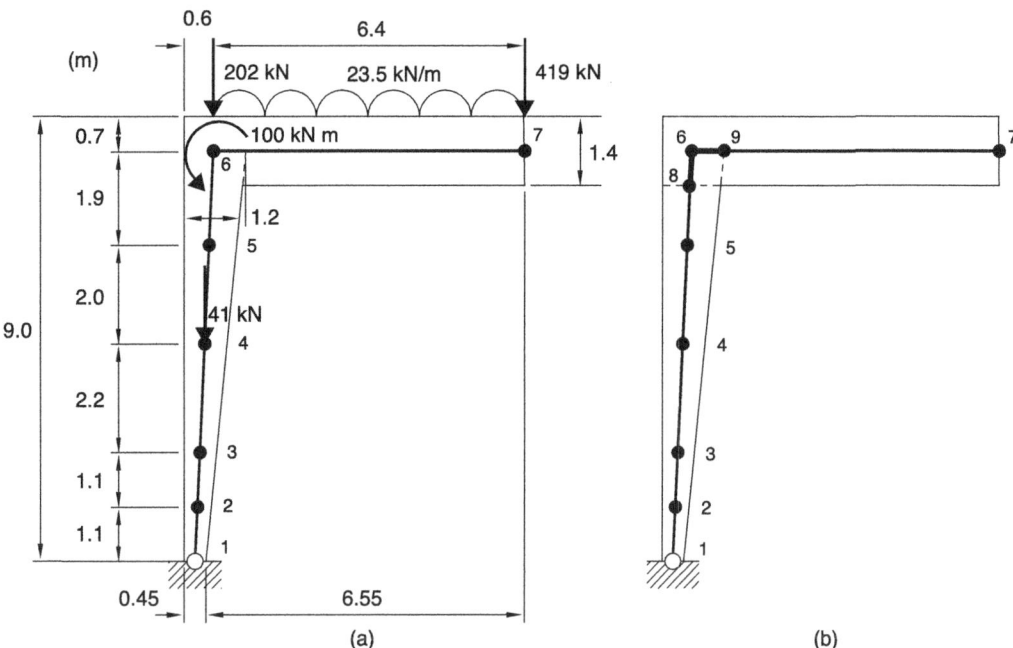

Figure 5.15   Computer model of half portal frame

load of intensity 23.5 kN/m along the beam member of the computer model (6–7). The portal self-weight was included, based on a unit concrete weight of 24 kN/m³.

The tapering legs of the portal were of rectangular section, with a thickness of 0.225 m. The uniform beam member was of T-section, having an effective I of 0.157 m⁴ and a cross-sectional area of 0.63 m². Section properties were based on the gross areas and the modulus of elasticity of the concrete was taken as 28 kN/mm².

The seven-node model shown in Figure 5.15(a) was used for the analysis, taking average properties for each of the members along the tapered legs. For comparison purposes, analyses will also be made based on (i) a uniform leg section based on its mean cross-section, and (ii) a tapering leg section, but with the connections from node 6 to the leg and the beam (members 8–6 and 6–9 in Figure 5.15(b)) made rigid to reflect the enhanced rigidity at the corner. Many software packages provide a facility for such rigid links to be specified directly. If this is not the case, then enhanced I or E values may be used for the links. However, this should be done with some care, since over-enhancement may cause numerical ill-conditioning. A factor of the order of a hundred will normally be satisfactory.

The pinned support at the foot of the half portal requires horizontal and vertical displacement restraints to be specified. However, the use of symmetry requires that restraints should also be applied at node 7. In accordance with the general precepts outlined in Section 1.6, the plane of symmetry is vertical and normal to the plane of the structure. Displacement normal to the symmetry plane (horizontally in this case) must therefore be restrained, as must rotation about lines contained by the symmetry plane (requiring a rotational restraint at node 7). These requirements

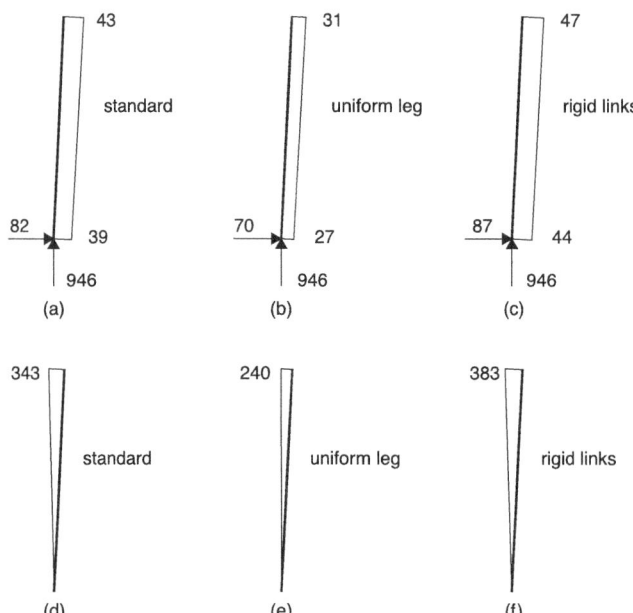

Figure 5.16   Example half portal: (a)–(c) pin reactions and leg SF diagrams (kN), (d)–(f) leg BM diagrams (kNm)

may alternatively be arrived at by noting that horizontal displacement or angular rotation at node 7 will result in a non-symmetric displaced shaped, but a symmetric shape is possible if vertical displacement only occurs at this node.

The results of the three analyses in respect of reaction components at the foot of the portal and the shear forces and bending moments in the leg are shown in Figure 5.16. The vertical component of reaction is identical for each of the analyses, since it simply balances the applied vertical loading. The horizontal thrust reaction varies considerably, however. The adoption of a uniform leg section reduces the relative stiffness of the leg at the leg–beam junction; hence leading to lower leg bending moments, shear forces and, hence, horizontal thrust. The use of rigid links, conversely, increases the relative stiffness and produces enhanced leg internal forces and base thrust reaction.

Worksheet 5.5   Computer analysis of plane frames

W5.5.1 The cycle wheel shown in Figure W5.5.1 is constructed from a carbon fibre material, which is of constant 3 mm thickness. The material has a modulus of elasticity of 480 kN/mm^2. The rim centreline diameter is 560 mm and the spokes are arranged at 120°. The self-weight of the complete cycle may be taken as 0.1 kN and the self-weight of the rider (distributed 40/60 between the front and rear wheels) may be taken as 1 kN.

Create an idealised computer model of the rear wheel, following the standard recommendation (Steel Construction Institute, 1995) that straight members representing a change of arc angle of 15° normally model curved members reasonably well. The model should be used to identify and analyse the worst cases for bending moment, shear force, axial load and displacement in the rim (University of Portsmouth).

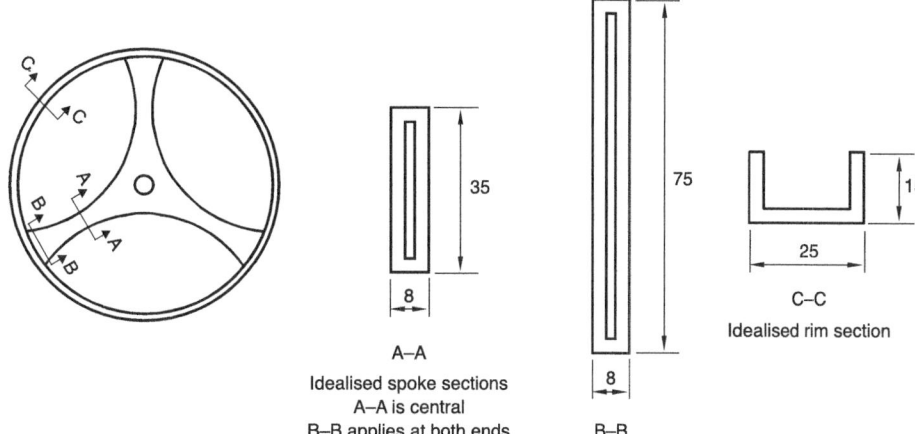

Figure W5.5.1   Worksheet 5.5.1

W5.5.2 Obtain a computer solution for the axial load, shear force and bending moment distributions in the rigid frame shown in Figure W5.5.2(a), due to the sway loads indicated. Take $E = 206\,\text{kN/mm}^2$ and the section properties as:

Section	$A\ (\times 10^3\,\text{mm}^2)$	$I\ (\times 10^6\,\text{mm}^4)$
Columns – L.H.S.	7.6	61
Columns – centre	12.3	222
Columns – R.H.S.	11.4	143
Beams – L.H.S.	7.58	254
Beams – R.H.S.	11.8	553

An approximate solution for a multi-storey frame under sway loading (Kennedy and Madugula, 1990) may be obtained by assuming internal hinges at the centres of each beam and each column length (column foot if column length is pinned at its base). The top storey, modified in this way, is shown in Figure W5.5.2(b). What is the degree of statical indeterminacy of this sub-structure? The statical indeterminacy may be solved by assuming the frame acts as a vertical cantilever and that the stresses in the columns may therefore be derived from simple bending theory as

$$\sigma = \frac{M(x - \bar{x})}{I} \quad \text{where } \bar{x} = \frac{\sum Ax}{\sum A} \quad \text{and} \quad I = \sum A(x - \bar{x})^2 \qquad \text{(Figure W5.5.2(c))}$$

The vertical components of reaction in the columns follow because $V = \sigma A$ and the horizontal components may then be determined by use of the zero bending moment conditions at the beam hinges. A similar procedure can be applied to the second and first storeys (Figure W5.5.2(d), (e)).

Determine approximating internal force distributions for the frame using the above model and compare with the computer solution. Re-run the computer solution with the beam I values decreased by a factor of 10 and comment on the changes produced in the bending moment distribution.

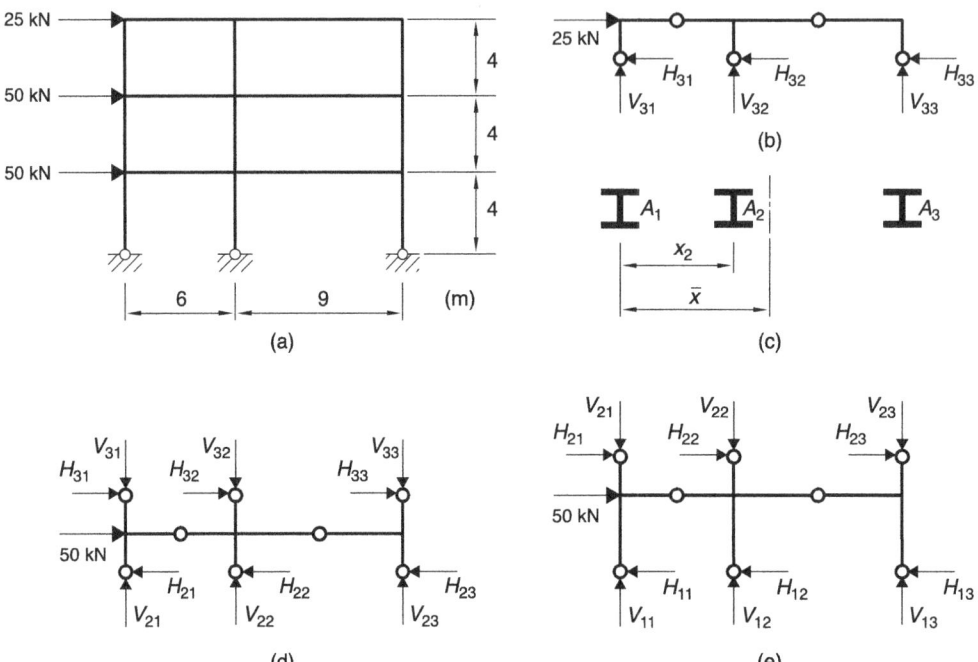

Figure W5.5.2   Worksheet 5.5.2

W5.5.3 The shear wall structure shown in Figure W5.5.3(a) is of constant thickness 0.3 m and is subjected to a lateral load of 20 kN/m applied to its left-hand side. In addition, $E = 30\,\text{kN/mm}^2$ and Poisson's ratio $= 0.15$. The wall may be modelled by a *wide-column* frame analogy (Stafford-Smith and Coull, 1991; Taranath, 1988) of the form shown in Figure W5.5.3(b). The analogy consists of beam and column members, whose cross-sectional properties are those of the corresponding wall sections, linked by rigid links. Use a frame analogy to obtain displacement and stress resultants throughout the structure.

Plot distribution diagrams of:

(a) shear forces in the connecting beams,
(b) lateral wall displacement,
(c) vertical direct stress at the base.

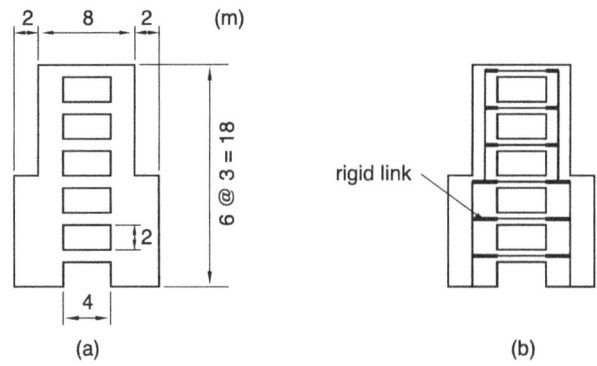

Figure W5.5.3   Worksheet 5.5.3

Explain the variation in connecting beam shear in relation to the lateral wall displacement. Relate the vertical base stress distribution to the overall structural behaviour of the shear wall structure. What features of the actual behaviour are unlikely to be modelled effectively by the frame analogy?

5.6   INFLUENCE LINES

There is relatively rarely a demand to produce quantitative influence lines for plane frames. Influence line sketches, however, are often helpful in identifying critical loading patterns. Whether determined quantitatively by computer analysis or qualitatively by the application of Müller-Breslau's principle, the procedures are similar to those employed for arch influence lines and will be demonstrated by finding representative influence lines for the frame shown in Figure W5.5.2(a).

Example 5.7   Influence lines for multi-storey frame

Figure 5.17 shows influence lines sketched for bending moments at two column and two beam positions for the multi-storey frame shown in Figure W5.5.2. The frame has been assumed to be laterally braced such that sway deformations are eliminated. The general form of the influence lines may be obtained by introducing a unit rotation at the relevant positions and sketching the resulting deflected shapes. If vertical, uniformly distributed, beam live loading is considered, then the sketches

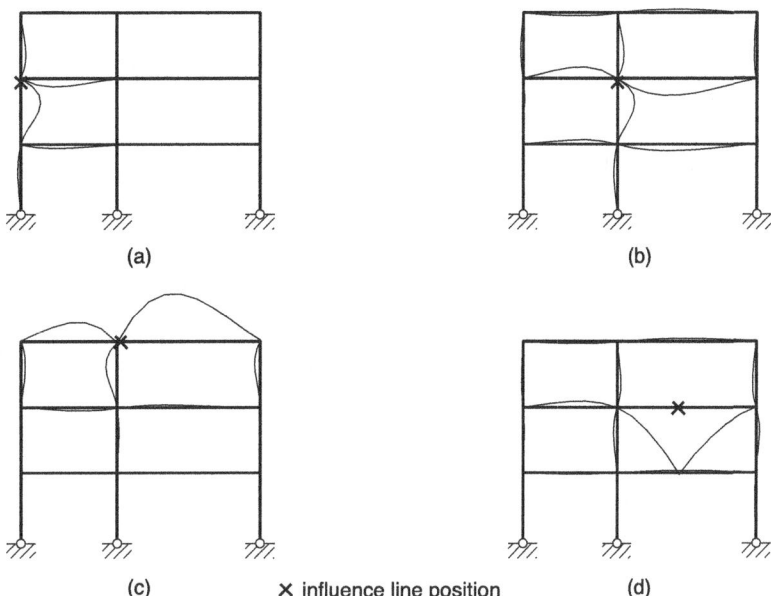

Figure 5.17   No-sway BM influence lines for example frame at (a) top of middle section left-hand column, (b) top of middle section of centre column, (c) left-hand end of right-hand beam, (d) within-span of middle right-hand beam

indicate that the greatest bending moment at the top of the middle section of the left-hand column (Figure 5.17(a)) will be achieved by loading the lowest two left-hand beams. Similarly the greatest bending moment at the top of the middle section of the centre column (Figure 5.17(b)) will be produced by loading the lower two right-hand beams and the upper left-hand beam. The beam influence lines (Figure 5.17(c), (d)) confirm the findings of the previous continuous beam investigation that end moments are maximised by loading adjacent spans and in-span moments by loading alternate spans.

It is conventionally considered (MacGinley, 1998) that maximal bending moments in external columns and beams will be approximately achieved if all beams are fully loaded, but that asymmetric loading patterns may need to be considered for internal columns. The sketches shown in Figure 5.17 are computer-derived validations of qualitative assessments and are therefore to scale. Figure 5.17(a), (c), (d) shows that the influence lines predominantly cause the same form of bending at the position under investigation, so that loading all the beams should produce a reasonable approximation to the maximal effect. In Figure 5.17(b), however, loading the lower beams to the left of the column position investigated, and the upper beam to the right, will significantly reduce the column moment and should be omitted from the critical loading pattern.

Should the example frame be free to sway, qualitative influence lines become somewhat more difficult to visualise, and it is perhaps more reliable to use computer analysis to generate the relevant deflected shapes. If a 0.5 m length member (I–J) is introduced at the top of the middle section of the left-hand column ($E = 206\,\text{kN/mm}^2$; $I = 61 \times 10^{-6}\,\text{m}^4$) and mm output units are presumed, then the bending moment influence line will be obtained from the displacements due to the introduction of a rotation, $\theta_I = 0.001$ rad, to the member. The end forces arising from the application of this rotation may be obtained from equation (4.6) as

$$
\begin{Bmatrix} N_I \\ F_I \\ M_I \\ N_J \\ F_J \\ M_J \end{Bmatrix} = E \begin{Bmatrix} 0 \\ -6I/L^2 \\ 4I/L \\ 0 \\ 6I/L^2 \\ 2I/L \end{Bmatrix} \{\theta_I\} = 206 \times 10^6 \times 61 \times 10^{-6} \begin{Bmatrix} 0 \\ -6/L^2 \\ 4/L \\ 0 \\ 6/L^2 \\ 2/L \end{Bmatrix} \{0.001\}
$$

$$
= \begin{Bmatrix} 0 \\ -301.6 \\ 100.5 \\ 0 \\ 301.6 \\ 50.3 \end{Bmatrix} \begin{matrix} \text{kN} \\ \text{kN} \\ \text{kN m} \\ \text{kN} \\ \text{kN} \\ \text{kN m} \end{matrix}
$$

In relation to a set of x(horizontal)- and z(vertical)-global axes, the nodal forces to be applied are therefore

$$\begin{Bmatrix} f_{xI} \\ f_{zI} \\ m_{yI} \\ f_{xJ} \\ f_{zJ} \\ m_{yJ} \end{Bmatrix} = \begin{Bmatrix} -301.6 \\ 0 \\ 100.5 \\ 301.6 \\ 0 \\ 50.3 \end{Bmatrix} \begin{matrix} \text{kN} \\ \text{kN} \\ \text{kN m} \\ \text{kN} \\ \text{kN} \\ \text{kN m} \end{matrix}$$

Similarly, if a 1 m member (I–J) is introduced at appropriate positions in the right-hand beam members ($E = 206\,\text{kN/mm}^2$; $I = 553 \times 10^{-6}\,\text{m}^4$), the relevant influence lines will be obtained by the application of member end-forces given, in local axes, by

$$\begin{Bmatrix} N_I \\ F_I \\ M_I \\ N_J \\ F_J \\ M_J \end{Bmatrix} = E \begin{Bmatrix} 0 \\ -6I/L^2 \\ 4I/L \\ 0 \\ 6I/L^2 \\ 2I/L \end{Bmatrix} \{\theta_I\} = 206 \times 10^6 \times 533 \times 10^{-6} \begin{Bmatrix} 0 \\ -6/L^2 \\ 4/L \\ 0 \\ 6/L^2 \\ 2/L \end{Bmatrix} \{0.001\}$$

$$= \begin{Bmatrix} 0 \\ -683.5 \\ 455.7 \\ 0 \\ 683.5 \\ 227.8 \end{Bmatrix} \begin{matrix} \text{kN} \\ \text{kN} \\ \text{kN m} \\ \text{kN} \\ \text{kN} \\ \text{kN m} \end{matrix}$$

Since the beam is horizontal, the nodal forces to be applied in the global axes are

$$\begin{Bmatrix} f_{xI} \\ f_{zI} \\ m_{yI} \\ f_{xJ} \\ f_{zJ} \\ m_{yJ} \end{Bmatrix} = \begin{Bmatrix} 0 \\ -683.5 \\ 455.7 \\ 0 \\ 683.5 \\ 227.8 \end{Bmatrix} \begin{matrix} \text{kN} \\ \text{kN} \\ \text{kN m} \\ \text{kN} \\ \text{kN} \\ \text{kN m} \end{matrix}$$

Applying these forms of nodal loads will provide the sway bending moment influence lines, which are shown in Figure 5.18 for the positions considered previously. In respect of sway forces, it may be noted from Figure 5.18 that, if the sway is applied in the form of horizontal **nodal** loading only, the loading produces bending moments of similar sign regardless of which node is loaded. This follows because, in any one case, the nodes all sway in the same direction. For combined sway/beam loadings, it follows that sway loading (presumed

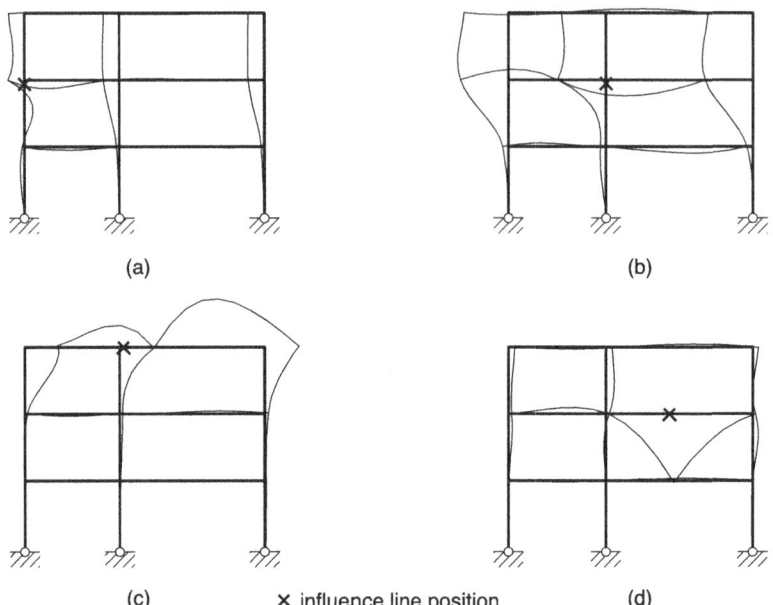

Figure 5.18   Sway BM influence lines for example frame at (a) top of middle section left-hand column, (b) top of middle section of centre column, (c) left-hand end of top right-hand beam, (d) within-span of middle right-hand beam

bi-directional) should be applied to all applicable nodes, in such a direction as to augment the moments caused by the beam loading.

If the bending moment at the top of the middle section of the centre column is considered, for example, then maximal effects from the beam loading will be achieved (Figure 5.18(b)) by applying the beam loading, as shown in Figure 5.19. These beam loadings produce a positive bending moment at the column position since the influence line displacement (downwards) is in the same sense the direction of the loading (Figure 5.18(b)). It follows that the sway loading should be applied from right to left, since this will also produce a positive

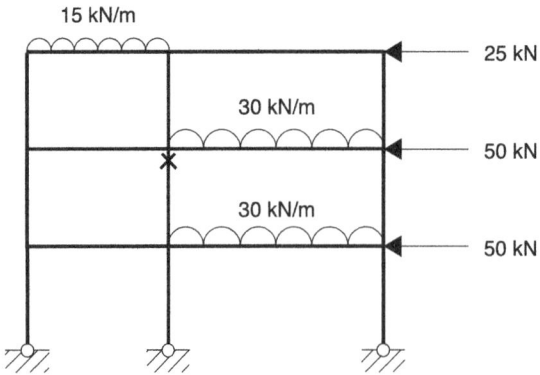

Figure 5.19   Example frame critical loading pattern

moment. If the loading values indicated in Figure 5.19 are used, then a computer analysis gives a column moment of 190 kN m. If the sway loading is taken with all the beams loaded at intensities of 15 kN/m at the roof level and 30 kN/m elsewhere, the column moment reduces to 142 kN m.

Worksheet 5.6 Influence lines for plane frames

W5.6.1 The frame shown in Figure W5.6.1 has an internal pin in the centre of its right-hand roof span. Assuming the frame is fully braced against sway, sketch influence lines for bending moment at the points indicated on the figure: (A) middle of the floor beam; (B) left-hand end lower roof beam; (C) top of centre column. Validate the sketches by obtaining the influence lines from appropriate computer analyses. Take $E = 206 \text{ kN/mm}^2$ and the section properties as:

Section	$A \ (\times 10^3 \text{ mm}^2)$	$I \ (\times 10^6 \text{ mm}^4)$
Beams – roofs	5.90	157
Beams – floor	7.58	254
Columns	5.88	45.6

Assuming uniformly distributed live loading of 10 kN/m for the two roof beams and 20 kN/m for the floor beam, use the influence lines to identify critical loading patterns for bending moment at the three positions. Use computer analyses to determine the maximal moments at the specified points and compare the values with those produced by loading all the beams simultaneously.

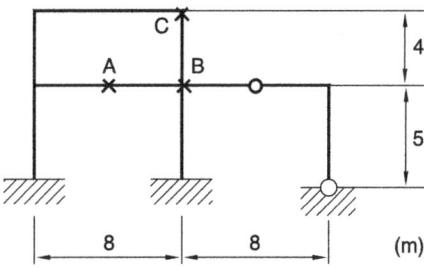

Figure W5.6.1   Worksheet 5.6.1

W5.6.2 The rigid frame shown in Figure W5.6.2(a) has the following section properties:

Section	$A \ (\times 10^3 \text{ mm}^2)$	$I \ (\times 10^6 \text{ mm}^4)$
External columns and rafters	9.51	334
Floors	7.58	254
Internal column	5.88	45.6

Take $E = 206 \text{ kN/mm}^2$ and use a computer package to derive bending moment influence lines for the points indicated in the figure (A eaves, B ridge, C left-hand end upper right-hand floor beam, and D centre lower right-hand floor beam).

Use the influence lines to determine critical loading combinations if the loading comprises live loading of 6 kN/m (on plan) for the rafters; 15 kN/m for the floors; and wind loading as shown in Figure W5.6.2(b). Analyse the frame for the critical loading combinations and compare the maximum moments obtained with values from a full loading case.

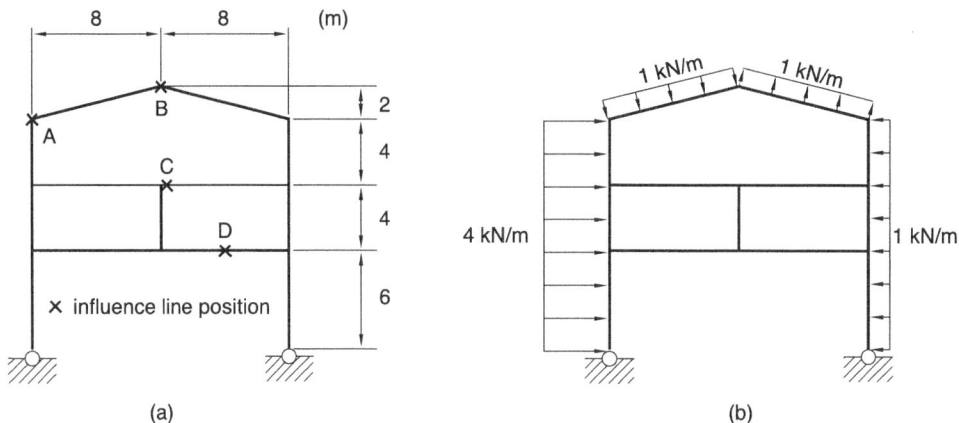

(a)   (b)

Figure W5.6.2   Worksheet 5.6.2

5.7   CASE STUDIES

C5.1.1   Figure C5.1.1 shows a simplified model of the Alamillo cable-stayed bridge in Seville (Aparcicio, 1997). There are 13 cables, which are spaced at equal 12 m intervals along the deck and inclined at $24°$ to the horizontal. At their upper ends, the cables are supported by a pylon, which is inclined at $32°$ to the vertical. The properties (in concrete units, $E_{conc} = 29 \, \text{kN/mm}^2$), the self-weight of the various portions of the pylon and deck and the measured forces in the cables under dead loading are given in Table C5.1.1.

Use a computer package to analyse the deck and pylon (omitting the cables) for the effects of cable loading (CL) and dead loading (DL). Produce comparative bending moment diagrams for the following combined loading conditions: (a) DL + CL; (b) 0.9DL + CL; (c) 1.1DL + CL.

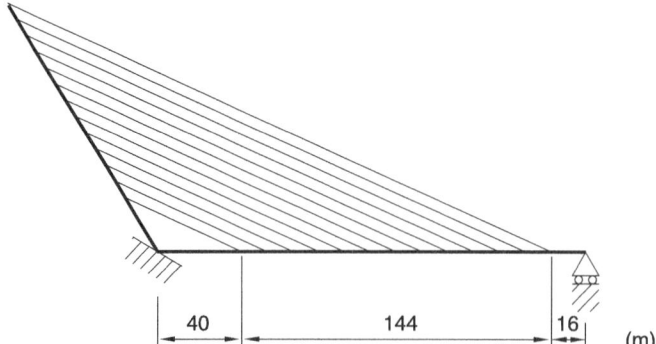

Figure C5.1.1   Case study 5.1.1

Table C5.1.1   Properties of Alamillo bridge

Member type	Position (before/cable no.)	A (m^2)	I (m^4)	Self-weight (kN/m)	Cable force (kN)
Deck	1	40	100	900	11 556
	2	14	37	420	11 694
	3	13	30	350	12 224
	4	12	24	313	10 614
	5	10	21	305	10 810
	6	9	21	305	10 164
	7	9	21	305	10 478
	8	9	21	305	10 242
	9	9	22	305	10 026
	10	10	22	305	9 280
	11	10	22	305	9 124
	12	10	22	305	8 398
	13 and after	10	22	305	8 730
Pylon	1	70	750	3200	11 556
	2	57	429	2800	11 694
	3	54	393	2700	12 224
	4	52	361	2600	10 614
	5	49	328	2500	10 810
	6	47	300	2500	10 164
	7	53	278	2300	10 478
	8	51	253	2200	10 242
	9	49	230	2300	10 026
	10	46	205	2400	9 280
	11	44	185	2300	9 124
	12	41	164	2100	8 398
	13	39	147	2000	8 730

Note: pylon loads are per unit horizontal distance

Determine the degrees of kinematic and statical indeterminacy, k and q, for the bridge. As a first approximation, it has been suggested (Pollalis, 1999) that the deck of the bridge can be considered to be a propped cantilever under live load. Comment on the reasonableness of this assumption. Both linear and non-linear elastic analyses of the bridge were undertaken for design purposes. What would be the main sources of the non-linear behaviour of the bridge? Would it be possible to omit the cables from the analysis if live loading were considered?

Compare the bending moment distributions for the combined analyses and relate each of them to the expected behaviour of the bridge. What implications do the bending moment diagrams have for the design, construction and maintenance of the bridge?

C5.1.2   For an interior fin frame of the Hines Meadow multi-storey car park (Figure C5.1.2), use published data (Thompson and Yates, 1994) to estimate the dimensions of the frame, the sizes of the frame members and the nature of its supports and joints. Estimate, also, the service dead and live gravity loads acting on the structure.

Analyse the displacements and internal forces in the frame, assuming that the fin frame is fully propped by the main car park structure. Re-analyse the frame on the basis that the fin frame is free of the main car park structure under dead loads, but is fully propped in respect of live loading.

Compare the analyses in respect of significant displacements and stress resultants. Why does the fin frame exert horizontal forces on the main car park structure, although it is subjected to vertical loading only? Why do these horizontal forces vary in magnitude and direction at the different attachment points?

Determine the kinematic and statical degrees of indeterminacy of the fin frame. If the frame were to be solved by hand, would it be more arduous to use the stiffness or the flexibility method of analysis? Most computer packages use the stiffness method. Why is this the case?

The car park was constructed using the fully propped solution. Why do you think this was the case? What special precautions would need to be taken for (a) the fully propped solution, and (b) the free under dead load solution?

Figure C5.1.2   Case study 5.1.2

C5.1.3   Figure C5.1.3 shows the dimensions and loadings for a proposed steel frame agricultural storage building. The frame is to be of constant section. Wind loads are reversible. In addition to the summer silage loads shown, winter silage loads should be considered. These may be taken to be 75% of the summer silage loads without a 6 kN impact load. If applied, silage loads always act on both stanchions simultaneously. The 6 kN impact loads can only act on one stanchion at a time.

Select loading cases that are considered appropriate for an investigation of the maximum bending moment that is likely to be sustained by the frame and undertake analyses for these cases. Determine the critical loading combination and produce computer output for this case.

State which loading cases were considered and why. Describe why the identified critical loading case gave rise to the greatest bending effects on the frame.

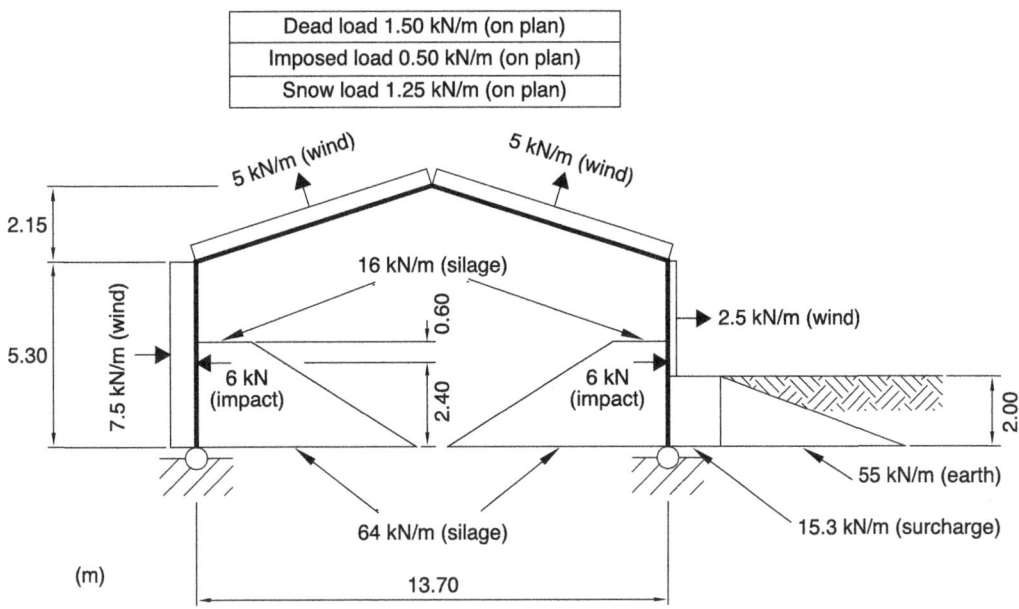

| Dead load 1.50 kN/m (on plan) |
| Imposed load 0.50 kN/m (on plan) |
| Snow load 1.25 kN/m (on plan) |

Figure C5.1.3   Case study 5.1.3

5.8   REFERENCES

Aparcicio, A. C. (1997) 'The Alamillo cable-stayed bridge: special issues faced in the analysis and construction'. *Proc. Instn. Civ. Engrs Structs and Bldgs*, **122**, Nov., 432–450.

Brohn, D. M. (1992) 'A new paradigm for structural engineering'. *The Structural Engineer*, **70**(13), 239–244.

Kennedy, J. B. and Madugula, M. K. S. (1990) *Elastic Analysis of Structures*. New York: Harper and Row – Chapter 12 of this reference gives transformation matrix details in connection with the stiffness method and its Chapter 14 deals with influence lines, including those for arches.

MacGinley, G. (1998) *Steel Structures – Practical Design Studies*. 2nd edn. London: E & F N Spon – Chapter 5 of this reference considers the analysis and design of multi-storey steel buildings.

Pollalis, S. N. (1999) *What is a Bridge? – the Making of Calatrava's Bridge in Seville*. MIT Press, Cambridge, MA.

Stafford-Smith, B. and Coull, A. (1991) *Tall Building Structures – Analysis and Design*. New York: Wiley.

Steel Construction Institute (1995) *Modelling of Steel Structures for Computer Analysis*. Ascot: Steel Construction Institute.

Taranath, B. S. (1988) *Structural Analysis and Design of Tall Buildings*. New York: McGraw-Hill.

Thompson, P. W. and Yates, A. S. (1994) 'The planning and design of Hines Meadow multi-storey carpark, Maidenhead, Berks'. *The Structural Engineer*, **72**(12), 187–191.

Vesey, D. (1987) 'A new floor at Fleet Building, Farringdon Street, London EC4'. *The Structural Engineer*, **65A**(6), 225–229.

6 Space trusses

6 Space trusses

6.1 INTRODUCTION

A space truss implies a three-dimensional arrangement of members that is capable of withstanding applied loads primarily by axial effects, any bending action being secondary. Such structures are a familiar sight either as tower cranes in town or as transmission towers in the country. Space trusses are inherently stiff and are therefore appropriate to long-span roofs and as supports for radar and radio telescope dishes, for which the deflection restrictions can be severe. The possible range of geometries is infinite, but regularity is often desirable. Irregularity, if nothing else, risks the inadvertent introduction of mechanisms, which can be difficult to detect in a complex spatial array without recourse to computer analysis. To generate regular space trusses that are guaranteed to be stable, a useful theorem (Timoshenko and Young, 1965) is that any closed polyhedron (a shape made up of plane sides) will be stable if its faces are triangular or are divided into a number of triangles. Thus, the square and triangular cross-section polyhedra shown in Figure 6.1(a), (b) are stable arrangements and form the basis of the tower and jib, respectively, of the tower cranes shown in this chapter's frontispiece.

The polyhedron of Figure 6.1(b) can alternatively be viewed as a number of interconnected pyramids (pentahedra). Rows of pentahedra such as those

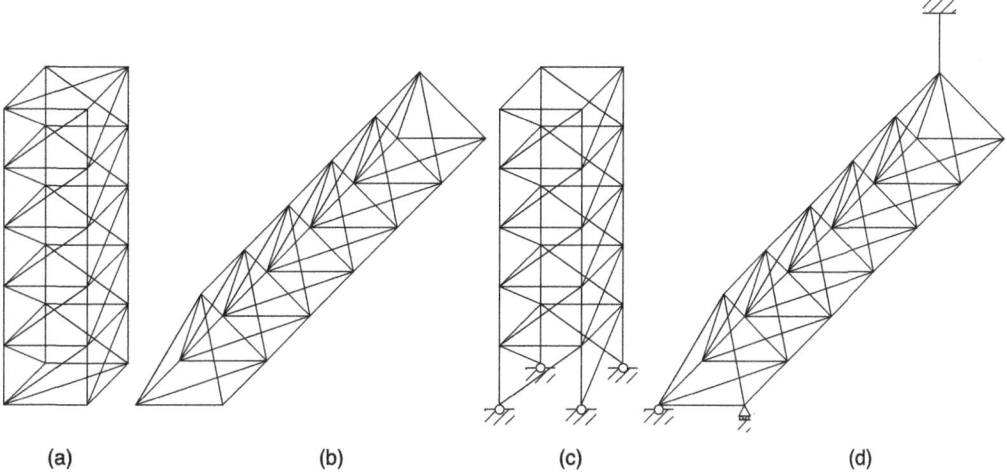

(a)   (b)   (c)   (d)

Figure 6.1   Simple triangulated polyhedra

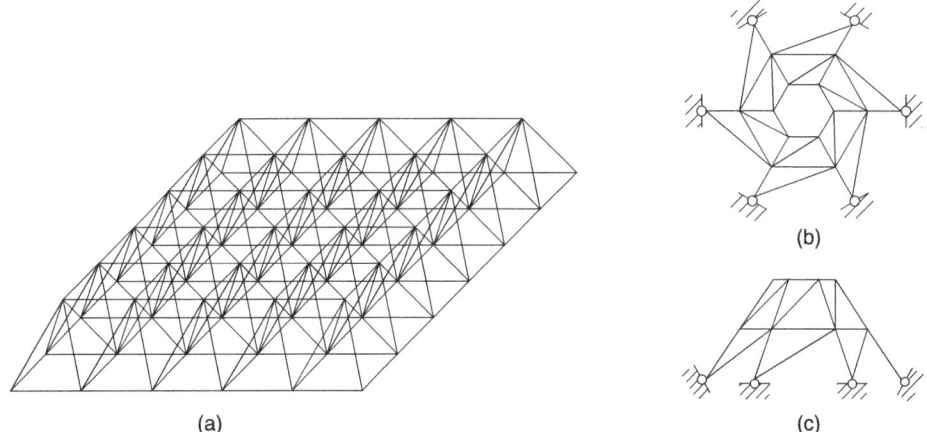

*Figure 6.2 (a) Double-layer grid, (b) Schwedler dome – plan, (c) Schwedler dome –
elevation*

shown in Figure 6.1(b), but without the base diagonals, can be placed alongside
each other. If the top nodes are then connected laterally, a double-layer grid
results (Figure 6.2). This form of space truss (MacGinley, 1998) is suitable for
long span roofing purposes and has the convenience that the pentahedra can be
stacked for transport to site. The tower shown in Figure 6.1(c) can of course be
tapered, and if this is carried out strongly or progressively then a determinate
Schwedler dome is produced (Figure 6.2(b), (c)).

6.2 KINEMATIC AND STATICAL INDETERMINACY

As shown, the truss in Figure 6.1(a) is unsupported and is therefore unsuitable for
terrestrial applications. If the base is supported by 'ball-and-socket' (universal or
spherical) joints that allow full rotation but prevent displacement completely, then
the truss becomes as shown in Figure 6.1(c). In Figure 6.1(c), the base members
have been removed, since these do not now carry load. The top face diagonal
has also been omitted, since the supports ensure the rigidity of the cross-section.
The kinematic indeterminacy of the truss may be found in a similar manner to the
procedure used for plane trusses. It is simply necessary to introduce an extra
degree of possible displacement degree of freedom at each node to include movement in the additional dimension. The structure has joints $j = 24$ and restrained
displacement components $d = 4 \times 3 = 12$. Its kinematic indeterminacy is
therefore $k = 3j - d = (3 \times 24) - 12 = 60$. Making similar assumptions as for
plane frames, in respect of pinned joints, nodal loading and no member joint
eccentricities, there will be only a single (axial) stress resultant in each of its m
members. Since the truss has 60 members, its statical indeterminacy, q, is therefore
$q = m - k = 60 - 60 = 0$, showing the structure to be determinate.

It may be shown (Timoshenko and Young, 1965) that six translational
constraints will ensure the rigidity of a body in space, provided the constraints

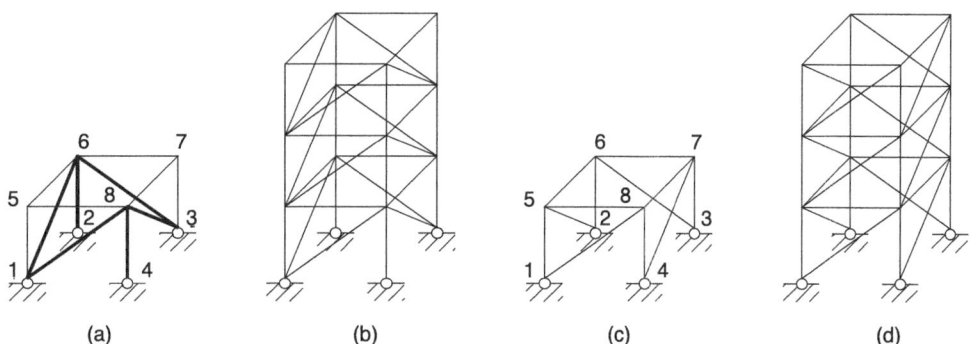

Figure 6.3   Space truss types: (a), (b) simple, (c), (d) complex

do not all intersect a unique line. The unsupported truss of Figure 6.1(b) may therefore be provided with satisfactory support if a universal joint and a transverse roller are provided at one end and a vertical support at the other (Figure 6.1(d)). In practice, the vertical support would probably be provided by the vertical component of a supporting tie as for the crane jibs shown in the frontispiece to this chapter. In this case, the number of restrained displacement components $d = 3 + 2 + 1 = 6$. The truss also has $j = 20$ and $m = 54$, so that $k = 3j - d = (3 \times 20) - 6 = 54$ and it is again determinate, since $q = m - k = 54 - 54 = 0$. In fact, all space trusses formed by triangulating the faces of polyhedra will be determinate if adequate, determinate supports are provided. This provides a possible alternative check on statical determinacy to the formulae used so far:

$$k = 3j - d \quad \text{and} \quad q = m - k \tag{6.1}$$

In general, the statical determinacy of space trusses, as found from equation (6.1), is not readily checked, but it is possible to do so if the truss is of a *simple* determinate type. Simple space trusses are such that it is possible to find a tripod starting sub-structure so that at least one node is connected to universal supports by three non-coplanar members. The tripods with apices at nodes 6 and 8 in Figure 6.3(a) are typical examples. These tripods are stable and will be determinate since they have $k = 3$, from the degrees of freedom at the unrestrained node, and $m = 3$ also. Nodes 6 and 8 are therefore stable nodes. Further stable nodes may be created by attachment either to supports or to existing stable positions by three non-coplanar members. Determinacy will be preserved since three extra degrees of displacement freedom are being created as well as the three extra members. Nodes 5 and 7 may be shown to be stable in this way. Using Figure 6.3(a) as a sub-structure to create a multi-storey tower, Figure 6.3(b), will thus produce a determinate, stable truss of the simple variety. It should be noted that bracing arrangements can be crucial in determining whether a truss is of a simple form or not. In Figure 6.3(c), for example, none of nodes 5–8 are connected to the supports by three members, so that no starting tripod exists. It follows that the truss, although stable and determinate, is not of a simple form and would be termed *complex*. The same would be true of any resulting forms (Figure 6.3(d)) for which it was used as a building block.

6.3 QUALITATIVE ANALYSIS

Many concepts relevant to the determination of the nature of the member forces in plane trusses are relevant to space trusses and, indeed, these will often behave in a similar manner to plane trusses, especially if they are not tapered, or have other geometric irregularities. Uniform lateral loading of the tower of Figure 6.1(c), for example, will be resisted by the faces parallel to the loading and the nature of the member forces induced can be readily established by use of the beam analogy type of procedure used for plane trusses. Similarly, the space truss shown in Figure 6.1(d), if acted on by uniform gravity load, will act in a beam type fashion, with the upper boom becoming a strut, the lower booms ties. The vertically sloping diagonals resist the shear, and the horizontal plane diagonals remain unloaded. Three-dimensional action of these truss forms is, therefore, principally caused by torsional loading. This is of significant practical concern, since torsional loading, caused by the absence of a single conductor, due to failure, for example, is often a critical case for transmission tower design (Open University, 1991). In assessing the nature of the forces in a space truss, the properties listed in Table 6.1 are often helpful.

If, as an example, the tower of Figure 6.1(c) is subjected to the torsional form of loading shown in Figure 6.4(a), then it may be conjectured that each load will be resisted by the side of the truss that is co-planar with the load and that the sides of the truss that are normal to the loads will be unaffected. The co-planar sides will act as plane truss cantilevers and the nature of the forces induced in the members will be as shown in Figure 6.4(a). This may be verified from the properties of Table 6.1 by noting that, at A in Figure 6.4(a), member AD is normal to the other members at A and, therefore, by Property 2 of Table 6.1, AD is unloaded. Similar results hold for BA, CB and DC at B, C and D respectively. By Property 3 at A and C, it then follows that the other two members at each of these nodes are unloaded. Consideration of the equilibrium conditions at B and D then provides the nature of the forces in the remaining members at these nodes and the argument may be continued down the lower storeys to confirm the original conjecture.

*Table 6.1   Member forces at **unloaded** joints*

	Type	Example	Result	Proof
1.	Three non-coplanar members only		All members unloaded	Repeated resolution normal to planes of pairs of members
2.	All members bar one are coplanar		Force in non-coplanar member is zero	Resolution normal to co-planar members
3.	All except two non-collinear members have zero force		Remaining two members also have zero force	Resolution normal to direction of each of the two members

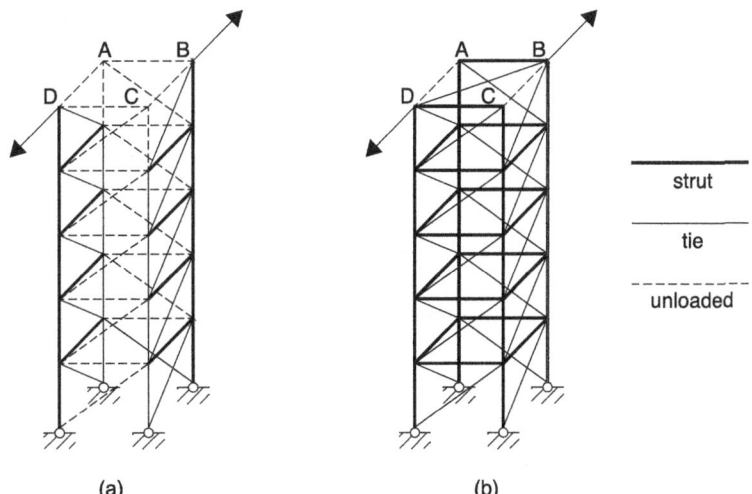

Figure 6.4   Tower under torsional loading: (a) horizontally unbraced, (b) horizontally braced

Should the cross-section of the tower be horizontally braced at the top (Figure 6.4(b)), then AD and BC are again unloaded by Property 2 at A and C, respectively. However, the loads will tension the brace BD, which in turn will compress AB and CD. The effect of these compressions, together with the loads, is to produce tension in each of the diagonal braces of the top panel. In turn, the diagonal brace tensions are balanced by compression in each of the tower legs. This argument may be continued down the tower to give the resultant force distribution shown in Figure 6.4(b), which shows that, in this case, the torsion is resisted equally by all four faces of the tower.

As a further example, the behaviour of the beam type of space truss shown in Figure 6.1(d) will be examined under the action of a central torsional effect (Figure 6.5). Probably the most important initial observation is that the vertical support at the remote end is unable to resist torsional effects since it is parallel and symmetrically placed to the forces that make up the torsional couple. It follows that this support is unloaded and the torque is balanced by equal and opposite vertical reactions at the near end. By starting at A (Figure 6.5) and applying Property 1 of Table 6.1 progressively at successive joints, it follows that all members beyond the torque application plane are unloaded, as indicated in Figure 6.5.

If the near half of the truss is examined in more detail (Figure 6.6(a)), then it may be observed that each of the two sloping side panels is subjected to a constant shear but are opposite in sense. The nature of the forces in the bracing members may therefore be deduced from the senses of the applied shear. Since the shears are constant, the bracing forces will also be constant and will be in equilibrium at the top boom nodes. It follows that the top boom is unloaded, which might alternatively have been deduced from its symmetric positioning.

The forces in the base panel may be established by initially considering equilibrium in a horizontal plane at nodes A and G (Figure 6.6(b)), from which

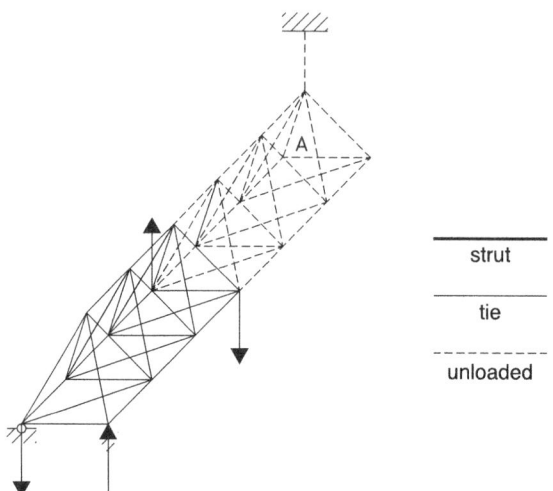

Figure 6.5   Beam space truss under torsional loading

it follows that AB, AD, GF and GH are all ties. Transverse equilibrium at B and H then shows that the diagonals BD and HF are struts. Since the nature of the respective types of member will not vary along the half-length, the assessment may be completed as shown in Figure 6.7.

As a final example of qualitative assessment, the Schwedler dome shown in Figure 6.8 will be considered. The dome has six planar sides and is subjected to the single vertical point load shown. By applying Property 2 of Table 6.1 (one non-coplanar member at an unloaded joint) to each of the nodes of the upper hexagon of members, it may be shown that all of these members except one must be unloaded. Applying Property 3 (all except two non-collinear members have zero force) to the same nodes shows that the indicated upper storey members (Figure 6.8(a)) must be unloaded. Moving down to the lower hexagon ring of members and applying Property 2 at each node, followed by Property 3, results

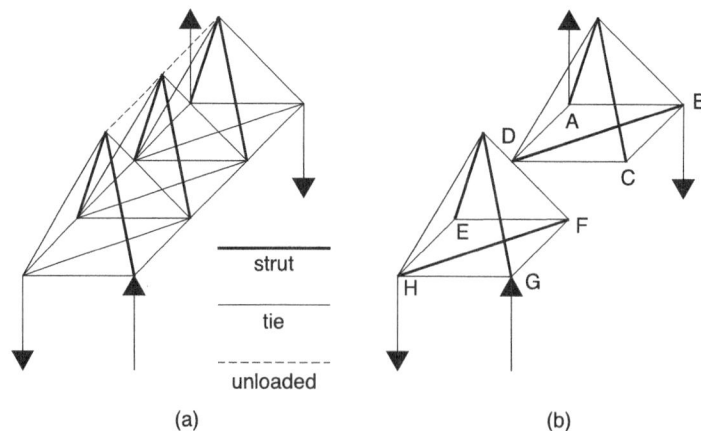

Figure 6.6   Beam space truss: (a) near half, (b) end bays of near half (central bay omitted for clarity)

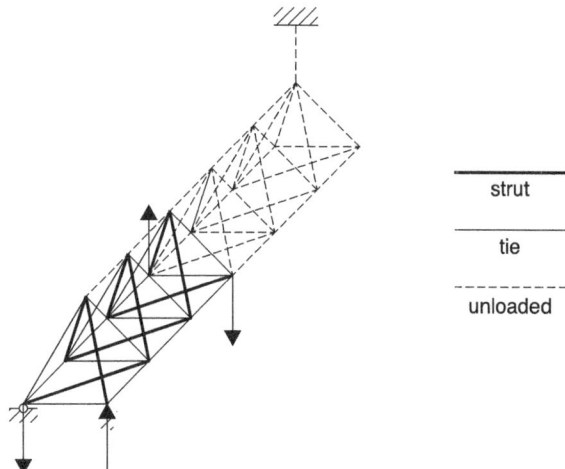

Figure 6.7   Qualitative analysis of a beam type space truss

in the indicated unloaded members at this level, with the exception of *ab* (Figure 6.8(a)). This member may, however, be shown to be unloaded by re-applying Property 2 at joint *a*. Figure 6.8(a) therefore shows the members that are known to be unloaded (dashed) and those that may sustain load (solid).

To determine the nature of the forces in the possibly loaded members, equilibrium at *a* (Figure 6.8(b)) normal to the plane of *abed*, indicates that *ac* is a strut. Resolution in horizontal plane normal to *ab* then shows *ad* also to be a strut. Similarly, resolution in horizontal plane parallel to *ab* shows *ab* to be a further strut. Resolution at *b* in the plane of the face containing *bd* and *be* can then be used to show these members to be a tie and a strut respectively. Similarly, resolution at *c* in the plane containing *cd* and *cf* shows these members also to be a tie and a strut.

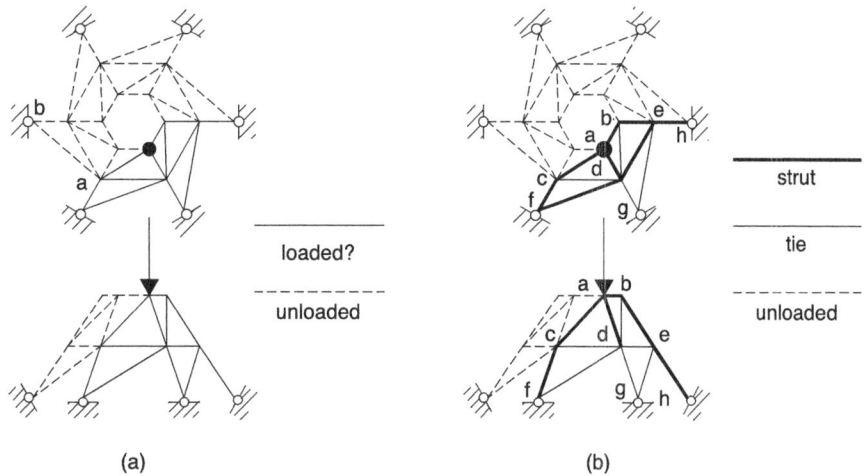

Figure 6.8   Schwedler dome plans and elevations: (a) unloaded and possibly loaded members, (b) qualitative assessment

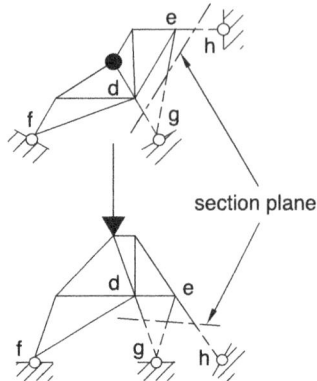

Figure 6.9   Schwedler dome plan and elevation (loaded members only for clarity)

To determine the senses of *eh* and *dg*, use may be made of a vertical section plane as shown in Figure 6.9. Since only the applied load and the internal force in *eh* have any moment about the line *fg*, *eh* must be a strut if it is to oppose the moment of the applied load. Finally, if the same section is used and moments taken about a horizontal line passing through *e* and *f*, then only the applied load and the internal force in *dg* have moments about this line, and *dg* must be a tie for equilibrium. The member *dg* might have been expected to be a strut in view of its position on the path from the load to the supported base, and the reason for it acting as a tie is perhaps best appreciated by considering the displacement of the truss. The load tends to displace its point of application, *a*, and the connected position, *d*, towards the centre of the dome, hence leading to a stretching of *dg*. Nodes *c* and *e*, conversely, are squeezed outwards by the load, resulting in *eh* and *cf* acting as struts.

Worksheet 6.1   Qualitative analysis of space trusses

W6.1.1   The space truss shown in Figure W6.1.1 is symmetric with respect to the *x–z* plane. Show that the truss is of a *simple* form and is therefore both statically determinate and stable. Identify the nature of the internal forces induced in the members of the truss by a point load at P, applied in (a) the positive *y* direction and (b) the negative *z* direction.

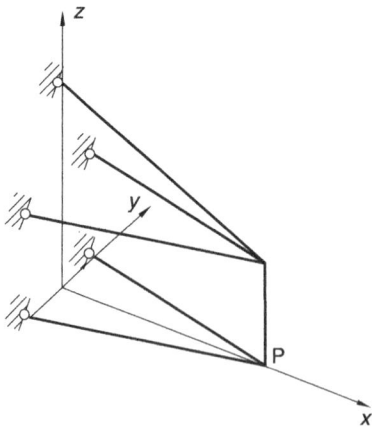

Figure W6.1.1   Worksheet 6.1.1

W6.1.2 The space truss shown in Figure W6.1.2 is such that QR and ST are symmetric with respect to the x–z plane and joint P lies in the x–z plane. Show that the truss is both statically determinate and stable. Why would the truss not be stable if P was in the same horizontal plane as QS? Identify the nature of the internal forces induced in the members of the truss by a point load at P applied in the negative z direction.

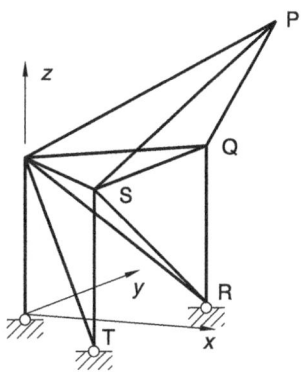

Figure W6.1.2   Worksheet 6.1.2

W6.1.3 Determine the kinematic and statical determinacies of the space truss shown in Figure W6.1.3. It may be shown (Timoshenko and Young, 1965) that trusses of the form shown in Figure W6.1.3 are stable if the upper and lower plane nodes lie on the vertices of a regular polygon having an odd number of sides, but will be unstable if, as in this case, the polygons have an even number of sides. Why will the introduction of a member QR make the truss stable? For the modified structure, identify the nature of the internal forces induced in the members of the truss by a point load at P applied in the negative z direction.

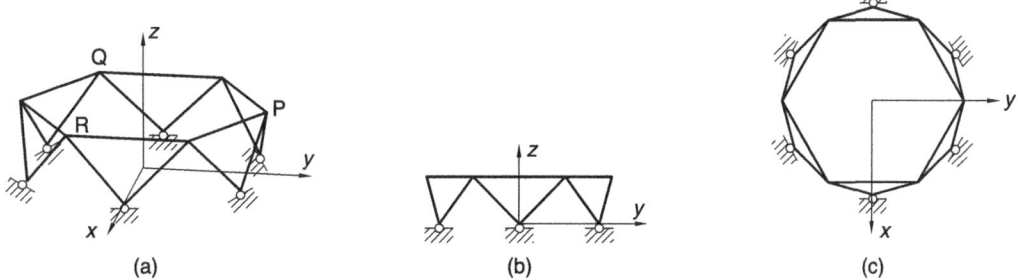

Figure W6.1.3   (a) Persepective, (b) elevation, (c) plan

6.4   COMPUTER ANALYSIS

Stiffness method theory

The analysis of space trusses by the stiffness method follows almost exactly the lines previously described for plane trusses (Chapter 1). In particular, the member stiffness matrix (equation 1.8) is identical, since pinned joints are again presumed and the only member stiffness involved is therefore axial. Dedicated space truss computer packages will therefore only require the specification of member cross-sectional areas as section properties. Should a dedicated space truss program not be available, then a general space frame package may be used and will give similar results if bending effects are minimised by the specification of reduced second moments of area (I values). Reduction by a factor of around one thousand is usually adequate.

The only difference between space and plane truss stiffness analysis lies in the transformation matrix employed. This matrix (equation 1.10) links the axial displacement at each end of a member to the displacements at the ends in the overall (global) axes. At both ends, each member still has a single axial displacement but there will be three global displacement components for the space member, as opposed to the two components for a plane member. The transformation matrix therefore still has two rows but will now have six columns (Ghali and Neville, 1997).

Restraints are purely displacement (translational) constraints and may be applied in any of the three co-ordinate directions. The commonest situations are the fully fixed pin, for which all three constraints are applied, and the smooth surface (usually horizontal), for which only normal constraint is applied. However, various constraint combinations may be needed to ensure that rigid body movements are eliminated or that symmetry conditions are correctly modelled. These forms of restraint are demonstrated in the example that follows.

Example 6.1 Computer analysis of a space grid

The space (double-layer) grid shown in Figure 6.10(a) will be analysed for the effects of a uniform load applied to the upper surface, which will be represented by nodal point loads. This results in non-corner edge nodes having loads one half of those applied to interior nodes and corner edge nodes having loads one

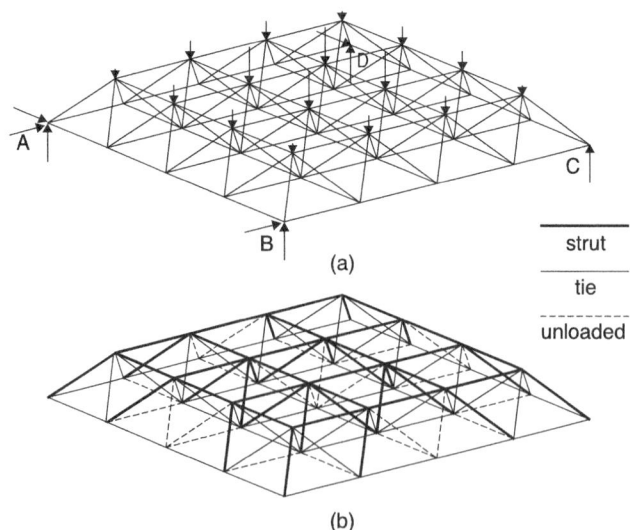

Figure 6.10   Space grid example: (a) layout, (b) axial load distribution

quarter of those applied to interior nodes. For convenience, uniform member cross-sectional areas are assumed. The grid is provided with vertical support only at points A, B, C and D. Although this support system is theoretically satisfactory in respect of the purely vertical loading applied, the structure is free to translate horizontally and to rotate about a vertical axis. These freedoms will be reported as errors by computer packages, and the analysis may be aborted. To eliminate the possible rigid-body movements, a suitable system of additional restraint is needed that preserves the symmetry of the structure. The double-layer grid may be viewed as a structure that bends in two orthogonal directions so that it is possible to follow the concept used in simply supported beam analysis, which uses a pin at one end and a roller at the other. In the span direction AD/BC, therefore, the pin is represented by constraining A and B in the span direction. Points A and D are similarly constrained in the direction of span AB/DC, providing the horizontal constraints shown in Figure 6.10(a). These additional constraints prevent translational displacement in a horizontal plane and also rotation about a vertical axis.

Based on a computer analysis, the natures of the axial loads in the truss members are as shown in Figure 6.10(b). The analysis may be validated by use of the orthogonal bending concept. Certainly, this would support all the upper boom members being in compression and all the lower boom members being in tension. Quantitatively, the magnitudes of the forces induced will increase towards the centre of any given gridline of members. However, the edge gridline members will carry significantly greater loads (between two and three times greater) than the internal lines, since the edges are better supported.

The sloping diagonal braces form a series of inclined Warren trusses and the nature of the forces induced in these members may be assessed from the shear analogy introduced in Chapter 3, although it should be stressed that this will not be relevant to internal bracing rows, which are not directly supported at their ends. Figure 6.11(b) illustrates the nature of the shear force distribution and Figure 6.11(a) shows the resulting alternating compression/tension system induced in the bracing members. The zero shear in the central panel indicates

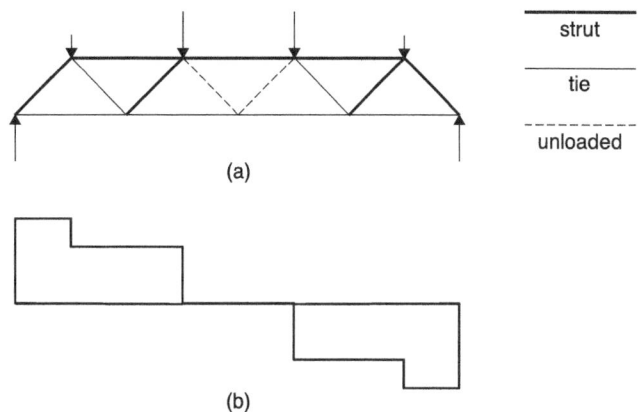

(a)

strut

tie

--------- unloaded

(b)

Figure 6.11 Equivalent Warren truss: (a) internal forces, (b) shear force diagram

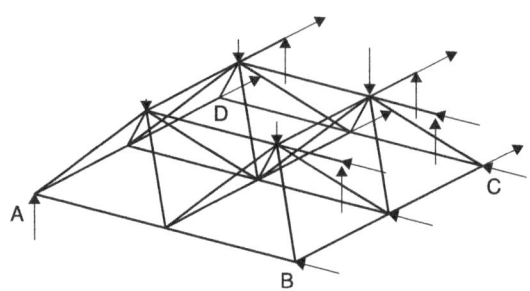

Figure 6.12   Space grid example – use of symmetry

that the two central braces will be unloaded. This conclusion may alternatively be arrived at by noting from vertical equilibrium at nodes along the lower edge boom (Figure 6.10(b)) that adjacent bracing members must have equal and opposite internal forces. However, at the centre, symmetry requires these forces to be equal. The only way these two requirements may be satisfied is by unloading the two members. From the equal and opposite forces in the edge sloping braces, it also follows, from a consideration of equilibrium in a horizontal plane, that the edge lower boom members must also be unloaded.

Although repetition features usually make it reasonably straightforward to generate even large space truss arrays, it may occasionally be desirable to analyse only a symmetric portion of a structure. As an example, the model needed for the analysis of a symmetric quarter of the space grid shown in Figure 6.10(a) will be developed. The member layout will be as shown in Figure 6.12, in which the cross-sectional areas of the members lying on the 'cuts' (BC and DC) should be halved, since these members are 'shared' between adjacent quarters. The only physical restraint on the quarter is the vertical support at A, but several other constraints are needed, both to prevent rigid body motion and to ensure that the symmetry conditions are fully modelled. Following general precepts, displacement normal to a line of symmetry must be prevented. This leads to the horizontal restraints along BC and CD, as well as the corresponding restraints applied to the cut upper boom members. The cut upper boom members are local mechanisms and it is desirable to stabilise these by applying vertical constraints at their cut ends. The resulting set of constraints shown in Figure 6.12 will both eliminate rigid body movement and will also effectively model the symmetry requirements.

Worksheet 6.2 Computer analysis

W6.2.1 The derrick crane shown in Figure W6.2.1 is supported purely vertically at C and has vertical supports at A and B, together with horizontal support such that it is stable and statically determinate. Select a set of constraints to model this support system. Why is it necessary to provide 'soft' spring supports in a horizontal plane at node E?

The jib length is fixed but the tie is a cable, whose length can be altered so that the elevation of the tie can be adjusted between 35° above the horizontal to 20° below (Figure W6.2.1(b)). In addition, the tie/jib system can be rotated about the post through 270° (Figure W6.2.1(c)).

Use computer analyses to determine the maximum tension and/or compression in each of the members of the crane and state the associated tie/jib orientations. What counterweights are needed at A and B to prevent uplift? Provide qualitative explanations for your results.

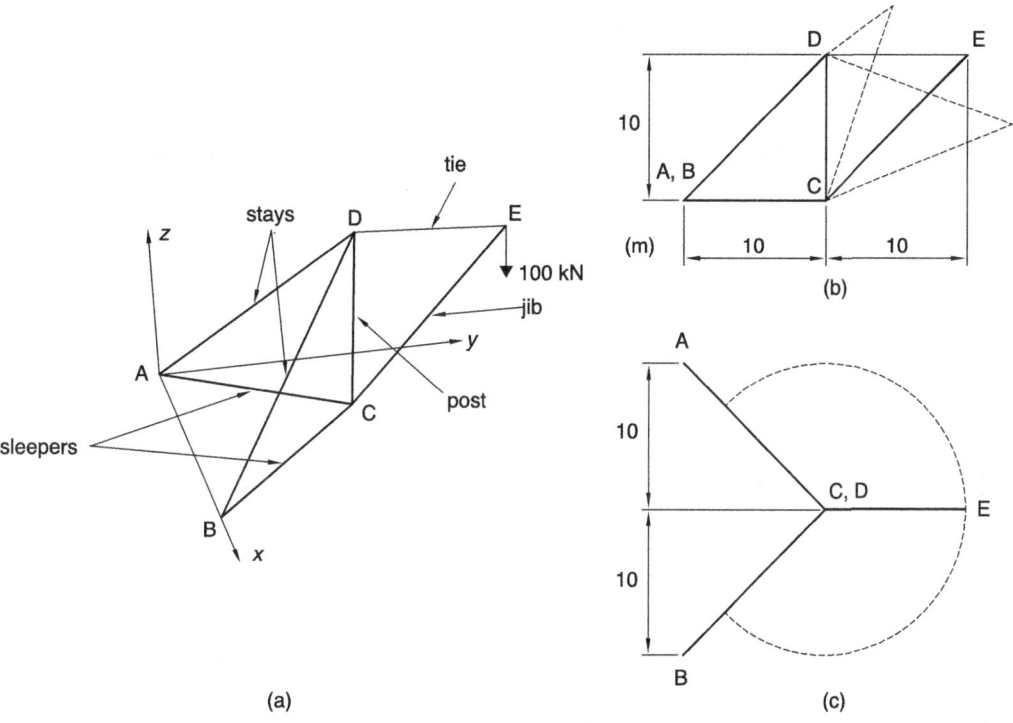

Figure W6.2.1   Derrick crane: (a) perspective, (b) elevation, (c) plan

W6.2.2 The double grid space truss shown in Figure W6.2.2 is similar to that of Figure 6.10(a), except the internal top boom members are arranged diagonally. These members are assumed not to be connected at their intersection points. The pyramidal units have a 2 m square base and are 1 m high. The structure is supported vertically at its corners and carries a uniform load of 5 kN/m^2 on its upper surface. The top boom members and inclined bracing members all have cross-sectional areas of 1000 mm^2. The bottom boom members all have cross-sectional areas of 500 mm^2. The structural material is mild steel, having $E = 207 \text{ kN/mm}^2$.

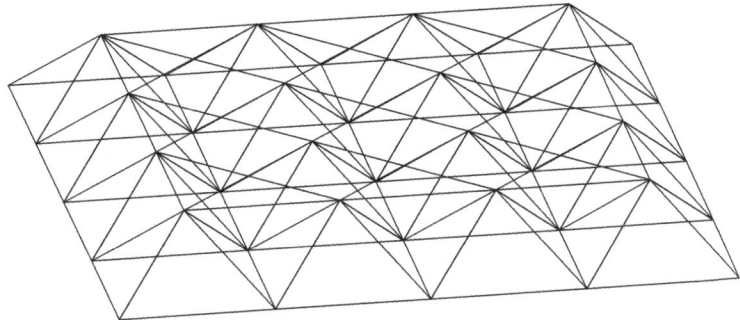

Figure W6.2.2   Worksheet 6.2.2

Use a computer package to analyse both the full structure and a symmetrical quarter of the structure. Compare and contrast the axial load distribution obtained with that of the truss considered in Example 6.1. Re-analyse the structure assuming a fully pinned support condition at each corner and explain the changes this produces in the axial load distribution.

W6.2.3 The transmission line tower shown in Figure W6.2.3 was designed to be economical and easily constructed in remote areas of Australia (Bull, 1958). The tower was designed for two loading cases: (a) three conductor cable dead loads of 10 kN each plus a transverse load of 60 kN (Figure W6.2.3(a)); (b) two conductor cable dead loads of 10 kN each plus a longitudinal, 'broken conductor' load of 25 kN (Figure W6.2.3(b)). The cross-sectional area of the eight principal truss members were all 1850 mm^2; for the cross-beam (AB), $A = 8000$ mm^2; for the cross-member (CD), $A = 1000$ mm^2; and for the guys, $A = 285$ mm^2. The structural members were constructed from steel with $E = 200$ kN/mm^2 and for the guy ropes $E = 80$ kN/mm^2. The overall dimensions of the tower are given in Figure W6.2.3(c), (d).

The structural members and the guys are both constructed from steel. Why, then, are significantly different values of Young's Modulus suggested for the two types of member? Explain why the tower as shown in Figure W6.2.3(a) cannot be directly analysed

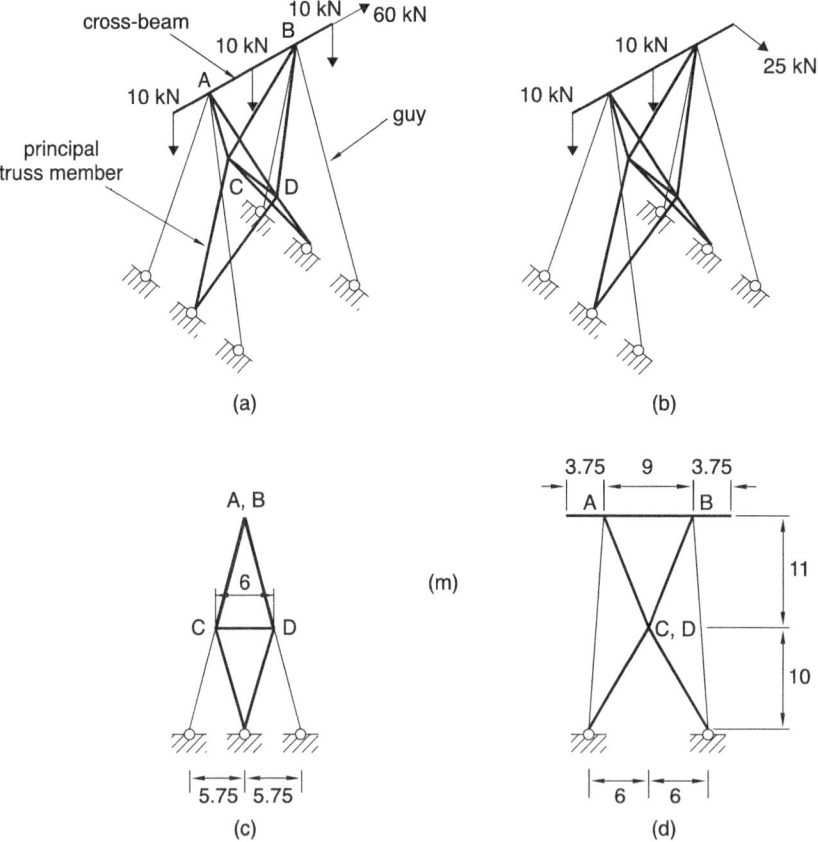

Figure W6.2.3 *Transmission line tower: (a) transverse load, (b) longitudinal load, (c) transverse elevation, (d) longitudinal elevation*

by a space truss package, but this is possible if the cross-beam is replaced by a single member AB and all loading is applied only to nodes A and B. Show that the tower has two mechanism degrees of freedom if the four guys are omitted and identify the two mechanisms.

Make the guys ineffective by reducing their cross-sectional areas by a nominal factor of 100. Use a computer analysis to obtain the forces produced in the truss by the pre-tension in the guys (35 kN in each guy). Why is it necessary to make the guys ineffective for this analysis but not to remove them completely? Determine equivalent nodal loads at A and B to represent the effects of the transverse and longitudinal loading cases. Using the actual guy areas, analyse the tower for the two loading cases and hence obtain resultant axial force distributions by combining these results with the pre-tension analysis.

If the maximum compression sustained by a tower member is to be the same under both loading cases, to what dimension should CD be changed?

W6.2.4 The nodes of the Schwedler dome shown in Figure W6.2.4 are situated at the vertices of one of three regular pentagons. The base nodes are at a radius of 10 m, the intermediate level nodes at a radius of 7.5 m and the top level nodes at a radius of 2.5 m. The dome is subjected to a horizontal load of 10 kN, as shown in Figure W6.2.4(a).

Make a qualitative assessment of the nature of the forces induced in the dome members if (a) the top is unbraced and (b) the top is triangularly braced, as indicated by the dashed members in Figure W6.2.4(a). Confirm your predictions by computer analyses and explain the differences in behaviour that result from the top bracing.

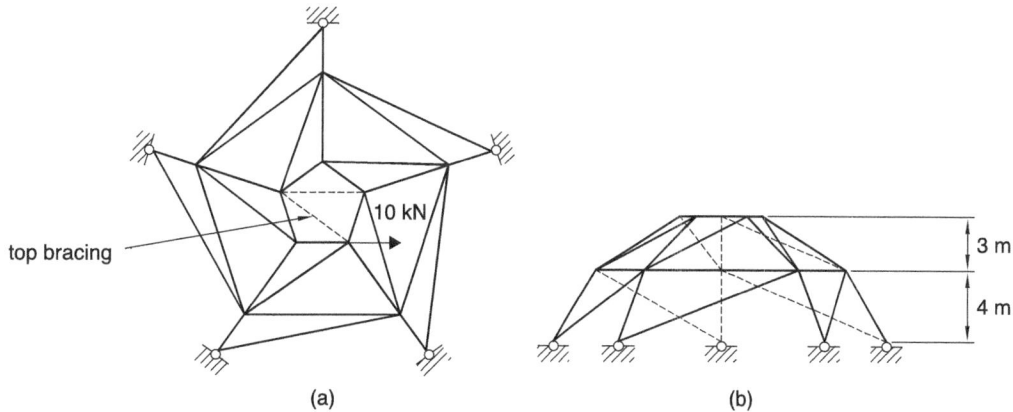

Figure W6.2.4   Schwedler dome: (a) plan, (b) elevation

6.5   INFLUENCE LINES

Influence lines for space trusses are generally not conveniently derived manually and computer generation using Müller-Breslau's principle is the usual approach. The principle is applied in exactly the same way as for plane trusses, but the geometrical interpretation is obviously somewhat more complex. The procedure is illustrated in the following example.

Example 6.2 Influence line application to a crane jib

The crane jib structure shown in Figure 6.13(a) consists of 11 similar, inter-connected pyramidal units. The units have a 2 m square base and are 1.414 m high. The jib is supported simply at one end and also by a central tie ($A_t = 1000$ mm^2), which is anchored 7 m above the base of the jib. All the members of the jib are initially taken to be similar ($A_j = 2000$ mm^2). Under the action of a travelling load of 100 kN, which is shared equally between the two lower booms, it is required to determine the maximum compression in a boom member and in an inclined bracing member.

To determine the maximum compression requires the identification of both a critical member and a critical load position. To narrow the range of possibilities, it is helpful to consider a two-dimensional 'stick-and-string' model of the jib, as shown in Figure 6.13(b). The boom members of the truss resist bending in the 'stick', and this is likely to be a maximum for negative bending at B and for positive bending at approximately D (the mid-point of AB). Introducing unit rotations at D and B provides the bending moment influence lines for these points as shown in Figure 6.13(c), (d). Clearly the maximum bending moment at D is obtained by placing the load at the free end, C. However, this causes negative bending at D, which results in compression on the underside of the jib, where it is shared between the two lower boom members. Although the bending moment resulting from positioning the load at D will only be a half of the end loading case, positive bending will be produced, which will be resisted by the single upper boom. It follows that loading at either position can be expected to produce rather similar compression values at point D.

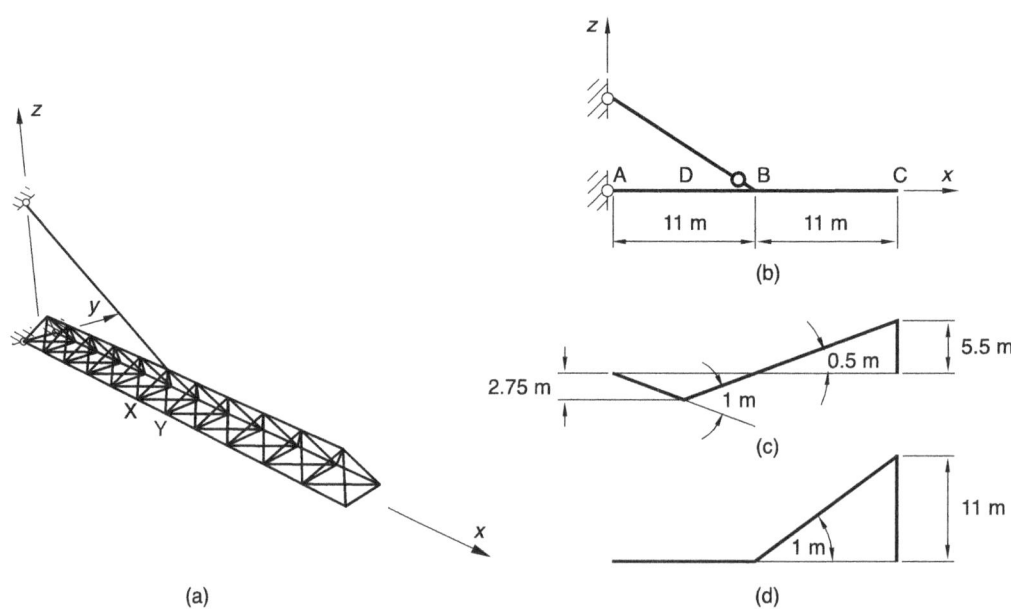

Figure 6.13   Crane jib influence line example: (a) layout, (b) stick-and-string model, (c) IL for BM$_D$, (d) IL for BM$_B$

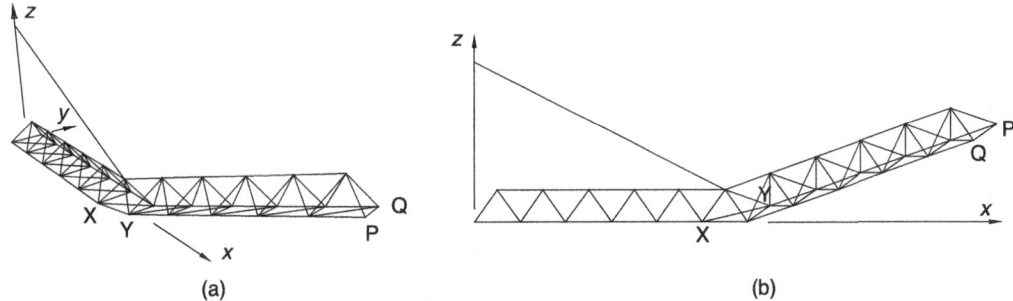

Figure 6.14   Influence lines for X–Y: (a) perspective, (b) elevation

If bending at B is considered, then end loading is clearly critical (Figure 6.13(d)) and will produce twice the maximum bending moment that was found at D. Negative bending is produced and it is therefore the lower boom members around point B that need to be investigated in further detail. An influence line for member X–Y (Figure 6.13(a)) will therefore be generated by computer. Following the procedure used for plane trusses, the end forces needed to generate the influence lines are those that produce unit extension of the member. Thus, taking $E = 200 \, \text{kN/mm}^2 = 200 \times 10^6 \, \text{kN/m}^2$, the forces required to produce a unit (1 mm) extension of a member may be obtained by taking $u_X = 0$ and $u_Y = 1 \times 10^{-3}$ m. In the case of XY, since it lies in the X direction

$$\left\{ \begin{array}{c} f_{xX} \\ f_{xY} \end{array} \right\} = \left\{ \begin{array}{c} N_X \\ N_Y \end{array} \right\} = \frac{EA_j}{L} \left[\begin{array}{cc} 1 & -1 \\ -1 & 1 \end{array} \right] \left\{ \begin{array}{c} u_X \\ u_Y \end{array} \right\}$$

$$= \frac{200 \times 10^6 \times 2 \times 10^{-3}}{2} \left[\begin{array}{cc} 1 & -1 \\ -1 & 1 \end{array} \right] \left\{ \begin{array}{c} 0 \\ 10^{-3} \end{array} \right\} = \left\{ \begin{array}{c} -200 \\ 200 \end{array} \right\} \text{kN} \qquad (6.2)$$

The displaced shape due to application of these nodal loads, which gives the axial load influence 'line' for X–Y, is shown in Figure 6.14. It will be seen from Figure 6.14(a) that the loading displaces the jib principally in the y and z directions. Under vertical loading, it is the z-displacements that provide the influence values, however, and the elevation shown in Figure 6.14(b) is therefore more helpful. Numerically, the vertical displacements of P and Q are $z_P = 4.24 \, \text{mm}$ and $z_Q = 3.54 \, \text{mm}$. When the specified loading of 100 kN is applied at the free end of the jib (divided equally between P and Q), the compression induced in X–Y is therefore $50(4.24 + 3.54) = 389 \, \text{kN}$.

If an inclined bracing member is considered, then shearing action on the stick-and-string model (Figure 6.15(b)) is relevant, and this is likely to be greatest either at the ends of the span AB or at the start of the overhang section BC. Shear force influence lines for positions A and B are therefore shown in Figure 6.15(c), (d). At A and B positive shear force is produced by downwards loading, which will produce compression (Figure 6.16(a)) in the diagonals that slope upwards in the positive x direction. Members X–Y and R–S (Figure 6.15(a)) are therefore selected for more detailed analysis since they are the closest relevant members to A and B, respectively.

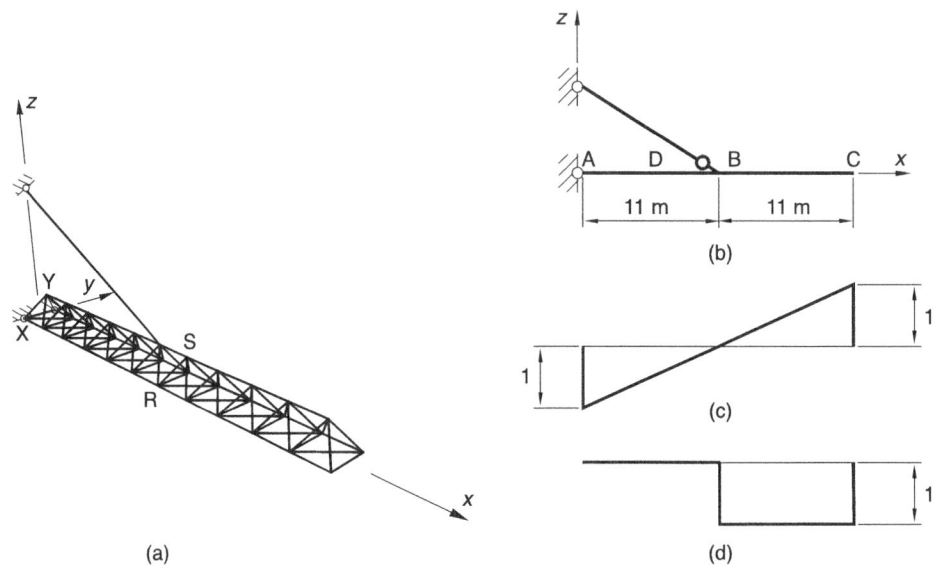

Figure 6.15   Crane jib influence line example: (a) layout, (b) stick-and-string model, (c) IL for SF_A, (d) IL for SF_B

To produce an influence line for a three-dimensional member, it is necessary to determine the force components in the three co-ordinate directions corresponding to the axial load created by unit elongation of the member. For the member shown in Figure 6.16(b) carrying an axial tension N, the components of N in the three co-ordinate directions are given by the product of N with the (direction) cosines of the angles between the direction of the members and the co-ordinate axes. Thus, if the member is of length L and has projections on the co-ordinate axes of X, Y, Z, then

$$\begin{Bmatrix} N_x \\ N_y \\ N_z \end{Bmatrix} = \begin{Bmatrix} X/L \\ Y/L \\ Z/L \end{Bmatrix} N \tag{6.3}$$

For either member X–Y or member R–S, $L = 2\,\text{m}$, $X = 1\,\text{m}$, $Y = 1\,\text{m}$ and $Z = 1.414\,\text{m}$, and, from equation (6.2), the axial load produced by a unit

Figure 6.16   (a) Diagonal bracing loads under positive shear, (b) three-dimensional force components

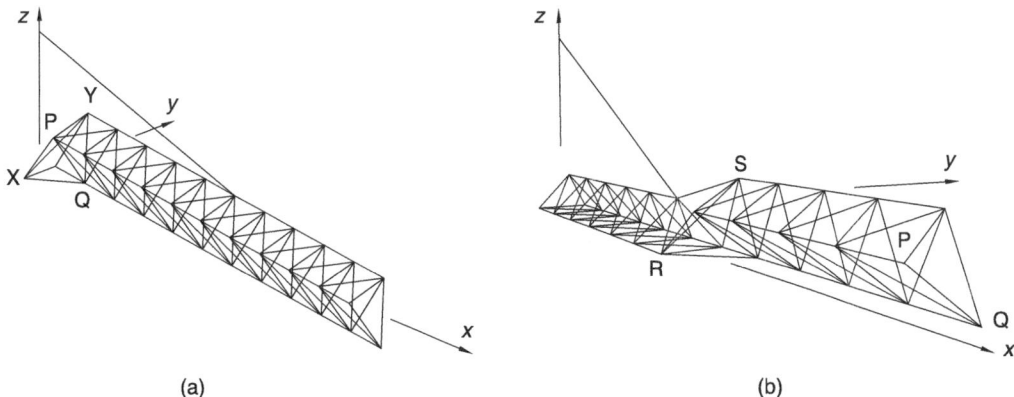

Figure 6.17   (a) Influence line for X–Y, (b) influence line for R–S

extension will be 200 kN. Hence the nodal loads needed to produce the relevant influence line for X–Y (and similarly for R–S) are

$$\begin{Bmatrix} f_{xX} \\ f_{yX} \\ f_{zX} \\ f_{xY} \\ f_{yY} \\ f_{zY} \end{Bmatrix} = \begin{Bmatrix} 0.5 & 0 \\ 0.5 & 0 \\ 0.707 & 0 \\ 0 & 0.5 \\ 0 & 0.5 \\ 0 & 0.707 \end{Bmatrix} \begin{Bmatrix} -200 \\ 200 \end{Bmatrix} = \begin{Bmatrix} -100 \\ -100 \\ -1.414 \\ 100 \\ 100 \\ 1.414 \end{Bmatrix} \text{kN} \tag{6.4}$$

From a computer analysis of the application of these nodal loads, the displaced shapes, and hence influence lines for the two members, are shown in Figure 6.17. From the figure, it may be seen that extension of the members principally produces upwards displacement, coupled with rotation about the longitudinal (x-) axis. The effect of the combined displacement is that the lower boom on the same side as the members moves upwards, and compression is therefore produced in the members by application of downward load on this side. On the opposite side of the lower boom, there is little resultant vertical movement so loading this side will not produce significant loads in the members considered. In addition, it may be noted that for member R–S (Figure 6.17(b)) the influence line agrees with that of the stick-and-string model (Figure 6.15(d)) that loading A–B does not affect the member, and that loading placed anywhere along B–C has a very similar effect.

In the case of member X–Y, the vertical displacement is greatest closest to the member, and the critical loading position will be to place the 50 kN loads at P and Q, the first available lower boom positions (Figure 6.17(a)). From the analysis, $z_P = 1.31$ mm and $z_Q = -0.1$ mm, and it follows that the compression induced in X–Y is $50(1.31 + (-0.1)) = 65$ kN. Using end loading points as P and Q for member R–S (Figure 6.17(b)), the displacements are $z_P = 1.41$ mm and $z_Q = 0.01$ mm, so that the compression in this case is $50(1.41 + 0.01) = 71$ kN. Member R–S is therefore the critical diagonal brace.

Worksheet 6.3   Influence lines for space trusses

W6.3.1 The supporting structure for the London Eye Ferris Wheel (Figure W6.3.1(a)) is shown diagrammatically in Figure W6.3.1(b). The structure consists of the wheel spindle ($A_s = 0.6\,\text{m}^2$) supported by an A-frame ($A_A = 0.2\,\text{m}^2$), both constructed in steel ($E = 206\,\text{kN/mm}^2$). Additional support is provided by sets of longitudinal ($A_{gl} = 0.2\,\text{m}^2$) and transverse ($A_{gt} = 0.15\,\text{m}^2$) guys, for which $E = 100\,\text{kN/mm}^2$. Approximate dimensions for the structure are given in Figure W6.3.1(c), (d). Use a computer package to generate an influence line for the axial force in the transverse guy AB (Figure W6.3.1(b)).

Explain how the influence line demonstrates the following properties of the guy AB:

(a) application of longitudinal (x direction) load at either B or C has an identical effect on the guy;
(b) application of transverse (y direction) load at B is most effective in loading the guy;
(c) application of transverse (y direction) load at C has no effect on the guy.

Explain, also, why the guy possesses the properties listed.

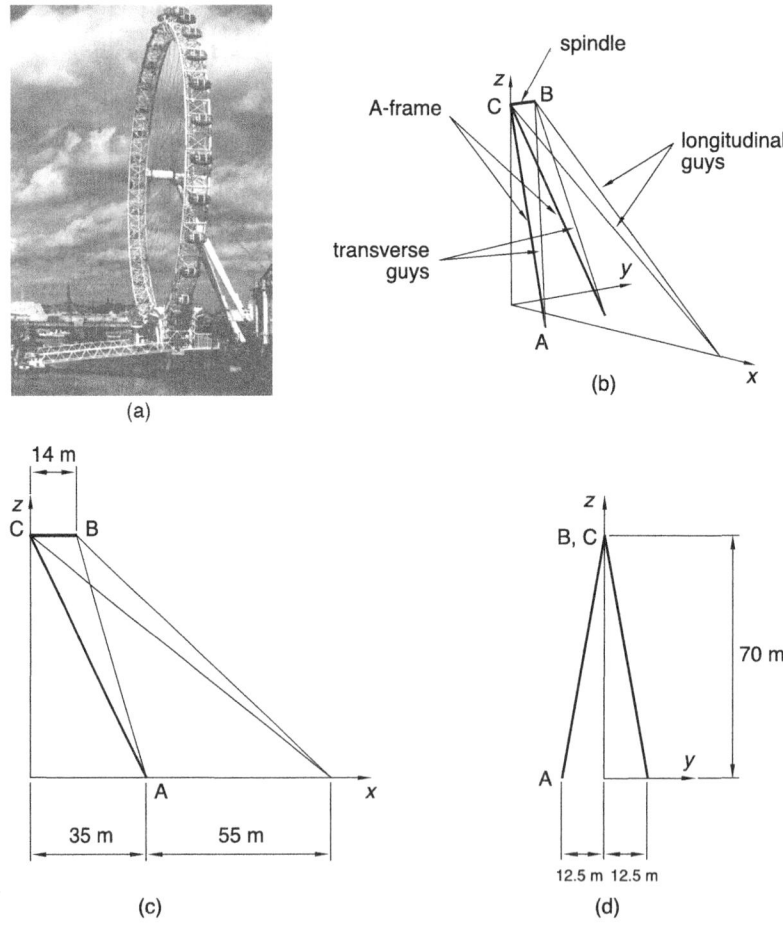

Figure W6.3.1   London Eye: (a) photograph, (b) perspective, (c) longitudinal elevation, (d) transverse elevation (reproduced with permission from Babtie Group. Independent Engineer for the London Eye)

It has been said of the structure that '*despite its lack of geometrical redundancy, it would take an enormous force to push the structure over*' (Institution of Structural Engineers, 1999), and '*For stability, the rear of the spindle is tailed down via one set of cables attached to the feet of the A-frame columns and by a second cable set anchored to the tension foundation. This dual system makes the force distribution between cable sets statically indeterminate*' (Berenbak *et al.*, 2001). Can these statements be reconciled? What is the degree of redundancy of the structure?

W6.3.2 The double length crane jib shown in Figure W6.3.2 is based on a pyramidal unit which has a base 2 m square and is of height 1 m. The jib is simply supported at one end and has additional support from the post and guy arrangement illustrated (see frontispiece to this chapter). The upper and lower boom members are to be of the same cross-section and all the inclined bracing members are to be similar. The jib is to be designed for a mobile point load, to be shared equally between any opposite pair of lower boom nodes.

Use a 'stick-and-string' hand calculation model to identify likely critical members and loading positions. Employ a computer package to generate influence lines for the identified critical members and use these to determine the maximum compressions sustained by both a boom and an inclined bracing member if the maximum load is to be 100 kN. Why does it not matter what sections sizes are assumed for this investigation?

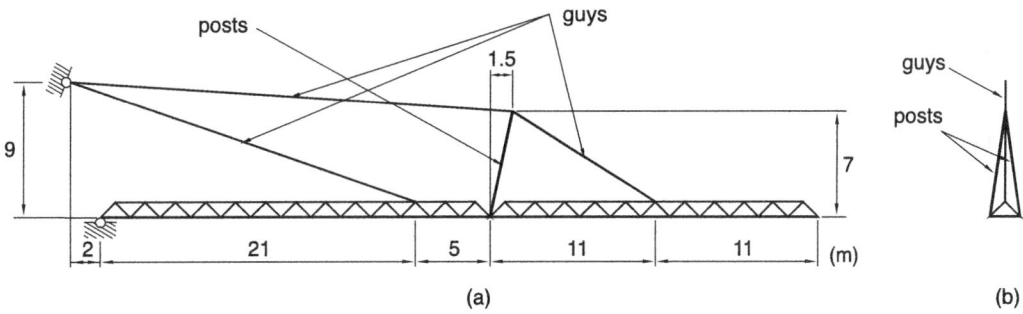

Figure W6.3.2   Double length crane jib: (a) side elevation, (b) end elevation

6.6   REFERENCES

Berenbak, J., Lanser, A. and Mann, A. P. (2001) 'The British Airways London Eye Part 2: Structure'. *The Structural Engineer*, **79**(2), 19.

Bull, F. B. (1958) 'The development and testing of a transmission line tower'. *Proc. of the Fiftieth Anniversary Conference*. The Institution of Structural Engineers. 247–253.

Ghali, A. and Neville, A. M. (1997) *Structural Analysis*. 4th edn. London: E & F N Spon – Chapter 6 of this reference deals with computer analysis in general and includes transformation matrices for different structural forms.

Institution of Structural Engineers (1999) Report: the London Eye. *The Structural Engineer*, **77**(23), 12.

MacGinley, G. (1998) *Steel Structures – Practical Design Studies*. 2nd edn. London: E & F N Spon – Chapter 5 of this reference considers the analysis and design of multi-storey steel buildings.

Open University (1991) *Structural Components*. Milton Keynes: Open University – a VHS format video (ref: T235/05V) that includes a discussion of electrical pylon design and shows a full-scale test to destruction of a prototype tower.

Timoshenko, S. P. and Young, D. H. (1965) *Theory of Structures*. 2nd edn. New York: McGraw-Hill – Chapter 4 of this reference includes a thorough description of space truss theory.

7 Grids

7 Grids

7.1 INTRODUCTION

Grids, alternatively known as grillages, comprise a planar system of members subjected to loading that acts normally to the plane of the members. Perhaps the simplest grid is the 'bent beam' (Figure 7.1(a)). As with pure beams, grid members bend about in-plane axes that are perpendicular to the directions of the individual members of the grid. Thus, in Figure 7.1(b), CB is essentially a cantilever beam and its bending moment diagram will be as shown in the figure. The axis of bending for any point along CB, about which the bending moment acts, is perpendicular to the member and so parallel to AB. Member AB will also bend, and, at any point, its axis of bending will be normal to the member and so parallel to CB. At B, therefore, the axis passes through C and there is no bending moment at this point. The bending moment will increase linearly along BA, however, again giving rise to a cantilever form of diagram (Figure 7.1(b)).

It is clear from Figure 7.1(a) that AB will twist as well as bend and this action distinguishes grid members from beams. Twisting moment, alternatively referred to as torque or torsion, is the moment *about the axis of the member* of the resultant force acting on a section. The sign convention used here will be a

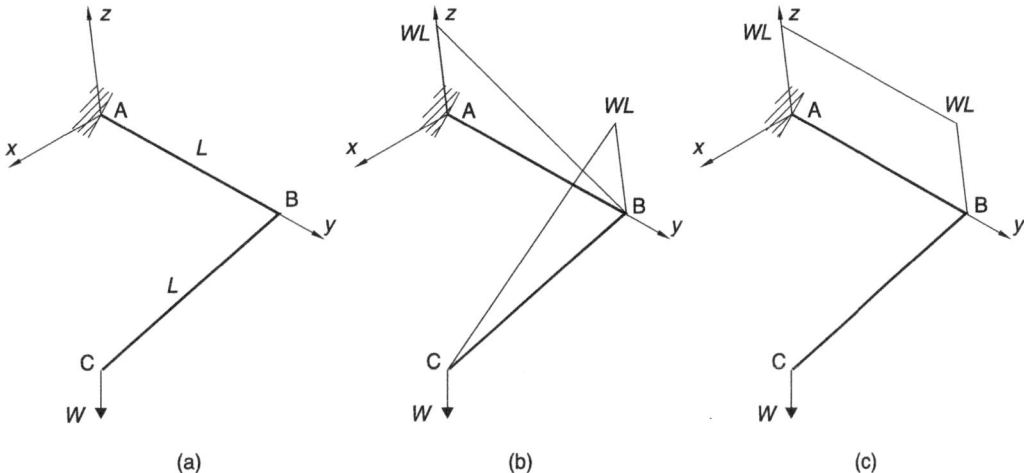

Figure 7.1   (a) Bent beam, (b) bending moment diagram, (c) torque diagram

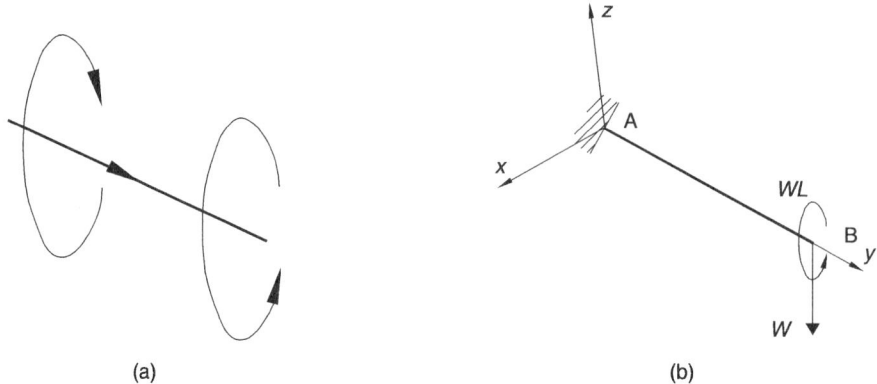

Figure 7.2 (a) Positive torsion, (b) end actions on AB

deformation-based convention that positive torque will create increasing clock-wise twist as progress is made along a member's axis (Figure 7.2(a)). It follows from these definitions that the torque action on AB is constant, since the moment about its axis (WL) does not change, and is also positive, since it is similar to Figure 7.2(a) in nature. The torque diagram for ABC is therefore as shown in Figure 7.1(c). By comparing Figure 7.1(b) and Figure 7.1(c), it may be noted that, at B, the torque value in BC (zero) becomes the bending moment value in AB and, conversely, the bending moment in BC (WL) becomes the torque value in AB. This occurs because the axis of AB is parallel to a normal to BC and will happen at the junction of any two perpendicular grid members. Finally, the bending moment and torque diagrams for AB could, alternatively, be readily developed if the load W at C is transferred to B, in which case it must be accompanied by its moment effect (WL) about AB (Figure 7.2(b)). The load, W, then produces the cantilever bending moment diagram and the torque, WL, the constant torsion.

Should the bent beam be curved, then similar principles will apply. It is simply necessary to be careful in the determination of moments about a member's axis (to find the torque) and about a normal to the axis (to find the bending moment). If the bent beam of Figure 7.1(a) is replaced by the quadrant beam, of radius L, as shown in Figure 7.3(a), then the torque at a general point B will be given by Wa, since a is the offset to the tangent (axis) direction at B. Also, Wb will give the bending moment, since b is the offset from the normal direction at B.

The bending moment and torque distribution diagrams (Figure 7.3(b), (c)) are now continuous, since a and b vary continuously. It may be observed from the figures that the bending moment initially increases at a faster rate, but subsequently the torque increases more quickly so that the identical values of WL are achieved at the support.

Like shear force, torque is an anti-symmetric quantity and will therefore be zero at points of symmetry. The circular beam shown in Figure 7.4(a), for example, is supported vertically at the four points indicated and is loaded uniformly. The

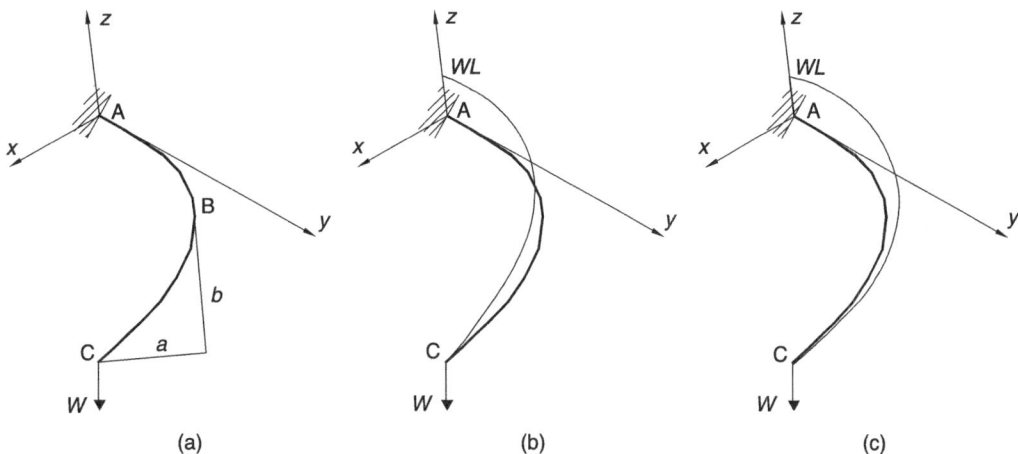

Figure 7.3   (a) Quadrant beam, (b) bending moment diagram, (c) torsion diagram

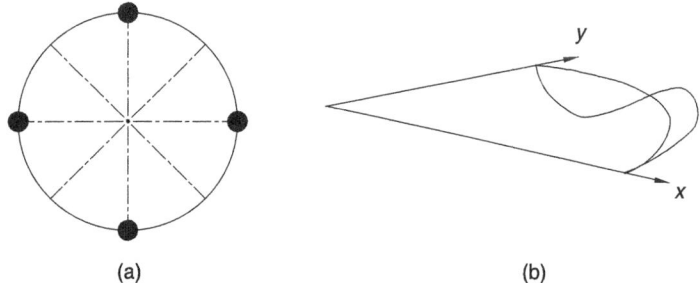

Figure 7.4   (a) Circular beam, (b) torque diagram for a quadrant

beam has points of symmetry at the supports and at positions orientated at 45° to the lines joining the supports. It follows that the torque must be zero at these points and will also alternate sign at these positions. This results in the form of a torque distribution diagram shown in Figure 7.4(b) for a quarter of the beam. The implication of this distribution for displaced shape is shown in Figure 7.5, in which out-riggers have been added to the beam to indicate its rotation. From the figure, it may be seen that the inwards rotation at a support gradually changes to an outwards twist at the 45° position and then, with the change of torque sign, turns back again after this point.

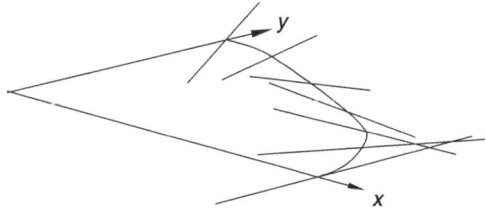

Figure 7.5   Displaced shape for a quadrant

7.2 KINEMATIC AND STATICAL DETERMINACY

A general node on a grid, such as B in Figure 7.1, will displace vertically and will also rotate about both the x- and the y-axes. As long as the loading is normal to the plane of the grid and small deflections are presumed, it may be assumed (as with 'pure' beams) that in-plane displacement does not occur, so that x- and y-displacements are taken to be negligible. Further, there will be no rotation about the vertical (z-) axis. There are therefore three displacement degrees of freedom at a general node (Figure 7.6(a)): z-displacement and x-, y-rotations. It follows that, if there are j joints and d restrained displacement components, the kinematic indeterminacy, k, will be given by

$$k = 3j - d \tag{7.1}$$

As already discussed, grid members both bend and twist. The bending moment will inevitably be accompanied by a shearing force, so that, at any point, each member will have three stress resultants: torque, shear force and bending moment (Figure 7.6(b)). It follows that, if there are m members and h stress resultant releases (normally moment or torsion hinges), the statical indeterminacy, q, will be given by the number of stress resultants that may not be found by statics as

$$q = (3m - h) - k \tag{7.2}$$

If, for example, it is required to find the kinematic and statical indeterminacies of the grid shown in Figure 7.7(a), which is supported vertically at the indicated points, then $d = 5$, $j = 12$ and $m = 17$, so that $k = (3 \times 12) - 5 = 31$ and $q = (3 \times 17) - 31 = 20$. The redundancy value of 20 may be checked in one of two ways. First, use may be made of the fact that a grid will be determinate if it has three non-collinear, vertical supports only and its members do not form any closed loops. The number of releases it is necessary to introduce to convert a grid to this form will therefore give its statical indeterminacy. For the example grid, two of the supports can be removed and six 'cuts' introduced to produce a statically determinate grid (Figure 7.7(b)). At each cut, the full set of stress resultants are released (Figure 7.6(b)), so that the degree of statical determinacy

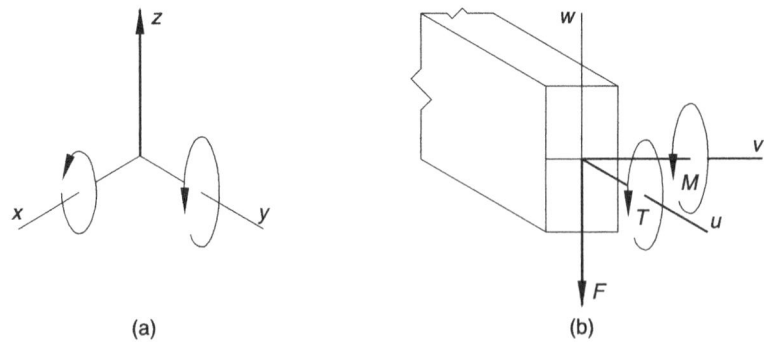

(a)   (b)

Figure 7.6   (a) Displacement degrees of freedom, (b) stress resultants

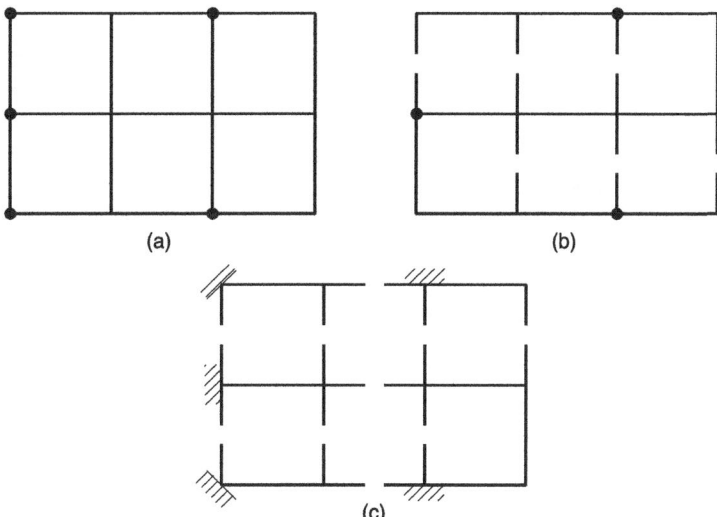

Figure 7.7 (a) Indeterminacy example grid, (b) determinate version with vertical supports, (c) determinate version with fixed supports

is checked as $2 + (3 \times 6) = 20$. An alternative approach is to make all the supports fixed, so that a 'new' structure is created, as shown in Figure 7.7(c) for the example grid. The redundancy of the actual grid may then be related to that of the new one by the number of extra constraints introduced. For the example, two rotational restraints must be introduced at each of the five vertical supports so that $q = q_{\text{new}} - 10$. The new grid may then be made determinate by introducing sufficient cuts to turn it into a set of branched, bent beams, each of which will be determinate. For the example grid, this requires ten cuts, which may be arranged as shown in Figure 7.7(c). It follows that $q_{\text{new}} = 3 \times 10 = 30$ and, hence, $q = 30 - 10 = 20$, as before.

7.3   STATICALLY DETERMINATE GRIDS

Stress resultant determination

Although not common, statically determinate grids are useful in becoming familiar with the inter-relationships between the stress resultants relevant to grids. As an example, the stress resultant distributions for the grid shown in Figure 7.8(a) will be determined. The grid is rectilinear, supported vertically at A, C and E, and carries a distributed load of 8 kN/m on DE. The three non-collinear vertical supports and the 'open' nature of the grid ensure it is determinate. However, as a check, it may be noted that $j = 5$, $d = 3$, $m = 4$, so that $k = (3 \times 5) - 3 = 12$ and $q = (3 \times 4) - 12 = 0$.

Reaction determination is an essential preliminary step and, as with beams, is most securely performed by use of moments for reaction evaluation and reserving

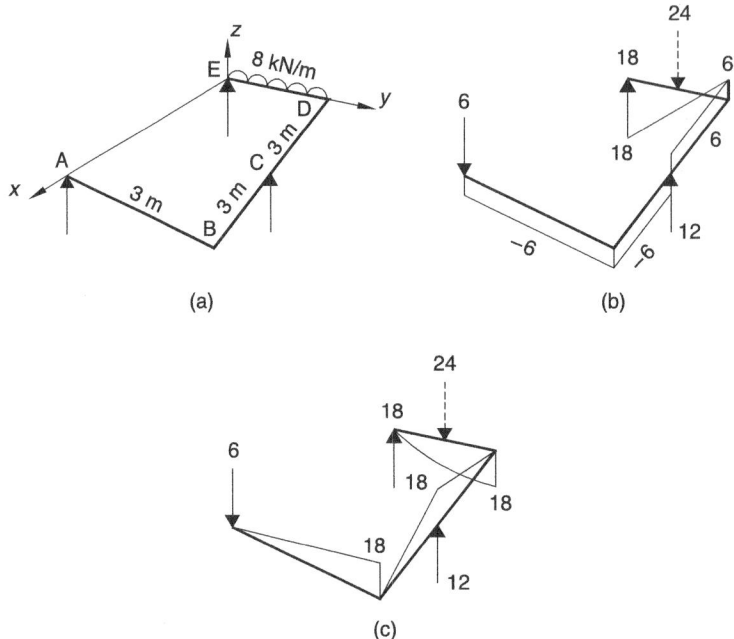

Figure 7.8 *(a) Statically determinate grid example, (b) SF and forces (kN), (c)* BM *(kNm) and forces (kN)*

vertical force equilibrium as a check. Thus, by moments about the *x*-axis in Figure 7.8(a)

$$3R_{\mathrm{C}} = (8 \times 3) \times 3/2 \Rightarrow R_{\mathrm{C}} = 12 \,\mathrm{kN}$$

And, by moments about the *y*-axis and a line parallel to the *y*-axis through A

$$6R_{\mathrm{A}} + 3R_{\mathrm{C}} = 0 \Rightarrow R_{\mathrm{A}} = -6 \,\mathrm{kN} \quad \text{and}$$

$$6R_{\mathrm{E}} + 3R_{\mathrm{C}} = (8 \times 3) \times 6 \Rightarrow R_{\mathrm{E}} = 18 \,\mathrm{kN}$$

Checking by resolving vertically gives

$$R_{\mathrm{A}} + R_{\mathrm{C}} + R_{\mathrm{E}} = -6 + 12 + 18 = 24 = (8 \times 3)$$

If the grid is 'opened-up' and viewed from the outside, the shear force convention used with beams may be applied and the shear force distribution will be obtained by 'following through' the loads to produce the diagram shown in Figure 7.8(b). If the beam bending moment convention is also followed, then the bending moments at the nodes are

$$BM_{\mathrm{B-}} = -6 \times 3 = -18 \,\mathrm{kN\,m}; \quad BM_{\mathrm{B+}} = 0 \,\mathrm{kN\,m}; \quad BM_{\mathrm{C}} = -6 \times 3 = -18 \,\mathrm{kN\,m};$$

$$BM_{\mathrm{D-}} = (-6 \times 6) + (12 \times 3) = 0 \,\mathrm{kN\,m}; \quad BM_{\mathrm{D+}} = +6 \times 3 = 18 \,\mathrm{kN\,m}$$

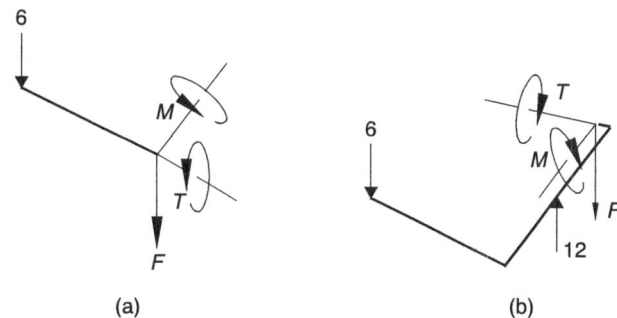

Figure 7.9 *(a)* BM_{B-}, *(b)* BM_{D+}

In determining these moments, it is particularly important to be careful regarding the axis the moment acts about (normal to the member under consideration), and to make sure that the sagging positive convention is correctly applied. The relevant sections for the determination of the bending moment in AB at B (BM_{B-}) and the bending moment in DE at D (BM_{D+}) are, for example, as shown in Figure 7.9(a), (b), respectively. From Figure 7.8(b), (c), it should be noted that the relationship between the change in bending moment and the area under the shear force diagram only holds within individual members and cannot be used around the full grid. Perhaps the most useful application of the property is in locating and evaluating the maximum bending moment in DE. The position of the maximum moment will be at the point of zero shear force, which is located at $18/8 = 2.25$ m from E. The value of the moment will be $0.5(18 \times 2.25) = 20.3$ kN m.

Along AB and DE, there is no offset loading at any point as progress is made from either of the support points (A and E), where the torque must be zero by the nature of the supports. It follows that AB and DE are free of torque (Figure 7.10(a)) and, although these members rotate (Figure 7.10(b)), there is no twist (**change** of axial rotation). From Figure 7.8, the torque in BD is constant and is of value $+18$ kN m. It follows that the axial rotation of BC will increase in a clockwise sense as progress is made from B to D (Figure 7.10(b)). By comparing

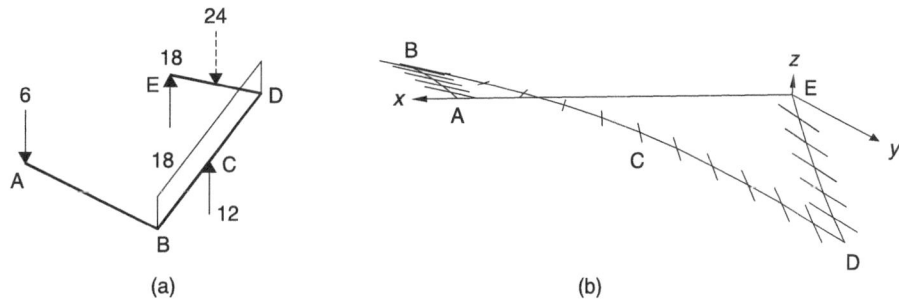

Figure 7.10 *(a) Torque distribution, (b) deflected shape*

Figures 7.8(c) and 7.10(a), the exchange of torque and bending moment values at the right-angled bends, B and D, may be observed.

Worksheet 7.1 Statically determinate grids

W7.1.1 Sketch shear force, bending moment and torque diagrams for the two grids shown in Figure W7.1.1, stating principal values. The grid shown in Figure W7.1.1(a) is fully fixed at its support. The grid shown in plan in Figure W7.1.1(b) is supported vertically at points A, B and E. The grid carries a uniformly distributed downward load of intensity 4 kN/m along BD.

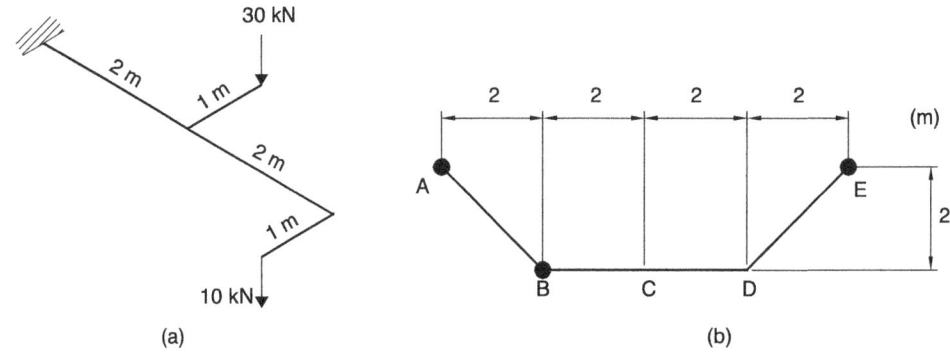

Figure W7.1.1   Worksheet 7.1.1

W7.1.2 If the connections at the joints A, B, C and D are such that torsion cannot be transmitted, prove that the grid shown in Figure W7.1.2 is statically determinate. By considering nodal equilibrium at nodes A, B, C and D, determine the reactions at E, F, G and H due to a vertical point load of 10 kN applied to node A. Each member of the grid is 1 m long. Hence, sketch a bending moment distribution for the grid. If torsional connections are provided (and EI = GJ), the decreasing rank order of the magnitudes of the reactions is E, F, H, G. Why is the rank order different for the torsion-free solution?

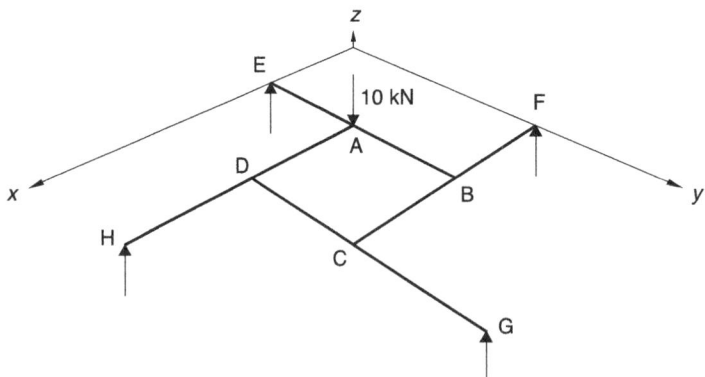

Figure W7.1.2   Worksheet 7.1.2

W7.1.3 Figure W7.1.3(a) shows the plan view of a cranked foundation which is supported by piles and carries a uniformly distributed load of 20 kN/m. The foundation is of square cross-section of side 1 m. As a first approximation, it was considered that the bending moment distribution for the foundation might be obtained from the continuous beam representation shown in Figure W7.1.3.

Determine the kinematic and statical degrees of indeterminacy, k and q, for the structures of Figure W7.1.3. Calculate the torque, bending moment and shear force distributions for the cranked foundation. Calculate, also, the shear force and bending moment distributions for the continuous beam, given that the end and central reactions for a uniform, two-span beam, under uniform loading of intensity w, are given by $3wL/8$ and $5wL/8$ respectively.

Why is it easier to determine the reactions for the cranked beam structure than for the continuous beam? Why are the bending moments at the ends of the cranked portion of the foundation the same and equal to the torque in this section? Why does the continuous beam not provide a reasonable estimate of the bending behaviour of the cranked foundation?

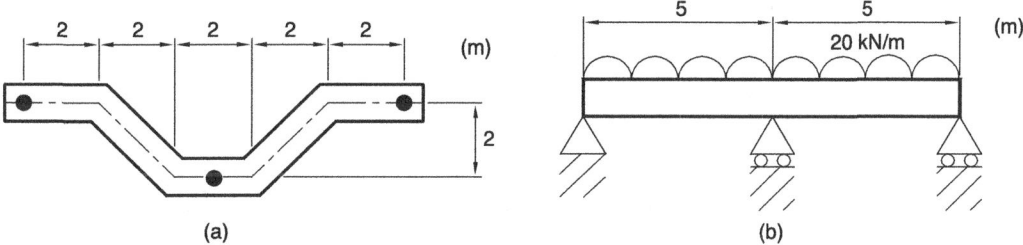

(a)   (b)

Figure W7.1.3   Worksheet 7.1.3

7.4   STATICALLY INDETERMINATE GRIDS

The behaviour of statically indeterminate grids will be examined by looking at the typical bridge deck structure shown in Figure 7.11(a). If a grid of this form is subjected to a set of concentrated loads, such as produced by a heavy vehicle, then the problem of 'load distribution' arises. In essence, load

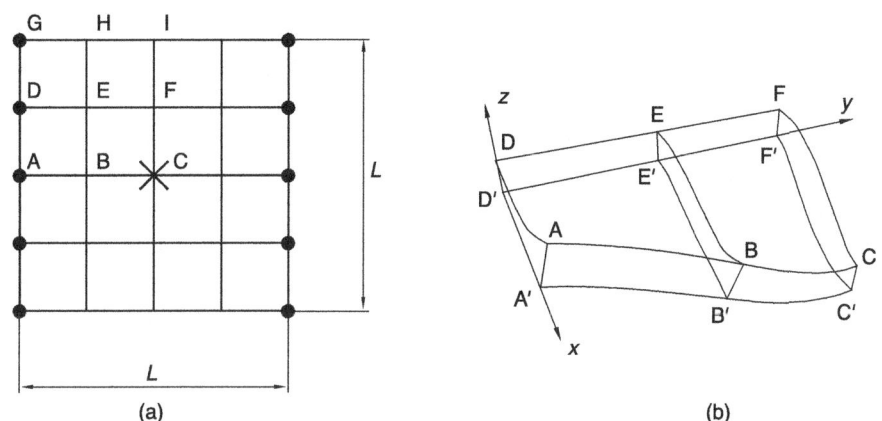

(a)   (b)

Figure 7.11   (a) Plan of example bridge grid, (b) effect of deformation of A–B–C

distribution refers to the proportions of the load that are sustained by each of the longitudinal beams, since this will clearly decrease with transverse distance from the loaded area of the bridge. For simplicity, the effect of a single, central, concentrated load will be examined, as illustrated in Figure 7.11(a). Clearly, the major impact of this load will be on the longitudinal beam that it is applied to (ABC), but some of its effect will be 'distributed' to the remaining beams. The manner in which this is accomplished will be investigated by considering the effects of the deformation of ABC on the neighbouring half-beam DEF, assuming, initially, that no distribution occurs so that DEF does not displace (Figure 7.11(b)).

By symmetry, ABC will not twist and CC′ will remain vertical. Node FF′ also remains vertical, since deflection along DEF has not been permitted. It follows that no twist is induced in CF by deflection along AC. Bending is, however, induced in CF, since C is now lower than F. Applying similar ideas to AD, it is deduced that there is no bending in this member. This occurs because there is no twist in AB (and hence no bending rotation applied to AD), while A′ and D′ remain at the same level, that of the support. There **is** a twist developed in AD because AA′ has rotated from its originally vertical orientation, while DD′ has been kept upright. For the more general member BE, both twisting and bending effects will be present due to the relative rotation of BB′ and EE′ and the relative vertical displacement of B and E.

The complete structural action may be now be described as follows. Grid members along ABC resist the load by bending alone and are stiffened by the transverse grid members, typically BE, which assist by a combination of bending and twisting actions. The transverse members, in turn, invoke resistance from the neighbouring longitudinal members by causing them to bend and twist. This process is repeated, with diminished effect, for the more distant systems of transverse and longitudinal members, hence resulting in distribution of the load throughout the bridge deck.

The effectiveness of the distribution process will, of course, depend on the stiffness of the transverse members provided, since a highly indeterminate ($q = 55$) system is being considered. The bending stiffness will depend on the 'I' values and the torsional stiffness on the torsional constant, 'J' values (Johnson, 2000). For open steel sections, such as 'I' beams, the torsional constant is very low, as compared with the corresponding second moment of area, I, value. Assuming the ratio J/I is 1/1000 for all members, the effect of varying the transverse bending stiffness may be examined by altering the second moment of area of the transverse members (I_t) in relation to the value for the longitudinal members (I_l).

The result of altering the transverse bending stiffness in this way is shown in Table 7.1 in the form of the ratio of the central longitudinal bending moment in the unloaded members to that at the centre of the loaded member. Clearly the load is less well distributed as the transverse stiffness reduces, this reduction being much more pronounced for the edge members than for the members adjacent to the central one. For equal longitudinal and transverse stiffness, it may be noted that the pair of members adjacent to the centre carry more load

Table 7.1 Effect of variation of transverse bending stiffness on load distribution $(J = I/1000)$

	I_t/I_l	I_t/I_l	I_t/I_l	I_t/I_l	I_t/I_l
	0.10	0.25	0.5	0.75	1.00
BM_F/BM_C	0.406	0.483	0.544	0.585	0.617
BM_1/BM_C	0.080	0.024	0.121	0.192	0.250

Table 7.2 Effect of variation of bending stiffness on load distribution $(J = I)$

	I_t/I_l	I_t/I_l	I_t/I_l	I_t/I_l	I_t/I_l
	0.10	0.25	0.5	0.75	1.00
BM_F/BM_C	0.408	0.496	0.563	0.603	0.634
BM_1/BM_C	0.106	0.188	0.262	0.310	0.347

than the central one, but the pair of edge members carry about a half that of the central member.

For the solid type of section used in concrete construction, torsional constants become of the same order as second moments of area. Equal J and I values are therefore assumed in Table 7.2, which provides similar results to Table 7.1. The increased torsional stiffness of the system may be seen to have relatively little effect on the distribution to the adjacent members, but does significantly improve the distribution to the edge members. In general, torsional effects in grids are less significant than bending effects and were often ignored in the past in order to simplify hand analysis. This is no longer necessary with computer analysis but 'torsion free' design is sometimes employed (Hambly, 1991), for which torsional effects need to be eliminated.

7.5 COMPUTER ANALYSIS

Stiffness method theory

Mathematically, the stiffness analysis of grids closely resembles that of plane frames, although the structural behaviour differs markedly. The mathematical similarity arises because each of the two forms is planar and involves three displacement components and three stress resultants at each node. The various matrices involved in the stiffness method are therefore the same size, and all that is required to convert a plane frame program to a grid one is to appropriately modify the plane frame local axis stiffness matrix (equation 4.6).

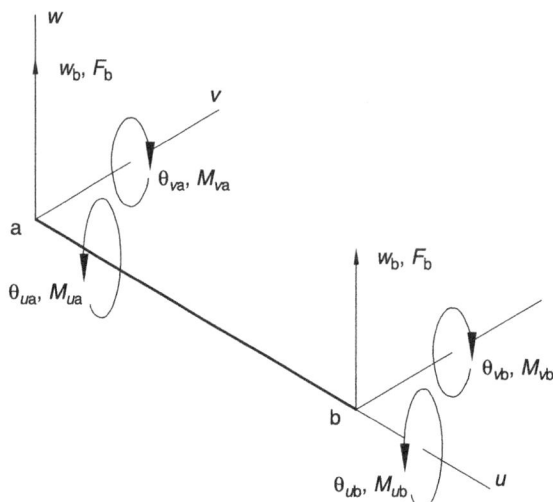

Figure 7.12   Typical grid member

Figure 7.12 shows the relevant displacement components and stress resultants at the ends of a typical grid member and these quantities are related by

$$
p = \left\{ \begin{array}{c} M_{ua} \\ F_a \\ M_{va} \\ M_{ub} \\ F_b \\ M_{vb} \end{array} \right\} = \left[\begin{array}{cccccc} GJ/L & 0 & 0 & -GJ/L & 0 & 0 \\ 0 & 12EI/L^3 & -6EI/L^2 & 0 & -12EI/L^3 & -6EI/L^2 \\ 0 & -6EI/L^2 & 4EI/L & 0 & 6EI/L^2 & 2EI/L \\ -GJ/L & 0 & 0 & GJ/L & 0 & 0 \\ 0 & -12EI/L^3 & 6EI/L^2 & 0 & 12EI/L^3 & 6EI/L^2 \\ 0 & -6EI/L^2 & 2EI/L & 0 & 6EI/L^2 & 4EI/L \end{array} \right]
$$

$$
\times \left\{ \begin{array}{c} \theta_{ua} \\ w_a \\ \theta_{va} \\ \theta_{ub} \\ w_b \\ \theta_{vb} \end{array} \right\} = \mathbf{K}_m e \tag{7.3}
$$

The bending terms in equation (7.3), which relate F and M_v to w and θ, are identical to the corresponding terms in the plane frame matrix. The only change needed to the plane frame matrix therefore is that the axial load terms are replaced by axial moment (torsion) terms. The expression used to relate torque (M_u) and axial rotation (θ_u) is based on Saint-Venant torsion theory (Johnson, 2000), and assumes that the torque in the member is uniform and that warping of the member's cross-section is unrestrained.

It follows from the member stiffness matrix that the member properties required will be the second moment of area (I) and the torsion constant (J). For non-standard cross-sections, torsion constants must be evaluated differently

Table 7.3   Torsional properties of thin-walled sections

Section	J	Example
Closed Variable thickness, t Area enclosed by contour, A	$\dfrac{4A^2 t}{\oint \dfrac{ds}{t}}$	
Closed Constant thickness, t Contour perimeter, S Area enclosed by contour, A	$\dfrac{4A^2 t}{S}$	
Unbranched open Variable thickness, t Contour perimeter, S	$\displaystyle\int_0^S \dfrac{t^3}{3}\,ds$	
Unbranched open Constant thickness, t Contour perimeter, S	$\dfrac{St^3}{3}$	
Unbranched open Consists of n unbranched, open, constant thickness sections	$\dfrac{1}{3}\displaystyle\sum_{i=1}^{n} B_i t_i^3$	

for various forms of section. Thin-walled sections, typical of metal construction, for example, are treated rather differently to solid sections that are typical of concrete and timber construction. Table 7.3 therefore provides data for thin-walled sections, while data for solid sections is given in Tables 7.4 and 7.5.

A further feature of torsional behaviour that it is sometimes necessary to take into account is the position on a cross-section from which torsional effects should be measured. For the general cross-section shown in Figure 7.13(a), the problem is to locate the *shear centre* (SC), about which the torque due to a lateral load F should be measured. The shear centre can alternatively be considered to be a point on a cross-section through which a lateral load must act if it is not to cause a section to twist.

Table 7.4   Torsional constants for solid sections

Section	J	Example
Circle $D =$ diameter	$0.098D^4$	
Square $s =$ side	$0.141s^4$	
Rectangle $B =$ breadth $t =$ thickness (for c see Table 7.5)	cBt^3	
Equilateral triangle $h =$ height	$0.0385h^4$	

Table 7.5   Rectangular section coefficients

B/t	1	1.2	1.5	2	2.5	3	4	5	10	∞
c	0.141	0.166	0.196	0.229	0.249	0.263	0.281	0.291	0.312	0.333

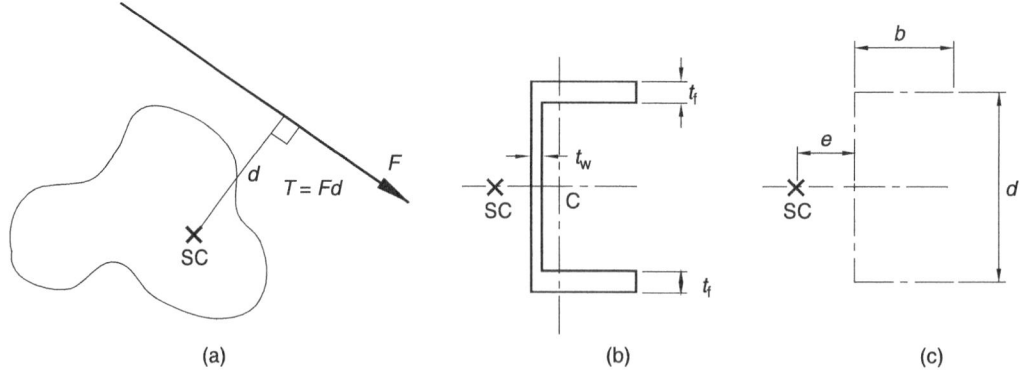

Figure 7.13   (a) General section, (b) channel section, (c) channel contour

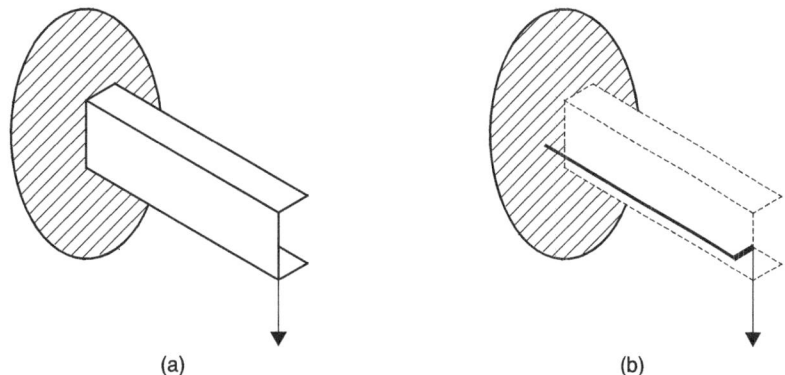

Figure 7.14   (a) Cantilever channel, (b) computer model

It may be shown (Gere and Timoshenko, 1991) that shear centres lie on lines of symmetry. Torsional effects are also most significant in the case of open thin-walled sections (Table 7.3). For grid action, in which in-plane displacements are ignored, it is necessary that sections possess a horizontal axis of symmetry, about which bending occurs, so that displacements due to bending are purely vertical. If the section also has a vertical line of symmetry, then the intersection of the two lines of symmetry will be at the centroid of the section and will also be the shear centre. In the case of thin-walled sections, the commonest section that possesses a horizontal line of symmetry but not a vertical one is the channel section. For a channel (Figure 7.13(b)), therefore, the shear centre will lie on the horizontal line of symmetry, but will not coincide with the centroid (C). The shear centre is, in fact, located at the opposite side of the web to the centroid at an eccentricity, e, from the centre-line (contour) of the web (Figure 7.13(c)) given by

$$e = \frac{3b^2 t_{\mathrm{f}}}{dt_{\mathrm{w}} + 6bt_{\mathrm{f}}} \tag{7.4}$$

It follows from the above that the simple cantilever shown in Figure 7.14(a) will display grid action under the influence of a vertical end load applied through its centroid, since it will twist as well as bend. To capture this behaviour, it will there-fore be necessary to model the beam as a linear element situated along the shear centre line (Figure 7.14(b)), and to employ a rigid off-set element to represent the torsional effect of the load correctly.

Example 7.1   Computer analysis of bent beam

The bent beam shown in plan in Figure 7.15(a) is constructed from a channel section shown in Figure 7.15(b). The channel has $I = 2.25 \times 10^6\,\mathrm{mm}^4$ and its torsion constant may be calculated from Table 7.3 as

$$J = 2\left(\frac{50 \times 10^3}{3}\right) + \left(\frac{80 \times 5^3}{3}\right) = 36.7 \times 10^3\,\mathrm{mm}^4$$

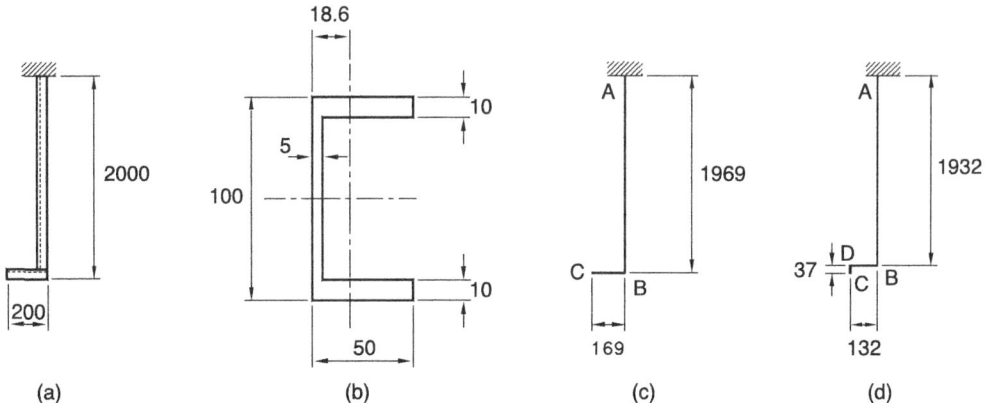

Figure 7.15 *(a) Bent beam, (b) cross-section, (c) centroid model, (d) shear centre model*

The eccentricity of the shear centre from the centre-line of the web is found from equation (7.4) as

$$e = \frac{3 \times 47.5^2 \times 10}{(90 \times 5) + (6 \times 47.5 \times 10)} = 20.5 \, \text{mm}$$

It follows that the shear centre line is $20.5 - (5/2) = 18.0$ mm from the outer edge of the web. The bent beam is to be analysed for the effects of a downward load of 0.5 kN applied at point C to its centroid position. Two possible computer models will be compared. In the first, the effects of shear centre displacement are ignored (Figure 7.15(c)) to produce a centroid model. For the second, shear centre effects are included to produce the shear centre model shown in Figure 7.15(d).

Table 7.6 *Comparison of centroid and shear centre line analyses*

Analysis	BM_A (kN m)	T_{AB} (kN m)	δ_C (mm)
Centroid line	−0.985	0.085	12.4
Shear centre line	−0.985	0.066	8.6
Percentage change	0%	−22%	−30%

Comparative results are presented in Table 7.6 from which it may be seen that the change in computer model does not alter the bending moment at A (BM_A), since the relevant lever arm is unchanged. However, the torque in AB (T_{AB}) is substantially decreased, due to a reduced lever arm. The displacement at C (δ_C) is, proportionately, most affected, since the smaller torsional rotation at B accompanies a reduced magnification distance, BC (Figure 7.15(c)), as compared to BD (Figure 7.15(d)).

Example 7.2 Computer analysis of a skew concrete bridge slab

The skew concrete bridge slab shown in Figure 7.16(a) is of depth 0.7 m and is to be analysed for the effects of a central, concentrated load of 1000 kN. It may be

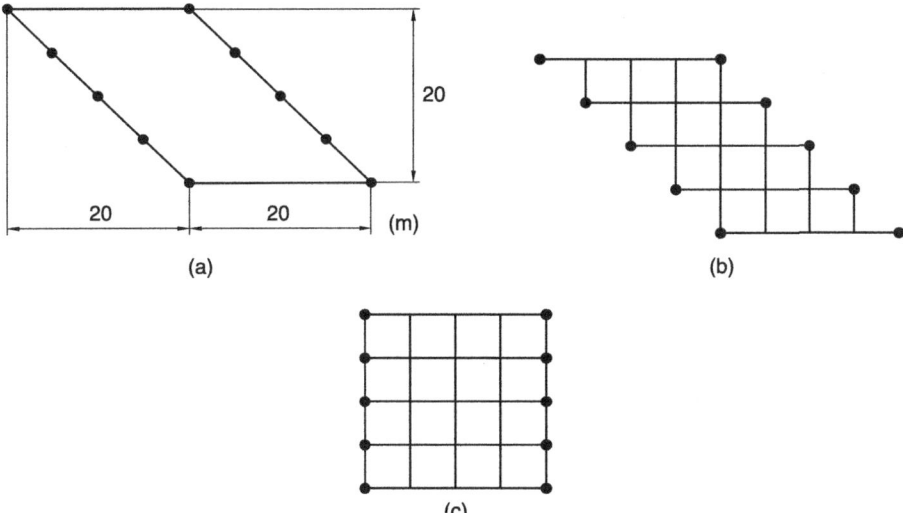

Figure 7.16   Computer analysis example: (a) skew slab, (b) grid analogy, (c) rectangular grid

assumed that the concrete is such that $E = 20\,\text{kN/mm}^2$ and Poisson's ratio, $v = 0$. Since $G = E/2(1 + v)$ (Johnson, 2000), the zero Poisson's ratio is alternatively equivalent to taking $G = 10\,\text{kN/mm}^2$. The slab is supported by discrete bearings, as indicated in the figure, which provide vertical support only.

Slabs may be analysed by a grid analogy (Hambly, 1991) in which a set of orthogonal longitudinal and transverse grid members are used to represent the slab. A suitable grid of members is shown in Figure 7.16(b). It is assumed that each grid member represents the bending and torsional stiffness of a width of slab extending between the centres of the spaces between neighbouring members. For the longitudinal members, this concept results in the interior grid members representing a 5 m width of slab, but for the edge members a width of only one half of this (Figure 7.17(a)). All the transverse members will have a 5 m width.

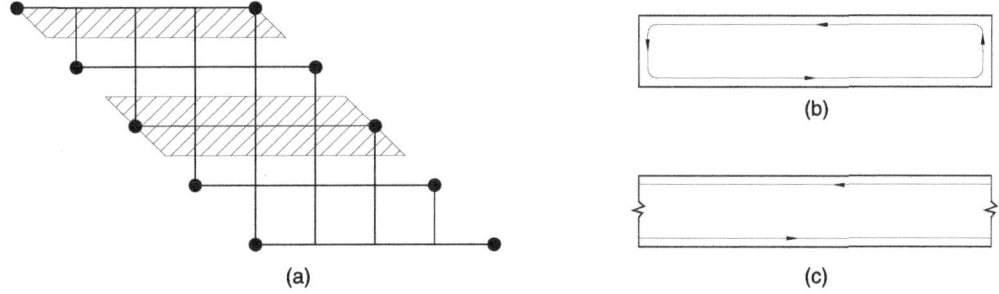

Figure 7.17   (a) Contributory widths for grid members, (b) torsional shear flow in solid rectangle, (c) shear flow in continuous slab

The second moments of area for the grillage members are readily evaluated as:

for the longitudinal members

$$I_{\text{edge}} = \frac{2.5 \times 0.7^3}{12} = 0.0715 \, \text{m}^4 \quad \text{and} \quad I_{\text{internal}} = \frac{5 \times 0.7^3}{12} = 0.1429 \, \text{m}^4$$

for all the transverse members

$$I = \frac{5 \times 0.7^3}{12} = 0.1429 \, \text{m}^4$$

The evaluation of the equivalent torsion constants needs rather more care. From Tables 7.4 and 7.5, it may be noted that the torsion constant of a thin, rectangular solid section is given by $J = Bt^3/3$. However, a solid section resists torsion by equal contributions from horizontal and vertical components of circulation shear stress (Figure 7.17(b)). In the case of a continuous slab, only horizontal shear stress components are present (Figure 7.17(c)) so that the torsional resistance is halved and, consequently, so is the torsion constant. Thus,

for the longitudinal members

$$J_{\text{edge}} = \frac{2.5 \times 0.7^3}{6} = 0.1429 \, \text{m}^4 \quad \text{and} \quad J_{\text{internal}} = \frac{5 \times 0.7^3}{6} = 0.2858 \, \text{m}^4$$

for all the transverse members

$$J = \frac{5 \times 0.7^3}{6} = 0.2858 \, \text{m}^4$$

A grillage computer analysis can readily be undertaken using the above member properties with the geometric, loading, material and support information provided, and results from such an analysis are presented in Figures 7.18 and 7.19. The reactions shown in Figure 7.18(a) are upwards positive. The figure therefore clearly demonstrates a fundamental feature of skew grids and slabs, namely the tendency to lift at the acute corners. The reaction values comply with the rotational symmetry of the grid considered, since the values are unaffected by a rotation of 180°. It may also be noted that the central 1000 kN load is primarily

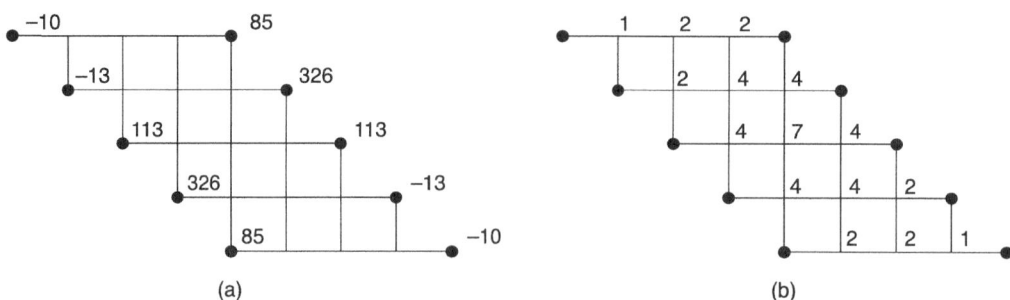

Figure 7.18   Skew grid analysis: (a) reactions (kN), (b) vertical nodal displacements (mm)

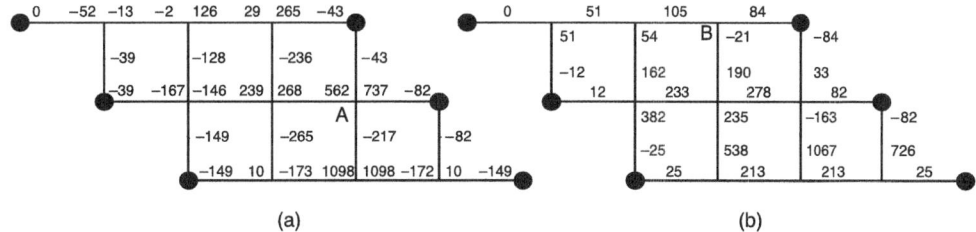

Figure 7.19 Skew grid analysis torques (kNm) and bending moments (kNm) for upper half grid: (a) bending moments in longitudinal members and torques in transverse members, (b) bending moments in transverse members and torques in longitudinal members

distributed to the nearest supports and this implies a load transfer system normal to the support lines. Again, the tendency to span perpendicularly to the supports is a fundamental feature of skew bridge behaviour.

Figure 7.19(a) shows longitudinal bending moments and transverse torques in the upper half of the grid only. The values have been converted from the 'member axis' conventions of the computer output to deformation sign conventions, such that positive bending moment indicates sagging and positive torque indicates deformation of the form shown in Figure 7.2(a). Figure 7.19(b) shows complementary values of transverse bending moments and longitudinal torques. These combinations of bending moment and torque are convenient for equilibrium checking purposes since all moments at any given node will be acting about the same axis: transverse in the case of Figure 7.19(a) and longitudinal in the case of Figure 7.19(b). With the deformation sign conventions used, the moments at any node will not necessarily sum to zero algebraically but equilibrium may be readily verified. At an internal node, for example, the resultant of the longitudinal bending moment will be balanced by the resultant of the two transverse torques.

At a typical node, A (Figure 7.19(a)), the bending moments and torques (in a joint to member action convention) are as shown in Figure 7.20 and, to the three figure accuracy adopted, the resultant bending moment ($737 - 562 = 175$ kNm) is balanced by the resultant torque ($217 - 43 = 174$ kNm). At the support nodes,

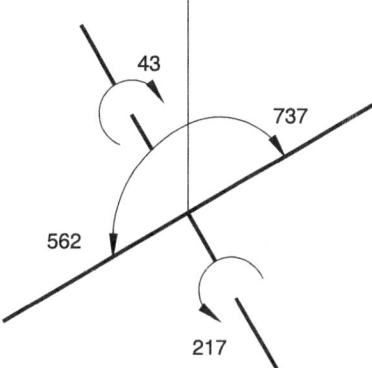

Figure 7.20 Moment equilibrium about a longitudinal axis at node A

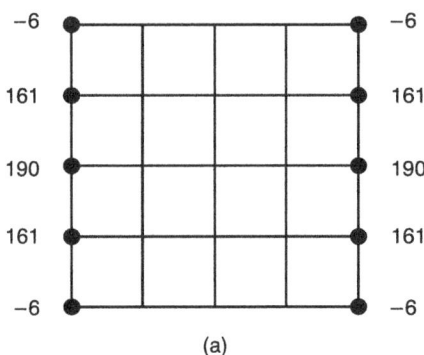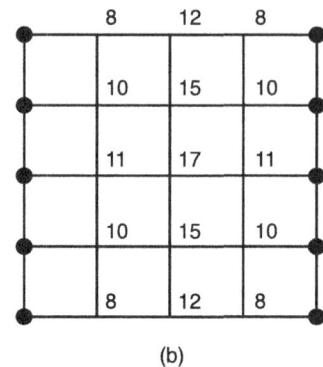

Figure 7.21 Rectangular grid analysis: (a) reactions (kN), (b) vertical nodal displacements (mm)

where two perpendicular members meet, it is clear that the concept of balancing moment/torque at such a node applies. It is also the torsional stiffnesses of the transverse members at the supports that allow the longitudinal members to develop hogging moments towards their ends. These members therefore tend to act as semi-fixed beams. Similar considerations explain the presence of hogging moments close to the support positions of the transverse beams (Figure 7.19(b)). However, at the top of the grid, it may be seen that the torsional stiffness of the edge longitudinal beam is only sufficient to generate a hogging moment at node B.

For comparison purposes, a 20 m span rectangular grid (Figure 7.16(c)) has also been analysed. Reactions and displacements for the comparison grid are shown in Figure 7.21. It is apparent from Figure 7.21 that this geometry is more effective in distributing the effects of the 1000 kN load, since the intermediate longitudinal members carry almost the same load as the central one. On the other hand (Figure 7.21(b)), the rectangular grid is significantly more flexible than the skew one (Figure 7.18(b)) due to its greater effective span.

Bending moment/torque distributions for the upper half of the rectangular grid are given in Figure 7.22. If these distributions are compared with those for the skew grid (Figure 7.19), it will be noted that bending moments, due to the

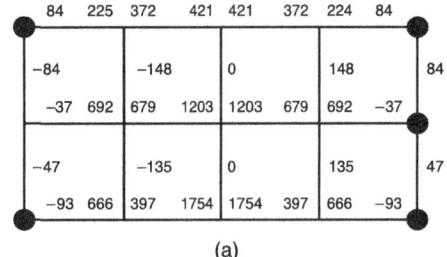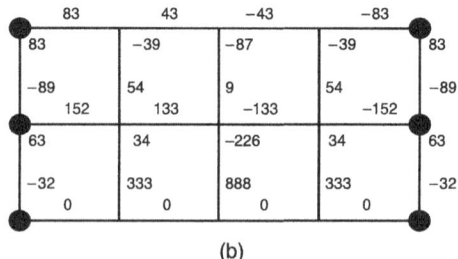

Figure 7.22 Rectangular grid analysis torques (kN m) and bending moments (kN m) for half grid: (a) bending moments in longitudinal members and torques in transverse members, (b) bending moments in transverse members and torques in longitudinal members

increased span, are significantly increased. Torques, on the other hand, are somewhat reduced, since the presence of symmetry about the two central axes ensures zero torque along these lines. Torsional effects are therefore limited to regions towards the corners of the slab. The torques also alternate sign in alternate quadrants, in accordance with the anti-symmetric nature of torsion. Maximal values occur in the neighbourhood of the quarter points of the grid. Transverse torsional stiffness again allows hogging moments to develop in the longitudinal beams at the supports. Furthermore, longitudinal torsional stiffness promotes transverse hogging moments at the free edge more consistently than in the skew case.

Worksheet 7.2 Computer analysis of grids

W7.2.1 Figure W7.2.1 shows a rectangular hollow and a channel section; both are of uniform thickness 10 mm and have the same external dimensions. The sections are to be analysed as cantilever beams of length 4 m. In each case, a vertical load of 8 kN is applied to the centre of the left-hand web at the end of the beam. The constructional material is steel for which $E = 205 \, \text{kN/mm}^2$ and $\nu = 0.3$.

Use a computer package to analyse the two beams, based on shear centre line models. Compare the displacements and rotations at the load point, A, and provide physical explanations for the comparative values obtained.

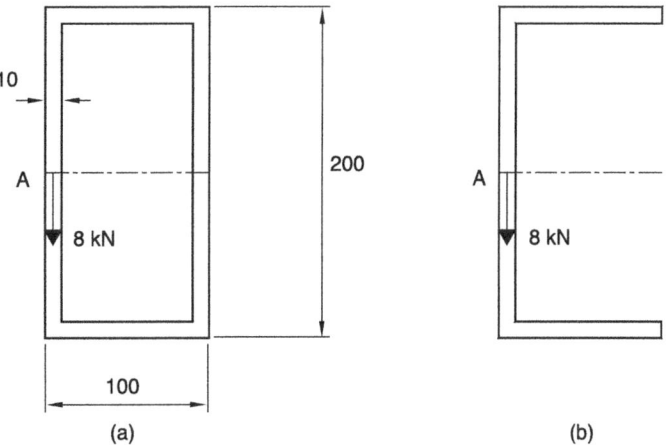

Figure W7.2.1   Worksheet 7.2.1

W7.2.2 Figure W7.2.2 shows the plan view of a curved bridge on a motorway slip road, for which the vertical gradient can be ignored so that it can be analysed as a plane grid. Section AB is straight, of length 25 m and has the cross-section shown in Figure W7.2.2(b). Curve BCDE is circular, of radius 25 m and has the cross-section shown in Figure W7.2.2(c). The thickness of the section is constant at 0.25 m and the elastic constants may be taken as: $E = 28 \, \text{kN/mm}^2$ and $G = 14 \, \text{kN/mm}^2$. The supports at ABCDE offer negligible rotational resistance.

Analyse the bridge in respect of both self-weight (weight density of concrete $= 25 \, \text{kN/}$ m^3) and also for the effects of a concentrated load of 1 MN, offset 3 m from the central position of span CD. Determine the kinematic and statical degrees of indeterminacy of the bridge. If the grid were to be solved by hand, would it be more arduous to use the stiffness or the flexibility method of analysis? Most computer packages use the stiffness method. Why is this?

Explain carefully how self-weight causes torsional effects, although the load is assumed to act along the axis of the bridge. Obtain bending moment and torque distribution diagrams for the two loading cases. Explain the changes in sign of the two quantities around ABCDE and relate these to the displacement of the bridge. Why has a different cross-section been adopted for AB? (Adapted from Hambly, 1991.)

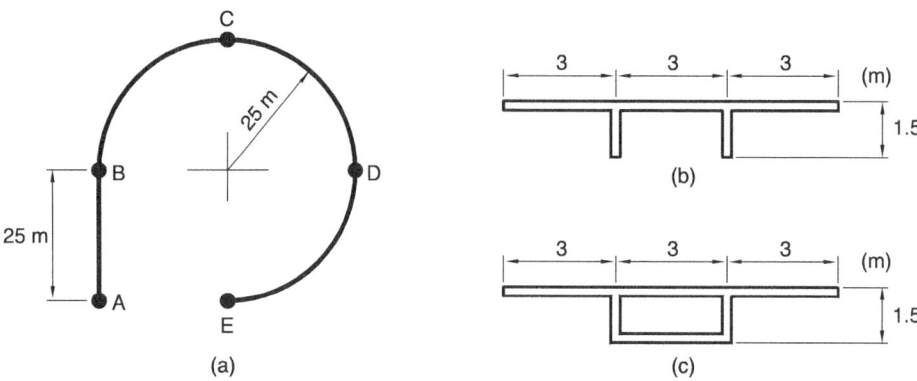

Figure W7.2.2   Worksheet 7.2.2

W7.2.3 The 45° skew grids shown in Figure W7.2.3 are such that the lengths of all the grid members are the same in the case of the orthogonal arrangement shown in Figure W7.2.3(a). Vertical support is provided at the indicated positions. Choose appropriate dimensions for the grid and suitable member properties on the assumption that the members are constructed of steel, and are such that the section properties of the transverse members are half those of the longitudinal members. Torsional constants should be taken as a quarter of the corresponding second moments of area. Analyse each of the three grid layouts shown for the effects of the same concentrated load placed at the centre of an edge. The load should be such as to produce a displacement of approximately span/300.

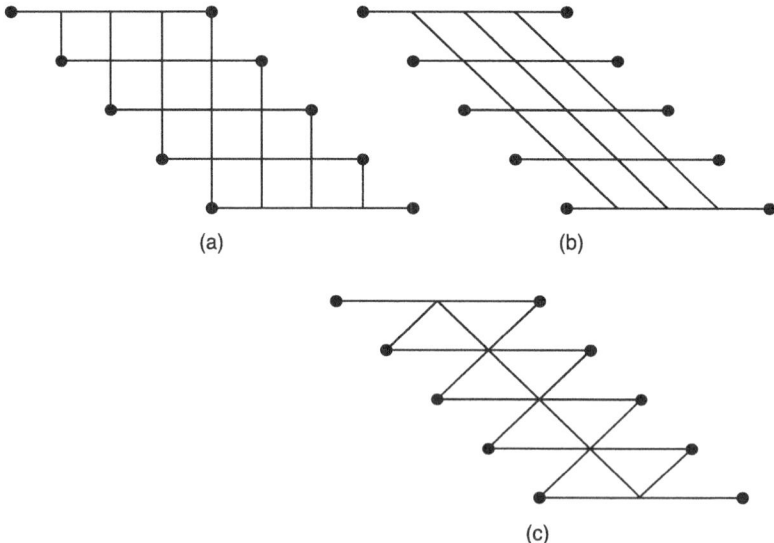

Figure W7.2.3   Worksheet 7.2.3

Compare the bending moment and torque distributions for the three grids and explain why they vary in their effectiveness in distributing the effects of the concentrated load away from the longitudinal edge member.

W7.2.4 Analyse the grid of Example 7.2 in both the skew and rectangular geometries, for the effects of a concentrated load of 1000 kN applied to the centre of the top edge. Compare the results of the analyses with each other and with the results from Example 7.2.

7.6   INFLUENCE LINES

Statically determinate grids

In the case of simple rectilinear bent beams, applying Müller-Breslau's principle will conveniently produce influence lines. Bending moment or torque influence lines will require the introduction of unit transverse rotations in members about a transverse or longitudinal axis, respectively. Should shear force influence lines be needed, these may be generated by the introduction of unit vertical displacements.

In the case of the bent beam shown in Figure 7.23(a), the influence line for bending moment at A can be produced by introducing a unit rotation at A about a transverse axis (Figure 7.23(b)), and the influence line for torque at A (Figure 7.23(c)) may be generated by introducing a unit axial rotation at A. Finally, the shear force influence line requires a unit vertical displacement at A (Figure 7.23(d)).

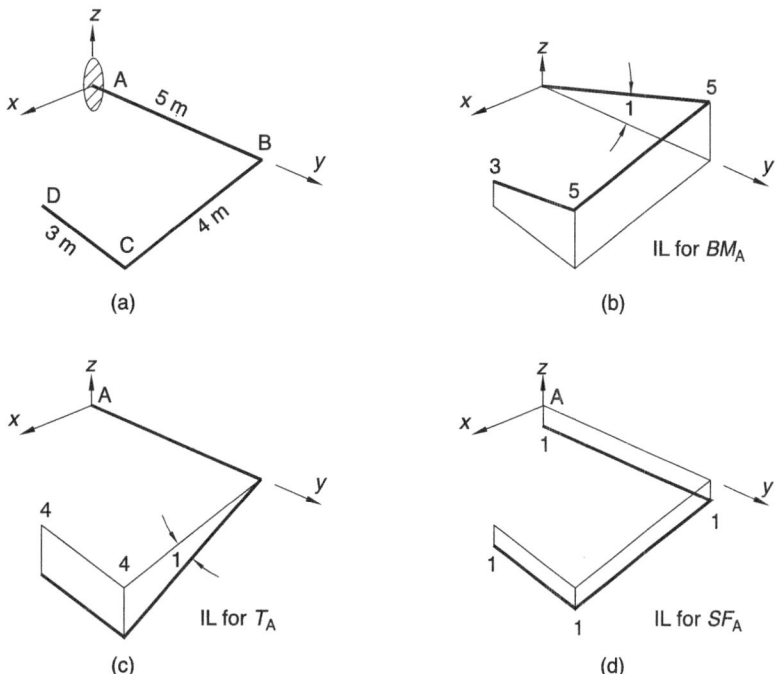

Figure 7.23   (a) Example bent beam and influence lines at A for (b) bending moment (m), (c) torque (m), (d) shear force

Example 7.3 Influence lines for a determinate grid

For more complex support conditions and geometries, it may be difficult to generate the appropriate displaced shapes to produce influence lines. In such cases it may be more convenient to calculate influence line values at the nodes of the grid and then to use the linear property of determinate influence lines to complete the diagrams. As an example of this type of approach, the grid shown in Figure 7.24 will be examined. An influence line for bending moment will be generated for a point B^+, which is on BC adjacent to node B. In respect of torque, an influence line will be found for a point C^+, which is on CD adjacent to point C. The lines will be used to determine maximal values of positive and negative bending moment and the maximum absolute value of torque produced by a uniform load of 15 kN/m applied to any length of the grid.

First, it is noted that, provided there is no load on AB, the bending moment at B^+ may be readily determined if the vertical reaction at A, R_A (upwards positive), is known. Thus, if AP and PB (Figure 7.24) are in the plane of the grid and are parallel and perpendicular, respectively, to BC, then, since $AP = \sqrt{2}$ m, it follows that

$$BM_{B^+} = -R_A \times AP = -1.41 R_A \tag{7.5}$$

To establish the required influence line, R_A therefore needs to be calculated for a unit load placed at each of the unsupported nodes B, C, E and F (Figure 7.24). Since the truss is determinate, moment equilibrium will be used to calculate the R_A values needed. Thus, for a unit load at B, moments about the x-axis gives $R_{DB} = 0$, and moments about axes parallel to the y-axis, through G and A, give $R_{AB} = 1.5$ and $R_{GB} = -0.5$. Since the three reactions sum to unity, vertical force equilibrium is satisfied, which acts as a check on the values. If the unit load is placed at C, then moments about the x-axis gives $R_{DC} = 1$, and moments about axes parallel to the y-axis, through G and A, give $R_{AC} = 0.5$ and $R_{GC} = -0.5$. Vertical force equilibrium is again satisfied. For the unit load placed at E and F, symmetry may be used to establish that $R_{AE} = R_{GC} = -0.5$ and $R_{AF} = R_{GB} = -0.5$. The required bending moments now follow from equation (7.5) and may be used to construct the influence line shown in Figure 7.25(a).

To obtain the maximum positive bending moment at B^+, it may be seen from Figure 7.25(a) that this will be achieved if the uniform 15 kN/m loading is applied to DEFG. If this is done, then

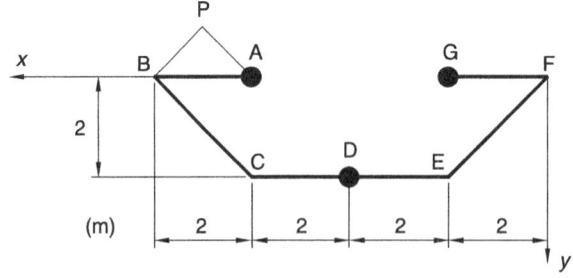

Figure 7.24 Determinate example grid

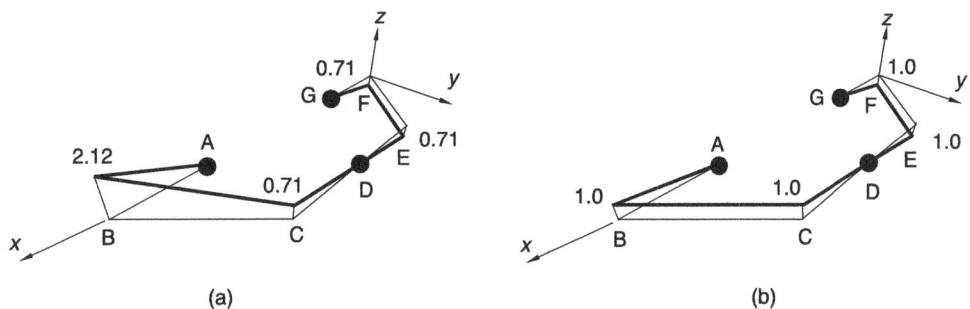

Figure 7.25   (a) Influence line for bending moment at B$^+$ (m), (b) influence line for torque at C$^+$ (m)

maximum positive bending moment at B$^+$

$$= 15[(\tfrac{1}{2} \times 2 \times 0.71) + (2\sqrt{2} \times 0.71) + (\tfrac{1}{2} \times 2 \times 0.71)] = 51\,\text{kN m}$$

Loading on ABCD will produce the worst negative bending moment, given by

maximum negative bending moment at B$^+$

$$= -15\left[\left(\frac{1}{2} \times 2 \times 2.12\right) + \left(2\sqrt{2} \times \frac{(2.12 + 0.71)}{2}\right) + \left(\frac{1}{2} \times 2 \times 0.71\right)\right]$$

$$= -103\,\text{kN m}$$

To create the torque influence line for C$^+$, it is necessary to consider two possible positions of a unit load

for unit load on AB: $T_{C^+} = (1 \times 2) - (R_A \times 2)$

and for unit load on CDEFG: $T_{C^+} = -R_A \times 2$

(7.6)

Using the values of R_A found previously, the nodal influence line values may be obtained by appropriate substitution in equation (7.6) and used to produce the complete diagram shown in Figure 7.25(b). Since this diagram is anti-symmetric, it follows that the maximum negative torque at C$^+$ is the same as the maximum positive value. Thus, placing the uniform load on either ABCD or DEFG gives

maximum absolute torque at C$^+$

$$= 15[(\tfrac{1}{2} \times 2 \times 1.0) + (2\sqrt{2} \times 1.0) + (\tfrac{1}{2} \times 2 \times 1.0)] = 72\,\text{kN m}$$

Statically indeterminate grids

The effectiveness of indeterminate grids in distributing loading can be conveniently examined by the use of influence lines, which, as for plane frames, can also be helpful in identifying critical loading cases. As in the case of previous indeterminate structures, Müller-Breslau's principle is used, in conjunction with computer analysis. For example purposes, the effects of central and edge loading on the behaviour of the skew bridge examined in Example 7.4 will be explored through the use of influence lines.

Example 7.4 Influence lines for skew grid

Influence lines for longitudinal bending moment and torque at both the centre of the grid and at the centre of an edge will be generated for the properties used previously. To produce the influence lines, 0.5 m length members (I–J) are introduced at the designated points (Figure 7.26).

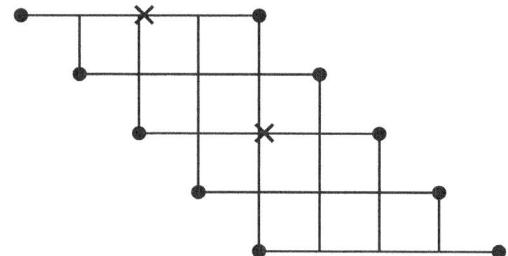

Figure 7.26 Skew grid influence line positions

If mm output units are assumed, the bending moment influence lines will be obtained from the displaced shape due to the introduction of a rotation, $\theta_{vI} = 0.001$ rad at the designated points. The end forces arising from the introduction of the rotation are given from equation (7.3) as

$$\begin{Bmatrix} M_{uI} \\ F_I \\ M_{vI} \\ M_{uJ} \\ F_J \\ M_{vJ} \end{Bmatrix} = EI \begin{Bmatrix} 0 \\ -6/L^2 \\ 4/L \\ 0 \\ 6/L^2 \\ 2/L \end{Bmatrix} \{\theta_{vI}\} \tag{7.7}$$

In relation to a set of x(horizontal)- and z(vertical)-global axes, the nodal forces to be applied for an edge member ($E = 20\,\text{kN/mm}^2$; $I = 0.0715\,\text{m}^4$) and a central member ($E = 20\,\text{kN/mm}^2$; $I = 0.1429\,\text{m}^4$), respectively, are

$$\begin{Bmatrix} f_{zI} \\ m_{xI} \\ m_{yI} \\ f_{xJ} \\ m_{xJ} \\ m_{yJ} \end{Bmatrix} = \begin{Bmatrix} -34\,320 \\ 11\,440 \\ 0 \\ 34\,320 \\ 0 \\ 5720 \end{Bmatrix} \begin{matrix} \text{kN} \\ \text{kN m} \\ \text{kN m} \\ \text{kN} \\ \text{kN m} \\ \text{kN m} \end{matrix} \quad \text{and} \quad \begin{Bmatrix} f_{zI} \\ m_{xI} \\ m_{yI} \\ f_{xJ} \\ m_{xJ} \\ m_{yJ} \end{Bmatrix} = \begin{Bmatrix} -68\,592 \\ 22\,864 \\ 0 \\ 68\,592 \\ 0 \\ 11\,296 \end{Bmatrix} \begin{matrix} \text{kN} \\ \text{kN m} \\ \text{kN m} \\ \text{kN} \\ \text{kN m} \\ \text{kN m} \end{matrix}$$

The resulting displaced surfaces that represent the influence lines are shown in Figure 7.27. Figure 7.27(a) shows that a central longitudinal bending moment is generated primarily by loading across a line joining the obtuse corners. This is therefore the principal direction of span at the centre. The influence line therefore confirms the basic behaviour discovered previously. Figure 7.27(b) indicates the

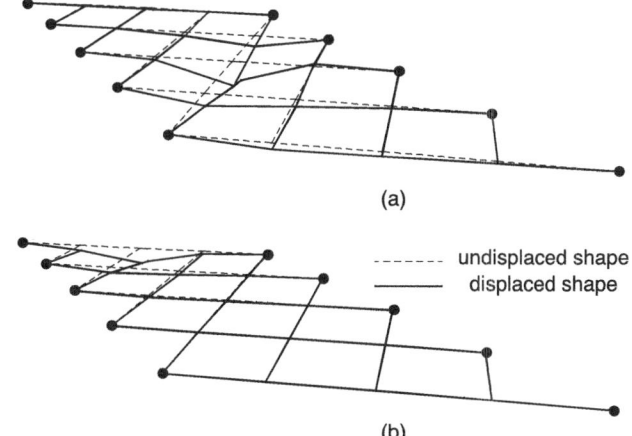

(a)

- - - - - undisplaced shape
———— displaced shape

(b)

*Figure 7.27   Influence lines for longitudinal bending moments at (a) central position,
(b) edge position*

converse tendency of edge longitudinal bending moment to be generated primarily
by loading along the edge.

The influence lines for torque will be obtained from the displaced shape due to the
introduction of a rotation, $\theta_{u\mathrm{I}} = 0.001$ rad at the designated points. The end forces
arising from the introduction of the rotation are given from equation (7.3) as

$$
\begin{Bmatrix}
M_{u\mathrm{I}} \\
F_{\mathrm{I}} \\
M_{v\mathrm{I}} \\
M_{u\mathrm{J}} \\
F_{\mathrm{J}} \\
M_{v\mathrm{J}}
\end{Bmatrix}
= GJ
\begin{Bmatrix}
1/L \\
0 \\
0 \\
1/L \\
0 \\
0
\end{Bmatrix}
\{\theta_{u\mathrm{I}}\}
\tag{7.8}
$$

In relation to a set of x(horizontal)- and z(vertical)-global axes, the nodal forces to
be applied for an edge member ($G = 10\,\mathrm{kN/mm}^2$; $J = 0.1429\,\mathrm{m}^4$) and a central
member ($G = 10\,\mathrm{kN/mm}^2$; $J = 0.2858\,\mathrm{m}^4$), respectively, are

$$
\begin{Bmatrix}
f_{z\mathrm{I}} \\
m_{x\mathrm{I}} \\
m_{y\mathrm{I}} \\
f_{x\mathrm{J}} \\
m_{x\mathrm{J}} \\
m_{y\mathrm{J}}
\end{Bmatrix}
=
\begin{Bmatrix}
0 \\
2858 \\
0 \\
0 \\
-2858 \\
0
\end{Bmatrix}
\begin{matrix}
\mathrm{kN} \\
\mathrm{kN\,m} \\
\mathrm{kN\,m} \\
\mathrm{kN} \\
\mathrm{kN\,m} \\
\mathrm{kN\,m}
\end{matrix}
\quad \text{and} \quad
\begin{Bmatrix}
f_{z\mathrm{I}} \\
m_{x\mathrm{I}} \\
m_{y\mathrm{I}} \\
f_{x\mathrm{J}} \\
m_{x\mathrm{J}} \\
m_{y\mathrm{J}}
\end{Bmatrix}
=
\begin{Bmatrix}
0 \\
5716 \\
0 \\
0 \\
-5716 \\
0
\end{Bmatrix}
\begin{matrix}
\mathrm{kN} \\
\mathrm{kN\,m} \\
\mathrm{kN\,m} \\
\mathrm{kN} \\
\mathrm{kN\,m} \\
\mathrm{kN\,m}
\end{matrix}
$$

The influence lines shown in Figure 7.28 indicate a significant difference between the
central and edge position in respect of torsional behaviour. The central position is
subjected to torque regardless of the position of the load, whereas edge torque is
only significant for loading local to the edge. This finding may again be related
to the basic tendency of the grid in tending to span perpendicularly to the supports.
The central position is thus involved for essentially all loading positions, but the

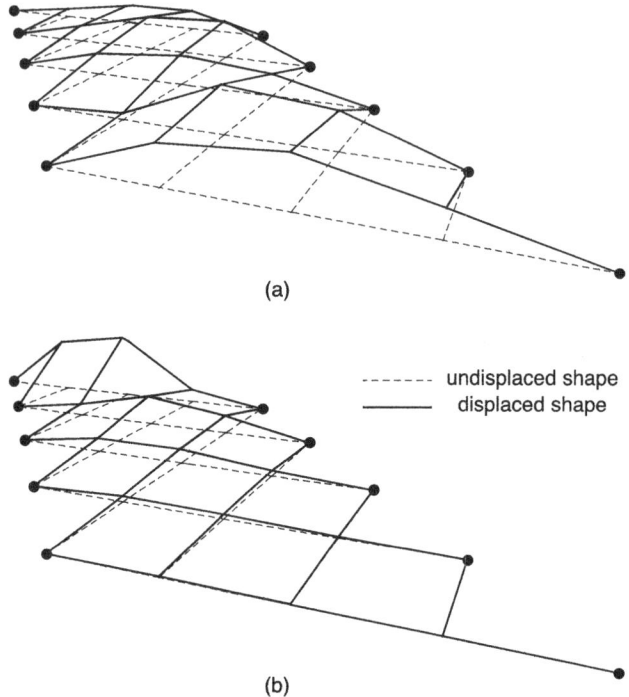

(a)

(b)

Figure 7.28   Influence lines for longitudinal torque at (a) central position, (b) edge position

edge is generally by-passed for other than loading in its immediate vicinity (where substantial torque is generated).

Worksheet 7.3   Influence lines for grids

W7.3.1   Figure W7.3.1 shows a determinate grid which has vertical support at points B, C and E. Construct influence lines for bending moment and torque at the mid-point of CD, as indicated in the figure. Why does loading on BC not produce any bending or torsion in CD? Use the diagrams to find the maximum positive and negative bending moment and the maximum absolute torque produced at the mid-point of CD, due to the action of a uniform live load of intensity 25 kN/m (applied to any lengths) and a point live load of 200 kN.

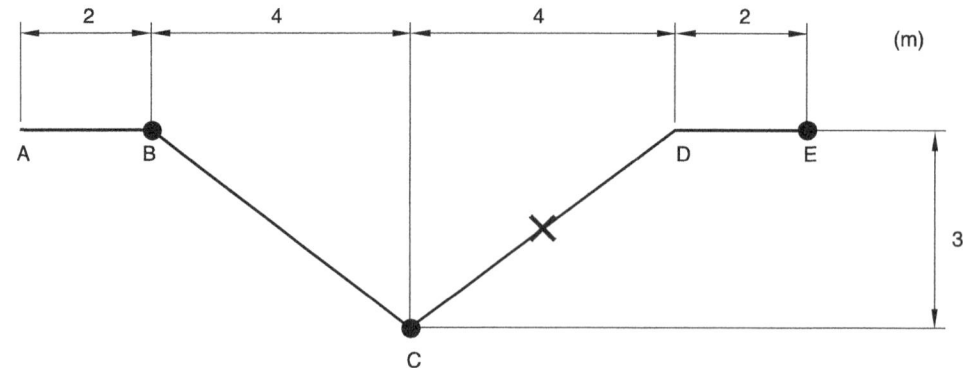

Figure W7.3.1   Worksheet 7.3.1

W7.3.2 Figure W7.3.2 shows the plan view of a concrete slab of thickness 200 mm, which has $E = 30\,kN/mm^2$ and $v = 0$. The slab is supported on columns, which may be presumed to offer vertical support only, at 5 m intervals around its perimeter. The slab is to be analysed by an approximating grillage, as shown in Figure W7.3.2.

Determine suitable sectional properties for the approximating grid and use a computer package to generate influence lines for bending moments and torques at the points A and B.

The slab is to be loaded by equal live point loads at grid intersection positions. Use the influence lines to determine which positions should be loaded to produce: (a) maximum positive bending moment at A; (b) maximum positive bending moment at B; (c) maximum absolute torque at A; (d) maximum absolute torque at B. Give explanations for the loading patterns determined in terms of the structural behaviour of the slab.

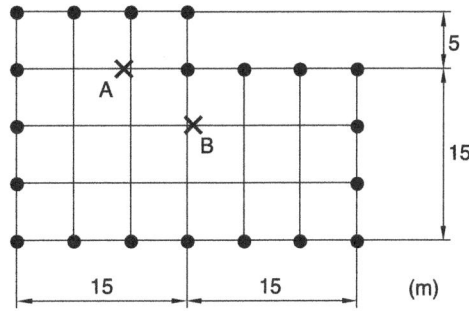

Figure W7.3.2   Worksheet 7.3.2

7.7   REFERENCES

Gere, J. M. and Timoshenko, S. P. (1999) *Mechanics of Materials*. 4th SI edn. London: Stanley Thornes – Section 6.8 of this reference covers shear centre theory.

Hambly, E. C. (1991) *Bridge Deck Behaviour*. 2nd edn. London: E & F N Spon – suggests appropriate grillage representations for a wide variety of practical bridge deck forms of construction.

Johnson, D. (2000) *Advanced Structural Mechanics*. 2nd edn. London: Thomas Telford – describes Saint-Venant torsion theory and the use of grillage analogies for the analysis of slabs.

8 Space frames

8 Space frames

8.1 INTRODUCTION

Space frames represent the most general form of skeletal structure and all the other types discussed so far can be considered to be space-frame subsets. Space frame members therefore possess the axial and bending stiffness of plane frames, for example, and also the torsional and bending stiffness of grids. Bending action in a space frame is, however, different to that in either a plane frame or a grid, since bending will occur in three-dimensional space. In a plane frame, bending occurs about a single axis for each member and all these axes are parallel (normal to the plane of the structure). Similarly, for a grid, each member has one axis of bending and all of these axes are in the same plane (the plane of the structure). In the case of a space frame, bending needs to be related to two axes. These axes should be the *principal* axes of the section (Gere and Timoshenko, 1999), since it is in respect of principal axes that simple bending theory is applicable. Axes of symmetry of a section are always principal axes, as are axes that are perpendicular to an axis of symmetry. Space frame members will often have an axis of symmetry in a vertical plane. This is so in the case of rectangular (Figure 8.1) or I-sections, for example. In such cases, this vertical plane axis is commonly taken as the local w-axis of the member and is defined such that positive w is in the direction of increasing z. For the ambiguous case of a vertical member, positive w needs to be separately defined. Here this will be taken to be in the negative x direction. As previously, the local u-axis is directed along the member. The v-axis will then complete the right-handed set and will be in a horizontal plane.

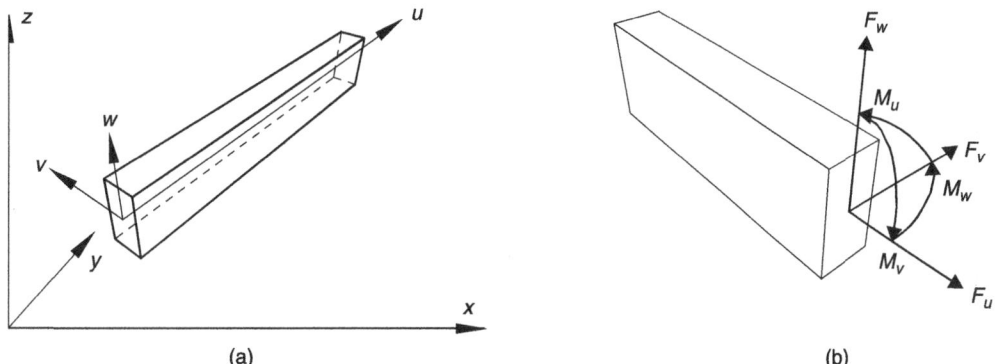

Figure 8.1   Space frame member: (a) local axes, (b) member forces

Bending moments in plane frames and grids relate to member v-axes. The difference for space frames, therefore, is that bending moments (and shear forces) will exist in relation to both the v- and w-axes of each member, as shown in Figure 8.1(b). Two bending moment/shear force diagrams will be needed, therefore, and it should be noted from Figure 8.1(b) that F_w accompanies M_v and F_v accompanies M_w. Diagrams for complete structures are also commonly shown in relation to v- and w-axes of the member, although these axes will, in general, be different for every member.

8.2 KINEMATIC AND STATICAL INDETERMINACY

Corresponding to the six possible stress resultant components for a space frame member, there will be six displacement components at a general node of a space frame (Figure 8.2). The six components will comprise three displacements and three rotations. It follows that, if there are j joints and d restrained displacement components, the kinematic indeterminacy will be given by

$$k = 6j - d \tag{8.1}$$

Further, if there are m members, having six unknown stress resultants each (Figure 8.1(b)), and h releases in the members (usually torsional or moment hinges), then the statical indeterminacy, q, is found from the number of stress resultants that cannot be determined by statics as

$$q = (6m - h) - k \tag{8.2}$$

The 'CUBIC' (Kubik, 1991) space frame shown in Figure 8.3(a) will be used as an indeterminacy example. The frame is supported vertically at its four lower corners A, B, C, D. To ensure the overall stability of the structure, additional support will be needed to prevent movement in a horizontal plane. This is provided by the horizontal components of reaction at A, which convert the support at this position into a ball-and-socket arrangement. Rotation about axes in the horizontal plane is prevented by the vertical supports (a minimum of three non-collinear supports is required for this, as for grids). Prevention of rotation about the vertical

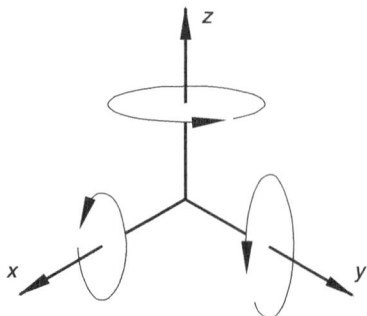

Figure 8.2   Displacement degrees of freedom

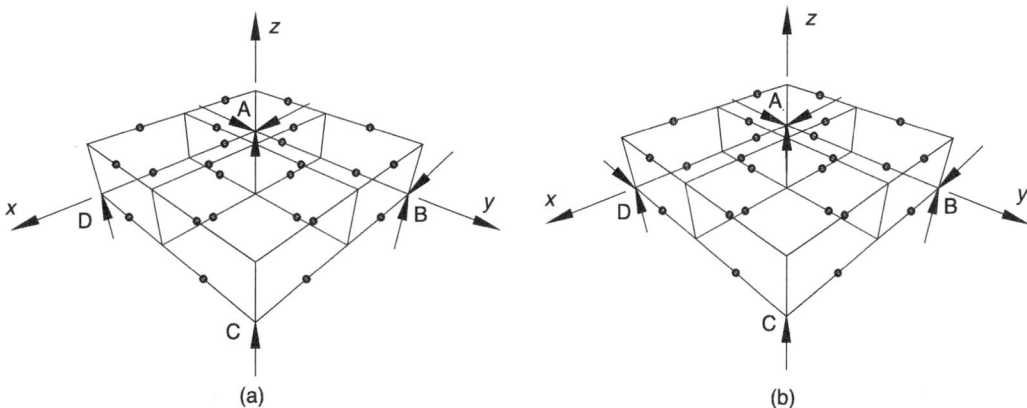

Figure 8.3   CUBIC space frame: (a) seven reaction components, (b) eight reaction components

axis requires an additional reaction in the horizontal plane, which is provided at joint B. Altogether there are seven reaction components, and therefore seven restrained displacement degrees of freedom, so that $d = 7$. The frame also has $j = 18$ and $m = 33$. Ball-and-socket type joints are provided at the centres of all the horizontal members. These member releases can transmit shears and axial load but no moment component. Each joint therefore provides three releases, giving a total of $h = 3 \times 24 = 72$. The kinematic and statical indeterminacies may now be obtained from equations (8.1) and (8.2) as

$$k = (6 \times 18) - 7 = 101 \quad \text{and} \quad q = ((6 \times 33) - 24) - 101 = 25 \tag{8.3}$$

Although this frame has a considerable number of internal releases, it is still highly statically indeterminate, which is typical of space frames. However, high indeterminacy does not ensure stability and, to prevent local rotation about a vertical axis, the frame needs either a further reaction component in the horizontal plane (such as at D in Figure 8.3(b)) or that z-rotation be suppressed at a convenient position. Identification of such local mechanisms will be detected by computer packages and will result in either an error message and/or the calculation of unreasonably large displacements.

Statically determinate space frames are rare and the only common form is a beam bent into three dimensions. The statical determinacy of redundant frames may be checked by introducing sufficient releases to convert the structure into a number of bent beams. The beams may be branched but must not incorporate closed loops. For the CUBIC frame just considered, it is first necessary to incorporate additional restraints to eliminate the internal releases and make the supports fully fixed (Figure 8.4(a)). This requires the addition of $(6 \times 4) - 7 = 17$ extra reaction components and 72 internal constraints, so that the indeterminacy of this 'new' structure is related to that of the original structure by

$$q_{\text{new}} = q + 89 \tag{8.4}$$

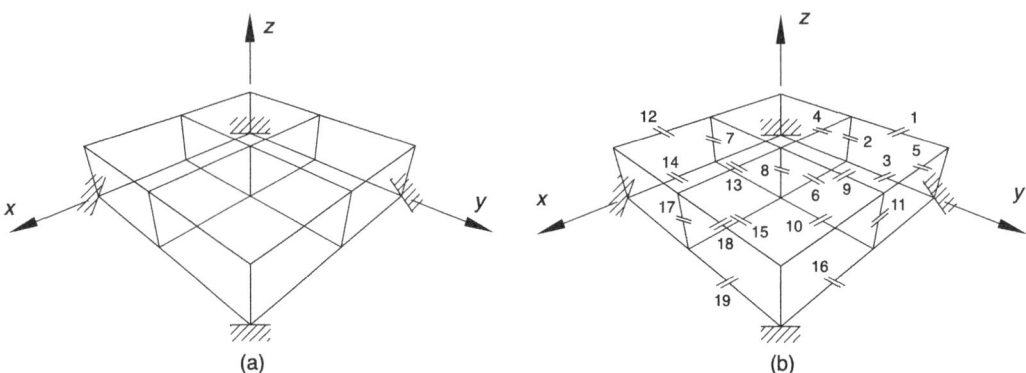

Figure 8.4   CUBIC space frame: (a) new structure, (b) released new structure

A system of bent beams may then be devised as shown in Figure 8.4(b) to produce a statically determinate system. Nineteen cuts are required to produce the bent beams and six internal stress resultants are released at each cut. It follows that

$$q_{\text{new}} = 6 \times 19 = 114 \quad \text{and} \quad q = 114 - 89 = 25 \tag{8.5}$$

The indeterminacy of the structure is therefore confirmed, but the conversion to a bent beam system is a cumbersome process and, for structures of any complexity, it is more convenient to use equations (8.1) and (8.2).

8.3   STATICALLY DETERMINATE SPACE FRAMES

Example 8.1   Bent beam example

Principally to illustrate the additional features introduced by three-dimensional action, stress resultant diagrams for the bent beam shown in Figure 8.5(a) will be developed. The members will be designated 1–2, 2–3 and 3–4, which define the positive u-directions for each member. The w and v directions then follow from the adopted convention and the three sets of local axes are shown in Figure 8.5(b).

The effects of the downward load, W, can most easily be assessed if they are converted into local member axes. By resolution, for example, it may be shown from Figure 8.6 that force components, F_x, F_y, F_z, in the global axes may be represented in the local axes by

$$\begin{Bmatrix} F_u \\ F_v \\ F_w \end{Bmatrix} = \begin{bmatrix} \cos\theta\cos\phi & \sin\theta\cos\phi & \sin\phi \\ -\sin\theta & \cos\theta & 0 \\ -\cos\theta\sin\phi & -\sin\theta\sin\phi & \cos\phi \end{bmatrix} \begin{Bmatrix} F_x \\ F_y \\ F_z \end{Bmatrix} \tag{8.6}$$

In equation (8.6), θ is the angle from the positive x-axis to the projection of u on the x–y plane, and ϕ is the angle from the projection of u on the x–y plane to u. For a member of length L (Figure 8.6(b)), which has $\bar{x}$, $\bar{y}$, $\bar{z}$ projections on the global

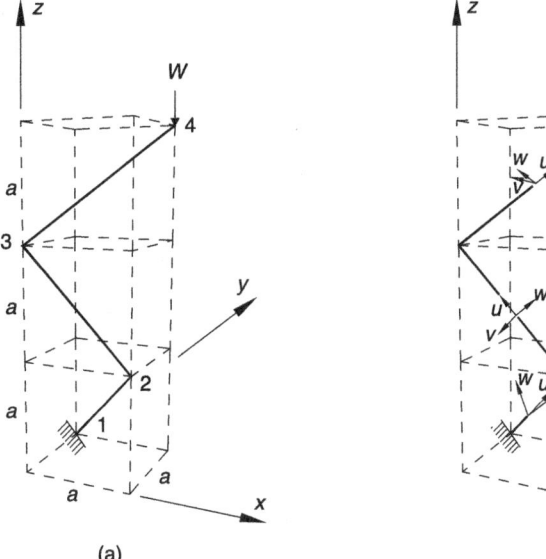

Figure 8.5 *(a) Bent beam, (b) local axes*

co-ordinate axes and projection H on the x–y plane, the trigonometric quantities in equation (8.7) may be calculated from

$$\cos\theta = \frac{\bar{x}}{H}, \quad \sin\theta = \frac{\bar{y}}{H} \quad \text{and} \quad \cos\phi = \frac{H}{L}, \quad \sin\phi = \frac{\bar{z}}{L} \tag{8.7}$$

For member 3 (3–4); $\bar{x} = a$, $\bar{y} = a$, $\bar{z} = a$, $H = \sqrt{2}a$, $L = \sqrt{3}a$ so that

$$\cos\theta = \frac{1}{\sqrt{2}}, \quad \sin\theta = \frac{1}{\sqrt{2}} \quad \text{and} \quad \cos\phi = \sqrt{\frac{2}{3}}, \quad \sin\phi = \frac{1}{\sqrt{3}}$$

The forces in the local axes at node 4 (member 3) due to the applied load $(F_z = -W)$ may therefore be obtained from equation (8.6) as

$$\left\{ \begin{array}{c} F_u \\ F_v \\ F_w \end{array} \right\}_4^3 = \left[\begin{array}{ccc} 1/\sqrt{3} & 1/\sqrt{3} & 1/\sqrt{3} \\ -1/\sqrt{2} & 1/\sqrt{2} & 0 \\ -1/\sqrt{6} & -1/\sqrt{6} & \sqrt{2/3} \end{array} \right] \left\{ \begin{array}{c} 0 \\ 0 \\ -1 \end{array} \right\} W = \left\{ \begin{array}{c} -1/\sqrt{3} \\ 0 \\ -\sqrt{2/3} \end{array} \right\} W \tag{8.8}$$

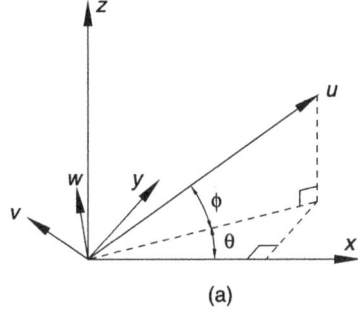

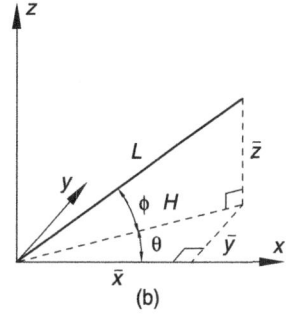

Figure 8.6 *Transformation of axes*

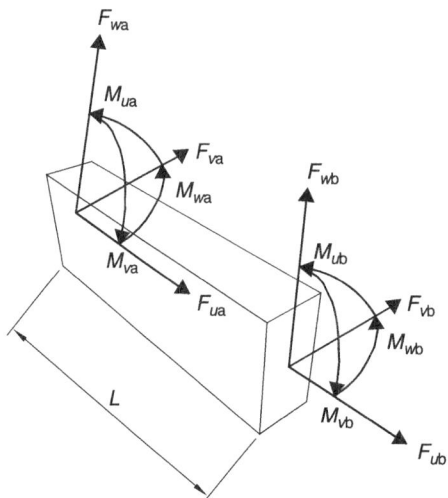

Figure 8.7   Equilibrium of space frame member

The moments at node 4 are all zero since it is a free end. The internal forces and moments at node 3 on member 3 may be obtained by the equilibrium conditions for the member, since, from Figure 8.7, it follows that

$$
\left\{
\begin{array}{c}
F_u \\
F_v \\
F_w \\
M_u \\
M_v \\
M_w
\end{array}
\right\}_a
=
\begin{bmatrix}
-1 & 0 & 0 & 0 & 0 & 0 \\
0 & -1 & 0 & 0 & 0 & 0 \\
0 & 0 & -1 & 0 & 0 & 0 \\
0 & 0 & 0 & -1 & 0 & 0 \\
0 & 0 & L & 0 & -1 & 0 \\
0 & -L & 0 & 0 & 0 & -1
\end{bmatrix}
\left\{
\begin{array}{c}
F_u \\
F_v \\
F_w \\
M_u \\
M_v \\
M_w
\end{array}
\right\}_b
\tag{8.9}
$$

Applying equation (8.9) gives

$$
\left\{
\begin{array}{c}
F_u \\
F_v \\
F_w
\end{array}
\right\}_3^3
=
\left\{
\begin{array}{c}
1/\sqrt{3} \\
0 \\
\sqrt{2/3}
\end{array}
\right\} W
\quad \text{and} \quad
\left\{
\begin{array}{c}
M_u \\
M_v \\
M_w
\end{array}
\right\}_3^3
=
\left\{
\begin{array}{c}
0 \\
-\sqrt{2} \\
0
\end{array}
\right\} Wa
\tag{8.10}
$$

The internal forces given by equations (8.8) and (8.10) are based on a right-handed, Cartesian sign convention, relative to the local axes of the member. Internal force diagrams, however, are usually related to the deformation sign conventions of Figures 1.26 and 7.2(a). To convert to deformation-type sign conventions, it is necessary to note from Figure 8.7 that, for positive values at end b:

F_{ub} represents positive axial load (tension);
F_{vb} and F_{wb} represent negative v and w shear forces;
M_{ub} represents a positive torque;
M_{vb} represents a negative bending moment;
M_{wb} represents a positive bending moment.

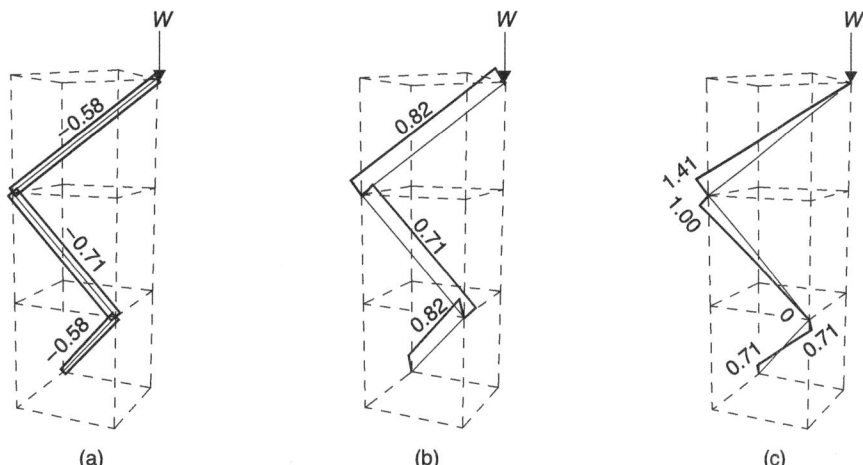

Figure 8.8   Example bent beam: (a) axial load (×W), (b) SF$_w$ (×W), (c) BM$_v$ (×Wa)

Reverse relationships apply at end a. In plotting deformation convention quantities, axial loads and torques will be centred on the member and the sign indicates the type. Shear forces will be plotted with positive values plotted in the positive v or w direction, as appropriate. Bending moments, however, will be plotted with negative values plotted in the relevant positive v or w direction, so that they appear on the tension side.

Using these conventions, the internal force diagrams for member 3–4 are shown in Figure 8.8. From the diagrams, it may be noted that this member acts as an inclined plane cantilever and could have been, more simply, treated as such. The benefits of a three-dimensional approach are, however, realised with the next member (2–3), which develops full space action. For member 2 (2–3), $\bar{x} = -a$; $\bar{y} = 0$; $\bar{z} = a$; $H = a$; $L = \sqrt{2}a$ so that

$$\cos\theta = -1, \quad \sin\theta = 0 \quad \text{and} \quad \cos\phi = \frac{1}{\sqrt{2}}, \quad \sin\phi = \frac{1}{\sqrt{2}}$$

Transferring the effect of the load to a set of global axes, with their origin at node 3 (Figure 8.5(a)), gives $M_x = -Wa$; $M_y = Wa$; $M_z = 0$. The forces in the local axes at node 3 due to these actions in global axes are found from equation (8.6) as

$$\begin{Bmatrix} F_u \\ F_v \\ F_w \end{Bmatrix}_3^2 = \begin{bmatrix} -1/\sqrt{2} & 0 & 1/\sqrt{2} \\ 0 & -1 & 0 \\ 1/\sqrt{2} & 0 & 1/\sqrt{2} \end{bmatrix} \begin{Bmatrix} 0 \\ 0 \\ -1 \end{Bmatrix} W = \begin{Bmatrix} -1/\sqrt{2} \\ 0 \\ -1/\sqrt{2} \end{Bmatrix} W \qquad (8.11)$$

Similarly, the moments at node 3 of this member are obtained as

$$\begin{Bmatrix} M_u \\ M_v \\ M_w \end{Bmatrix}_3^2 = \begin{bmatrix} -1/\sqrt{2} & 0 & 1/\sqrt{2} \\ 0 & -1 & 0 \\ 1/\sqrt{2} & 0 & 1/\sqrt{2} \end{bmatrix} \begin{Bmatrix} -1 \\ 1 \\ 0 \end{Bmatrix} Wa = \begin{Bmatrix} 1/\sqrt{2} \\ -1 \\ -1/\sqrt{2} \end{Bmatrix} Wa \qquad (8.12)$$

It will be noted from equation (8.12) that this member is subject to also torque and bending about both axes. The forces and moments at its first node (2) may be found from member equilibrium (equation (8.9)) as

$$\begin{Bmatrix} F_u \\ F_v \\ F_w \end{Bmatrix}_2^2 = \begin{Bmatrix} 1/\sqrt{2} \\ 0 \\ 1/\sqrt{2} \end{Bmatrix} W \quad \text{and} \quad \begin{Bmatrix} M_u \\ M_v \\ M_w \end{Bmatrix}_2^2 = \begin{Bmatrix} -1/\sqrt{2} \\ 0 \\ 1/\sqrt{2} \end{Bmatrix} Wa \tag{8.13}$$

Continuing the same procedure for member 1 (1–2), this has: $\bar{x} = a$; $\bar{y} = -a$; $\bar{z} = a$; $H = \sqrt{2}a$; $L = \sqrt{3}a$ so that

$$\cos\theta = \frac{1}{\sqrt{2}}; \quad \sin\theta = -\frac{1}{\sqrt{2}} \quad \text{and} \quad \cos\phi = \sqrt{\frac{2}{3}}; \quad \sin\phi = \frac{1}{\sqrt{3}}$$

In relation to a set of global axes at node 2, the effect of the load is: $F_z = -W$; $M_x = -Wa$; $M_y = 0$; $M_z = 0$. The forces in the local axes at node 2 due to these actions are found from equation (8.6) as

$$\begin{Bmatrix} F_u \\ F_v \\ F_w \end{Bmatrix}_2^1 = \begin{bmatrix} 1/\sqrt{3} & -1/\sqrt{3} & 1/\sqrt{3} \\ 1/\sqrt{2} & 1/\sqrt{2} & 0 \\ -1/\sqrt{6} & 1/\sqrt{6} & \sqrt{2/3} \end{bmatrix} \begin{Bmatrix} 0 \\ 0 \\ -1 \end{Bmatrix} W = \begin{Bmatrix} -1/\sqrt{3} \\ 0 \\ -\sqrt{2/3} \end{Bmatrix} W \tag{8.14}$$

Similarly, the moments at node 2 of this member are obtained as

$$\begin{Bmatrix} M_u \\ M_v \\ M_w \end{Bmatrix}_2^1 = \begin{bmatrix} 1/\sqrt{3} & -1/\sqrt{3} & 1/\sqrt{3} \\ 1/\sqrt{2} & 1/\sqrt{2} & 0 \\ -1/\sqrt{6} & 1/\sqrt{6} & \sqrt{2/3} \end{bmatrix} \begin{Bmatrix} -1 \\ 0 \\ 0 \end{Bmatrix} Wa = \begin{Bmatrix} -1/\sqrt{3} \\ -1/\sqrt{2} \\ 1/\sqrt{6} \end{Bmatrix} Wa \tag{8.15}$$

The forces and moments at the member's first node (1) are again found by equilibrium (equation (8.9)) as

$$\begin{Bmatrix} F_u \\ F_v \\ F_w \end{Bmatrix}_1^1 = \begin{Bmatrix} 1/\sqrt{3} \\ 0 \\ \sqrt{2/3} \end{Bmatrix} W \quad \text{and} \quad \begin{Bmatrix} M_u \\ M_v \\ M_w \end{Bmatrix}_1^1 = \begin{Bmatrix} 1/\sqrt{3} \\ -1/\sqrt{2} \\ -1/\sqrt{6} \end{Bmatrix} Wa \tag{8.16}$$

The complete sets of internal forces for the bent beam are shown in Figures 8.8 and 8.9. The downward vertical load does, of course, produce compression in each of the three members. The torque in the top member is zero since the load is applied on its axis. In the lower two members, the torque changes sign since the load is offset to different sides of the u-axes of the members. The w-shear is constant in all members due to the absence of local loading within the member lengths. The top two members are bent in single curvature in relation to their v-axes, but of differing signs, while the lowest member is bent in double curvature. There is no shear in the v directions, since these are all in horizontal planes and there is no horizontal plane loading. The w-axis of the top member is in the plane of the load and hence w-bending does not occur in this member. The load does, however, cause w-bending, of similar curvature, in the lower two members.

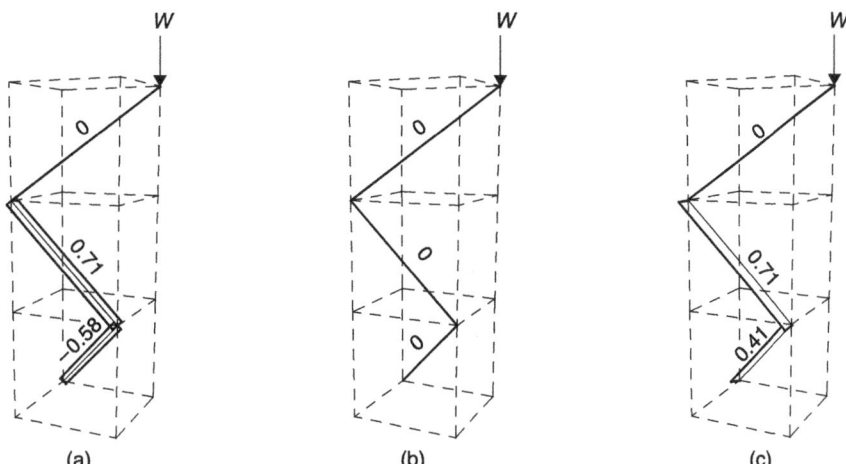

Figure 8.9   Example bent beam: (a) torque (×Wa), (b) SF_v (×W), (c) BM_w (×Wa)

Worksheet 8.1   Statically determinate space frames

W8.1.1   The members of the bent beam shown in Figure W8.1.1 are designated 1–2, 2–3 and 3–4. The structure is subjected to a downward load of W and a horizontal (x-direction) load of $W/2$ at node 4. It is fully fixed at node 1. Determine the internal forces in the members and hence draw internal force diagrams for the structure. What differences would it make to (a) the values calculated and (b) the diagrams drawn, if member 1–2 were to be designated 2–1?

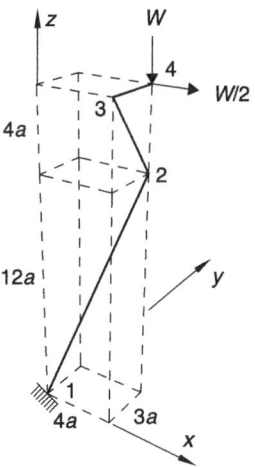

Figure W8.1.1   Worksheet 8.1.1

8.4   COMPUTER ANALYSIS

Orientation of principal axes

In Chapter 8, it was noted that a cantilever channel of the form shown in Figure 8.10(a) would rotate under the action of a vertical load applied to its web, since

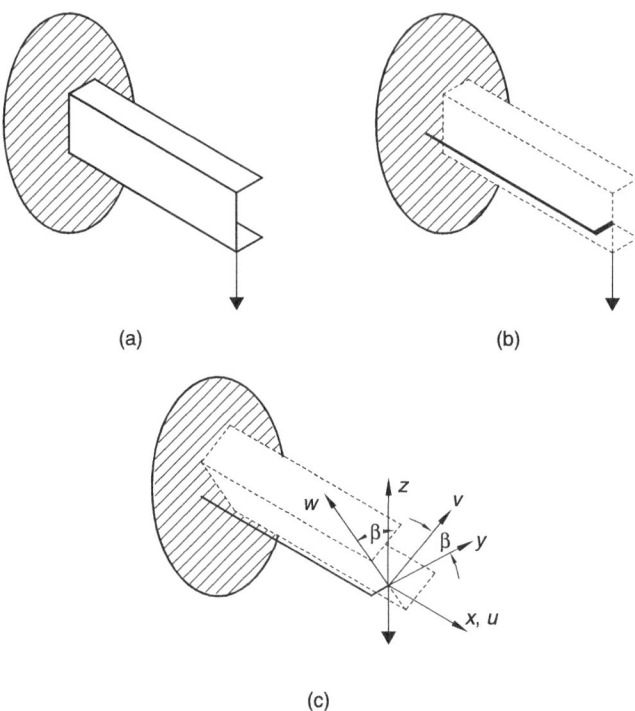

Figure 8.10 Cantilever channel: (a) physical, (b) non-rotated computer model, (c) rotated computer model

torques act about the shear centre of sections. If it was desired to include the shear centre effect, then the cantilever had to be modelled along the shear centre line and the load located on a rigid offset (Figure 8.10(b)). If the channel section is rotated about its axis, as in Figure 8.10(c), then its principal axes, v-, w-, will no longer co-incide with the global y-, z-axes. In these circumstances, a downward load will have components in both the v- and the w-axes and will cause displacements in both directions. In the case shown (Figure 8.10(c)) the displacements will be in the negative v, w directions and these movements will be equivalent to displacements in the negative y-, z-axes. It is therefore possible for a vertical load to cause lateral displacement, even to a straight member. To model such a member correctly for computer analysis purposes, it is necessary to specify the relationship of the local member axes to the global axes. This can usually either be done directly by specifying the angle β (Figure 8.10(c)) or, often more conveniently, by specifying a point in the u, w plane.

Sections with axes of symmetry are normally such that an axis of symmetry is parallel to the y- or z-global axis and β is therefore zero. This is unlikely to be the case for an angle section, however, since, for an equal angle the line of symmetry is inclined at 45° to the horizontal (Figure 8.11(a)). For unequal angles and sections such as zeds (Figure 8.11(b), (c)), the orientation of the principal axes needs to be determined from appropriate theory (Gere and Timoshenko, 1999) or from standard reference data. Angles, in common with all sections consisting of branches that intersect in a single point, have their shear centre at the intersection point. In the case of zeds the shear centre coincides with the centroid due to the rotational symmetry of the section.

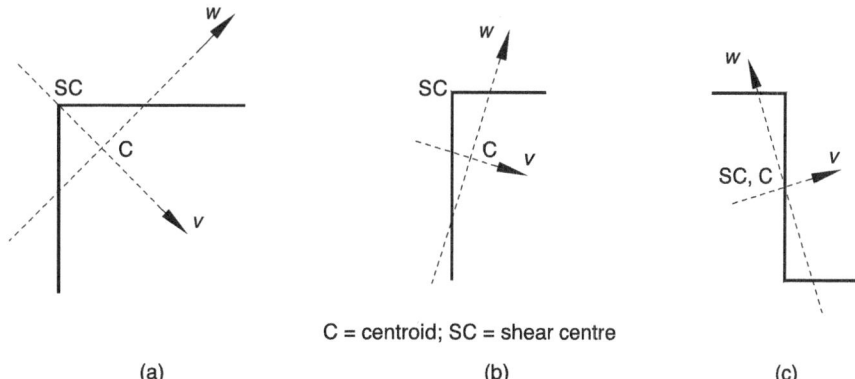

C = centroid; SC = shear centre

(a)   (b)   (c)

Figure 8.11   Principal axes and shear centres: (a) equal angle, (b) unequal angle, (c) zed section

Stiffness method theory

To adapt the general stiffness method to space frames, it is necessary to appropriately alter the transformation matrix $\mathbf{T}$ and the member stiffness matrix $\mathbf{K}_m$ to three dimensions. The member stiffness matrix is generated by incorporating axial load and torsional effects to bending effects about the two principal axes. Following the procedures adopted for plane frames and grids, the stiffness matrix is given by

$$p = \mathbf{K}_m e$$

where

$$\mathbf{K}_m =$$

$$
\begin{bmatrix}
\frac{EA}{L} & 0 & 0 & 0 & 0 & 0 & \frac{-EA}{L} & 0 & 0 & 0 & 0 & 0 \\
0 & \frac{12EI_w}{L^3} & 0 & 0 & 0 & \frac{6EI_w}{L^2} & 0 & \frac{-12EI_w}{L^3} & 0 & 0 & 0 & \frac{6EI_w}{L^2} \\
0 & 0 & \frac{12EI_v}{L^3} & 0 & \frac{-6EI_v}{L^2} & 0 & 0 & 0 & \frac{-12EI_v}{L^3} & 0 & \frac{-6EI_v}{L^2} & 0 \\
0 & 0 & 0 & \frac{GJ}{L} & 0 & 0 & 0 & 0 & 0 & \frac{-GJ}{L} & 0 & 0 \\
0 & 0 & \frac{-6EI_v}{L^2} & 0 & \frac{4EI_v}{L} & 0 & 0 & 0 & \frac{6EI_v}{L^2} & 0 & \frac{2EI_v}{L} & 0 \\
0 & \frac{6EI_w}{L^2} & 0 & 0 & 0 & \frac{4EI_w}{L} & 0 & \frac{-6EI_w}{L^2} & 0 & 0 & 0 & \frac{2EI_w}{L} \\
\frac{-EA}{L} & 0 & 0 & 0 & 0 & 0 & \frac{EA}{L} & 0 & 0 & 0 & 0 & 0 \\
0 & \frac{-12EI_w}{L^3} & 0 & 0 & 0 & \frac{-6EI_w}{L^2} & 0 & \frac{12EI_w}{L^3} & 0 & 0 & 0 & \frac{-6EI_w}{L^2} \\
0 & 0 & \frac{-12EI_v}{L^3} & 0 & \frac{6EI_v}{L^2} & 0 & 0 & 0 & \frac{12EI_v}{L^3} & 0 & \frac{-6EI_v}{L^2} & 0 \\
0 & 0 & 0 & \frac{-GJ}{L} & 0 & 0 & 0 & 0 & 0 & \frac{GJ}{L} & 0 & 0 \\
0 & 0 & \frac{-6EI_v}{L^2} & 0 & \frac{2EI_v}{L} & 0 & 0 & 0 & \frac{-6EI_v}{L^2} & 0 & \frac{4EI_v}{L} & 0 \\
0 & \frac{6EI_w}{L^2} & 0 & 0 & 0 & \frac{2EI_w}{L} & 0 & \frac{-6EI_w}{L^2} & 0 & 0 & 0 & \frac{4EI_w}{L}
\end{bmatrix}
$$

$$(8.17)$$

$$p = \left\{ F_{ua},\ F_{va},\ F_{wa},\ M_{ua},\ M_{va},\ M_{wa},\ F_{ub},\ F_{vb},\ F_{wb},\ M_{ub},\ M_{vb},\ M_{wb} \right\}^{T}$$

and

$$e = \left\{ u_{a},\ v_{a},\ w_{a},\ \theta_{ua},\ \theta_{va},\ \theta_{wa},\ u_{b},\ v_{b},\ w_{b},\ \theta_{ub},\ \theta_{vb},\ \theta_{wb} \right\}^{T}$$

To obtain the transformation matrix, it is necessary to account for the effects of the angle, β, as well as for the local/global transformation given by equation (8.6) (Figure 8.6). By resolution of force components in Figure 8.10(c), it follows that

$$
\begin{Bmatrix} F_u \\ F_v \\ F_w \end{Bmatrix} =
\begin{bmatrix}
1 & 0 & 0 \\
0 & \cos\beta & \sin\beta \\
0 & -\sin\beta & \cos\beta
\end{bmatrix}
\begin{Bmatrix} F_x \\ F_y \\ F_z \end{Bmatrix}
\tag{8.18}
$$

A combined transformation of the forms of equations (8.6) and (8.18) can therefore be represented by

$$
\begin{Bmatrix} F_u \\ F_v \\ F_w \end{Bmatrix} =
\begin{bmatrix}
1 & 0 & 0 \\
0 & \cos\beta & \sin\beta \\
0 & -\sin\beta & \cos\beta
\end{bmatrix}
\begin{bmatrix}
\cos\theta\cos\phi & \sin\theta\cos\phi & \sin\phi \\
-\sin\theta & \cos\theta & 0 \\
-\cos\theta\sin\phi & -\sin\theta\sin\phi & \cos\phi
\end{bmatrix}
\begin{Bmatrix} F_x \\ F_y \\ F_z \end{Bmatrix}
$$

or

$$
\begin{Bmatrix} F_u \\ F_v \\ F_w \end{Bmatrix}
$$

$$
=
\begin{bmatrix}
\cos\theta\cos\phi & \sin\theta\cos\phi & \sin\phi \\
-\cos\beta\sin\theta - \sin\beta\cos\theta\sin\phi & \cos\beta\cos\theta - \sin\beta\sin\theta\sin\phi & \sin\beta\cos\phi \\
\sin\beta\sin\theta - \cos\beta\cos\theta\sin\phi & -\sin\beta\cos\theta - \cos\beta\sin\theta\sin\phi & \cos\beta\cos\phi
\end{bmatrix}
$$

$$
\times \begin{Bmatrix} F_x \\ F_y \\ F_z \end{Bmatrix} = \mathbf{T}_0 \begin{Bmatrix} F_x \\ F_y \\ F_z \end{Bmatrix}
\tag{8.19}
$$

This transformation will apply both to forces and to moments and is relevant to both ends of a member. The required complete transformation matrix is therefore given by

$$
\mathbf{T} =
\begin{bmatrix}
\mathbf{T}_0 & 0 & 0 & 0 \\
0 & \mathbf{T}_0 & 0 & 0 \\
0 & 0 & \mathbf{T}_0 & 0 \\
0 & 0 & 0 & \mathbf{T}_0
\end{bmatrix}
\tag{8.20}
$$

It follows from equations (8.17) and (8.20) that the member properties required for a space frame analysis will be cross-sectional areas, torsional constants and principal second moments of area. The inclination of the latter to a vertical plane will also be needed, where relevant, and care is needed to ensure that the second moments of area are prescribed to their appropriate axes. Particular care is needed with vertical members, since conventions for the designation of member axes vary for this singular case. The use of 'full 3-D' member graphics (see the later Figure W8.2.3) is particularly helpful in checking member orientations.

Example 8.2 Channel cantilever beam

A channel cantilever beam of the form shown in Figure 8.10 will be analysed for the effects of a 0.5 kN downward load applied through the shear centre at the end of its 2 m length. The angle β will be taken as 30°. Comparative analyses will be undertaken in which the effects of the β angle are ignored and included, respectively. The horizontal and vertical components of displacement will be derived analytically and used to validate the computer solution. The analytical solution will also be used to illustrate the displacement response to varying β. The following properties will be assumed:

$$I_v = 2.25 \times 10^6 \, \text{mm}^4; \quad I_w = 0.354 \times 10^6 \, \text{mm}^4; \quad E = 206 \, \text{kN/mm}^2$$

Appropriate computer analyses result in:

Table 8.1   Channel cantilever displacement results (mm)

	δ_y	δ_z
β effects ignored	0	−2.88
β effects included	−6.63	−6.73

From Table 8.1 it will be noticed that including the β angle effect results in the vertical displacement more than doubling and the introduction of an approximately equal horizontal displacement. Both of these results are, of course, due to the vertical load producing a bending component about the more flexible w-axis when the β effect is included.

The analytical solution will be worked in general terms, assuming a vertical load, W, is applied through the shear centre and that the length of the beam is L. The load may be expressed as components in the u, v directions as shown in Figure 8.12. Since u and v are principal axes, the general formula for end deflection of a cantilever may be used (Figure 2.1) to produce the displacements in these directions:

$$\delta_v = -\frac{W \sin \beta L^3}{3EI_w} \quad \text{and} \quad \delta_w = -\frac{W \cos \beta L^3}{3EI_v}$$

The vertical and horizontal components of displacement are then given by

$$\delta_y = \delta_v \cos \beta - \delta_w \sin \beta \quad \text{and} \quad \delta_z = \delta_v \sin \beta + \delta_w \cos \beta$$

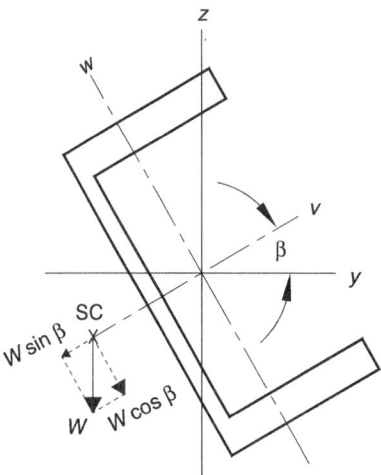

Figure 8.12 Cantilever channel beam

So that

$$\delta_y = -\frac{WL^3}{3E}\frac{\sin\beta\cos\beta}{I_w} + \frac{WL^3}{3E}\frac{\sin\beta\cos\beta}{I_v}$$

$$= -\frac{WL^3}{3E}\frac{\sin 2\beta}{2}\left(\frac{1}{I_w}-\frac{1}{I_v}\right) = -\frac{C_W}{2}C_I\sin 2\beta \qquad (8.21)$$

where

$$C_W = \frac{WL^3}{3E} \quad \text{and} \quad C_I = \left(\frac{1}{I_w}-\frac{1}{I_v}\right)$$

and

$$\delta_z = -\frac{WL^3}{3E}\frac{\sin^2\beta}{I_w} - \frac{WL^3}{3E}\frac{\cos^2\beta}{I_v}$$

$$= -\frac{WL^3}{3E}\left(\sin^2\beta\left(\frac{1}{I_w}-\frac{1}{I_v}\right)-\frac{1}{I_v}\right) = -C_W\left(C_I\sin^2\beta+\frac{1}{I_v}\right) \qquad (8.22)$$

For the values used in the computer analyses, $C_W = 6.472 + 10^6\,\text{mm}^5$; $C_I = 2.38 + 10^6\,\text{mm}^{-4}$ and substitution in the above equations for $\beta = 30°$ and $\beta = 0°$ produces checks on the computer results.

From equations (8.21) and (8.22), it may be observed that the displacement components vary with β according to the trigonometrical functions $\sin 2\beta$ and $\sin^2\beta$. Using the same numerical values as previously, plots of the y- and z-displacement components are shown in Figure 8.13. From Figure 8.13, it may be observed, and may be checked analytically, that the vertical displacement component naturally achieves its minimum and maximum values when bending occurs about the smaller or larger principal axis, respectively, and that there is no lateral displacement in these circumstances. Lateral displacement is greatest

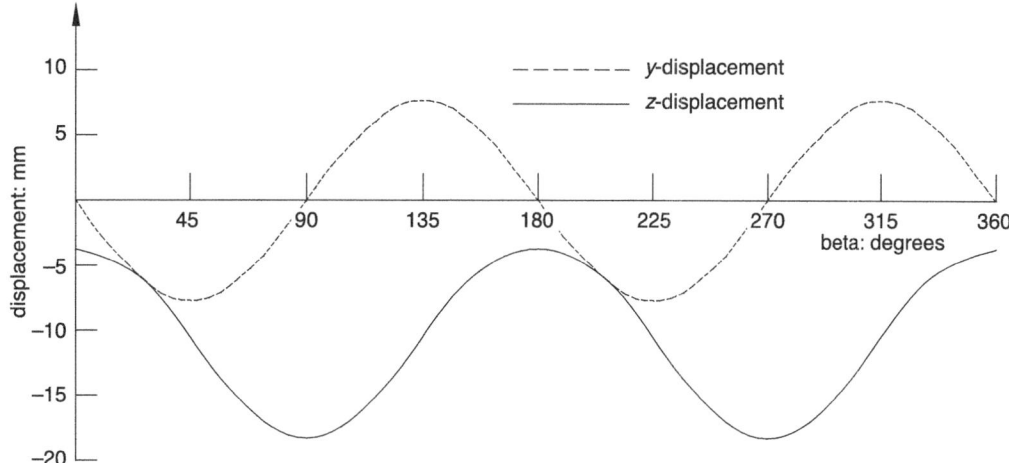

Figure 8.13   Displacement variation with beta for cantilever channel beam

when the principal axes are orientated at 45° to the x and y directions and alternates in sign.

Example 8.3   4 × 4 CUBIC space frame

A CUBIC space frame that has four 2 m square bays in each direction, a construction depth of 0.5 m, and is vertically supported at its lower corner nodes, will be analysed for the effects of a uniformly distributed load of 5 kN/m² and a central point load of 80 kN. The following properties will be assumed:

for the structural material: $E = 206\,\mathrm{kN/mm^2}$; $\nu = 0.3$

for the beams: $A = 4 \times 10^3\,\mathrm{mm^2}$; $I_v = 30 \times 10^6\,\mathrm{mm^4}$;

$$I_w = 4 \times 10^6\,\mathrm{mm^4}; \quad J = 100 \times 10^3\,\mathrm{mm^4}$$

for the columns: $A = 10 \times 10^3\,\mathrm{mm^2}$; $I_v = 100 \times 10^6\,\mathrm{mm^4}$;

$$I_w = 100 \times 10^6\,\mathrm{mm^4}; \quad J = 150 \times 10^6\,\mathrm{mm^4}$$

Since the structure and loading are both symmetric, only a quarter of the frame will be analysed, as shown in Figure 8.14(a). To model the symmetry requirements, the general precepts of suppressing displacements normal to the planes of symmetry and rotations about axes within planes of symmetry need to be observed. Constraints must therefore be applied such that, for all nodes in the plane BCGF, $\delta_y = \theta_x = 0$, and for all nodes in the plane CDHG, $\delta_x = \theta_y = 0$. Both sets of constraints will apply along the central vertical, CG. These symmetry constraints are sufficient to position the structure and the only further restraint to be applied is the vertical support at node A. Cross-sectional properties will need to be halved for members lying in the symmetry planes, resulting in a quartering of the properties for the post CG. The uniform loading will be represented by equivalent nodal loading. This again needs to be halved along the symmetry

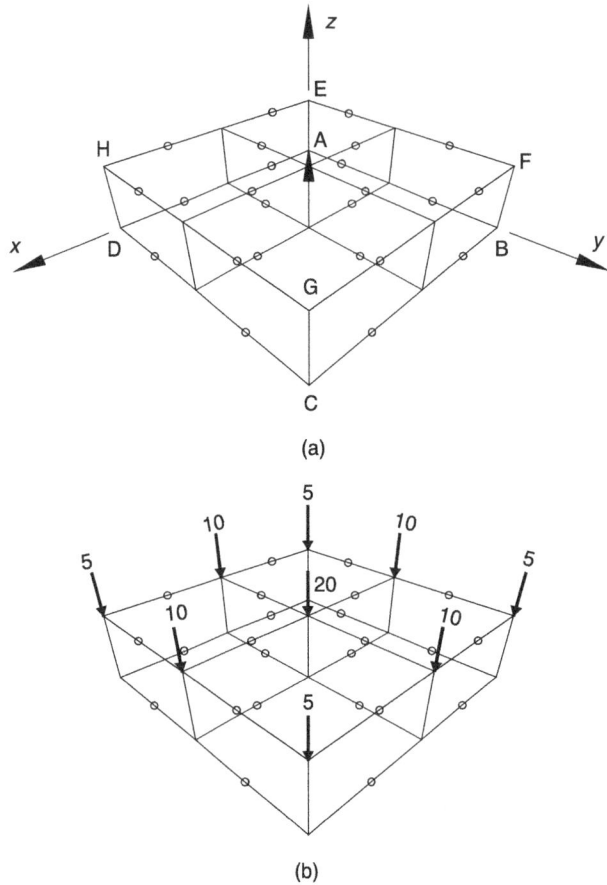

Figure 8.14   4 × 4 CUBIC space frame example: (a) layout, (b) equivalent uniform loading (kN)

lines and quartered at G, resulting in the nodal loads shown in Figure 8.14(b). The point loading case will need a load of 20 kN applied to node G.

The axial loads in the frame under the two loading cases are shown in Figure 8.15. The loads shown are 'true' loads in the sense that computer values for members in the planes of symmetry have been scaled to allow for the reduced section sizes employed in the analysis. The posts share load between the two layers and therefore simply sustain compressions equal to a half the load applied to the relevant upper node. The axial loads in the beam members provide the primary resistance to bending. Clearly the structure is basically in sagging bending and all the top beams are therefore in compression and all the lower beams in tension.

The distributions of the beam axial loads vary for the two loading cases. In both cases, the axial loads in individual two-bay frames increase towards the centre, as the bending effect increases. However, for the uniform loading, Figure 8.15(a), axial loading in the sets of two-bay frames increases as progress is made towards the edge of the structure. The reason for this lies in the load distribution paths followed. For the uniform loading, load applied to A (Figure 8.15(a)) will be

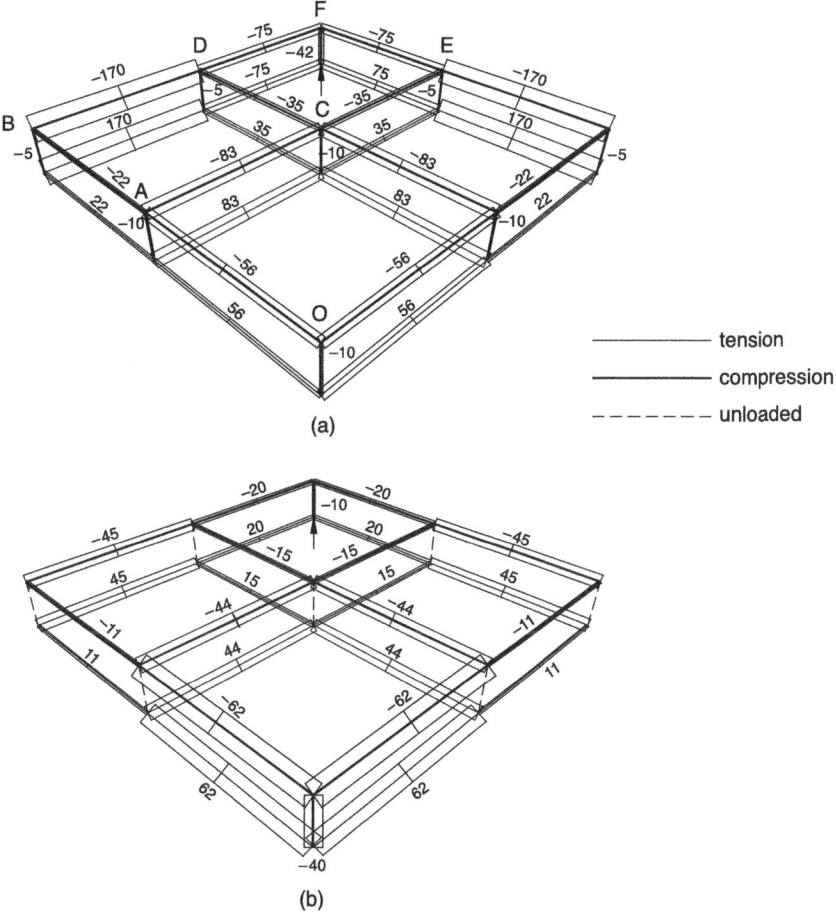

Figure 8.15   4 × 4 CUBIC space frame example axial loads (kN): (a) uniform loading, (b) point loading

preferentially transmitted along the paths AC(D/E)F to the corner support rather than along the more flexible route ABDF, thus increasing the effective load on the intermediate two-bay frame (ACE) and relieving the central one (OAB). The edge frame has no intermediate stiffening support of this type and must therefore bear the full effect of the loading applied to it, leading to it sustaining the greatest axial loads. For the central point load case (Figure 8.15(b)), it will be noticed that the greatest axial loads now appear in the central bays of the middle two-bay frames since these bays are the only ones to be directly loaded. The tendency to increasing axial loads in outer bays then reasserts itself beyond the central area.

Figure 8.16 shows a bending moment diagram for bending about the v-axes of the members and the corresponding shear force diagram for shear forces in member w directions. Again, diagrams in the symmetry planes have been scaled to produce 'true' diagrams. For simplicity, values are shown for only the posts and upper beams, the values on the lower beams being similar to those of the upper beams. Under vertical loading, there is no w-axis bending in the horizontal beam members, but, in general, there is w-axis bending in the posts. A diagram

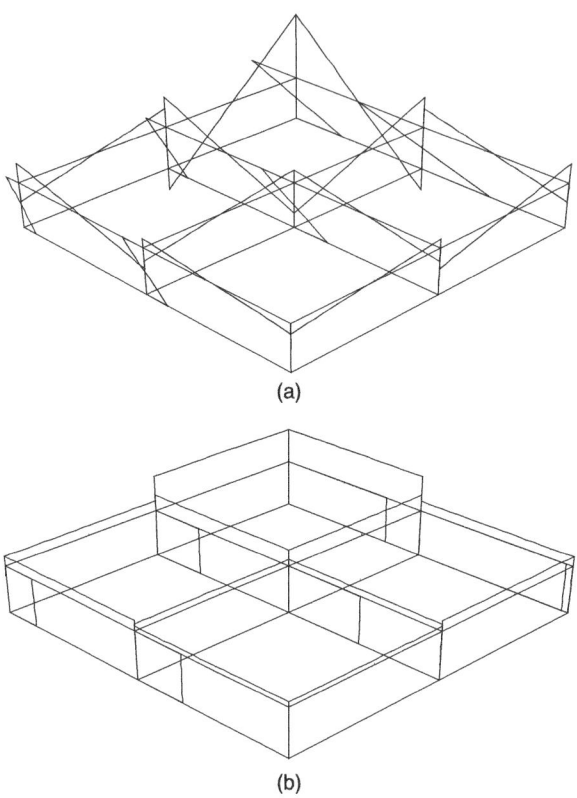

(a)

(b)

Figure 8.16   4 × 4 CUBIC space frame example: (a) BM$_v$ *diagram, (b)* SF$_w$ *diagram*

such as Figure 8.16(a) therefore has to be treated with care, since it needs to be combined with the corresponding w-axis bending moment diagram for members subject to biaxial bending.

As discussed previously (Chapter 5), Vierendeel girders resist shearing action by the development of local bending in the horizontal booms to generate matching internal shear forces. For the structure considered, the shear forces are largest towards the support position and will tend to zero as lines of symmetry are crossed. The shear force distribution shown in Figure 8.16(b) confirms these expectations. It follows that the bending moments will also increase towards the support, from near zero values at the centre. Each beam member is bent into double curvature, so producing characteristic shear deformation within each bay (Figure 5.13(b)).

Worksheet 8.2   Computer analysis of space frames

W8.2.1   The two sections shown in Figure W8.2.1 are of constant thickness, 10 mm, and have the same cross-sectional area, $A = 1900\,\mathrm{mm}^2$. The sections have principal axes orientated as shown in the figure. The angle section has principal second moments of area of 2.866×10^6 and $0.734 \times 10^6\,\mathrm{mm}^4$, while the zed section has values of 3.467×10^6 and $0.265 \times 10^6\,\mathrm{mm}^4$. Assume $E = 206\,\mathrm{kN/mm}^2$ for the sections.

Allocate principal second moment of area values appropriately to the axes v and w shown in Figure W8.2.1. Each of the sections is used as a cantilever beam of length

2 m, and a downward load of 250 N is applied to the ends of the beams through their shear centres. Employ a suitable computer package to determine the y-, z-components of deflection. Use the displacement components to determine the resultant displacement and its direction in each case. Justify the differences in the resultant displacements determined for the two sections. Repeat the analysis when each of the sections is given a $45°$ anti-clockwise rotation about its positive u-axis and compare the results with those from the first analyses.

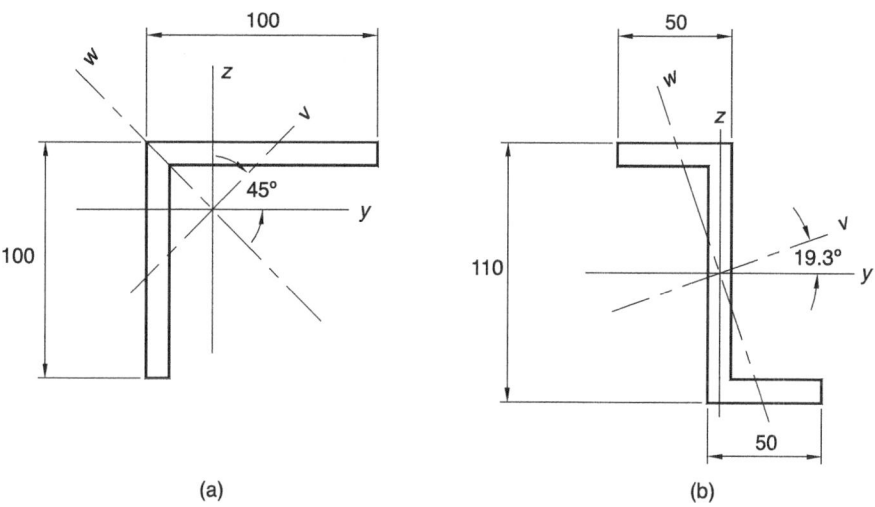

(a)   (b)

Figure W8.2.1   Worksheet 8.2.1

W8.2.2 The central spine support for a spiral staircase is shown in Figure W8.2.2(a)–(c). A general point P on the spiral has co-ordinates $(\cos\varphi, \sin\varphi, \phi/\sqrt{3})$ m (Figure W8.2.2(c)), where φ is measured in radians. At the ends of the staircase (A and B), the support beam is hinged

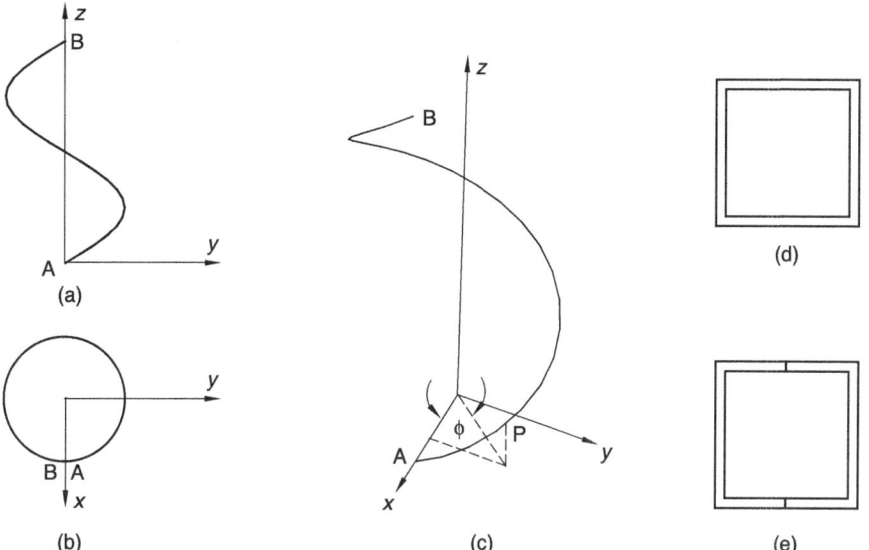

(a)   (b)   (c)   (d)

(e)

Figure W8.2.2   Spiral staircase beam: (a) elevation, (b) plan, (c) perspective, (d) box section, (e) double channel section

about the x-axis, but is constrained in all other directions. Two forms of section are to be considered for the beam. The first (Figure W8.2.2(d)) is a square box section of external side dimension 150 mm. The second (Figure W8.2.2(e)) comprises two channel sections placed so as to form a square box of external side dimension 150 mm, but with no structural interconnection. Both sections have constant thickness of 10 mm. The beam is to be subjected to a uniform, downward load of 2 kN/m on plan. Assume $E = 206\,\mathrm{kN/mm^2}$ and $\nu = 0.3$ for the sections.

Using a suitable division into linear segments, analyse the staircase beam for both sections. Relate the internal force distributions obtained to the expected structural behaviour of the beam. Explain the differences in response obtained from use of the two different sections.

W8.2.3 The heavy machinery hall shown in Figure W8.2.3(a) is 10 m high and has transverse and longitudinal bay widths of 5 m. All beams and columns are similar I-sections and are orientated as shown in the figure. The following properties should be assumed:

for the material: $E = 206\,\mathrm{kN/mm^2}$; $\nu = 0.3$

for the sections: $A = 10 \times 10^3\,\mathrm{mm^2}$; $I_v = 1000 \times 10^6\,\mathrm{mm^4}$; $I_w = 30 \times 10^6\,\mathrm{mm^4}$;

$$J = 1 \times 10^6\,\mathrm{mm^4}$$

The feet of the columns may be presumed fully fixed.

Analyse the frame for a uniform, downward, distributed load of $3\,\mathrm{kN/m^2}$, combined with a point load of 1000 kN due to hoisted heavy machinery, which is applied to point P (Figure W8.2.3(a)). Compare the axial loads and bending moments resulting from this analysis with analyses for the frame if the two central columns on the end transverse bays are omitted (Figure W8.2.3(b)), and if the central longitudinal beams are also omitted (Figure W8.2.3(c)).

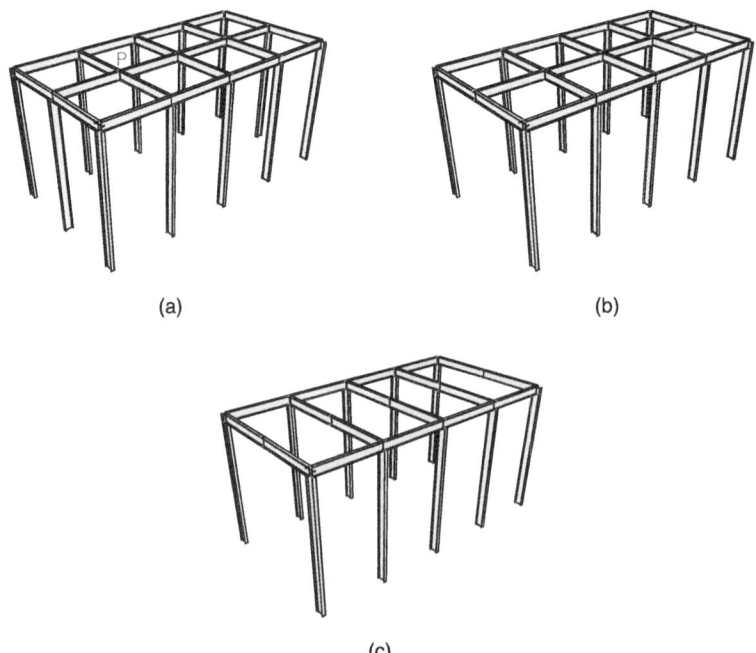

(a)   (b)

(c)

Figure W8.2.3   Worksheet 8.2.3

8.5 INFLUENCE LINES

The general strategies applied previously for the generation of influence lines all apply to space frames. However, space structures inevitably tend to be more complex and are rarely determinate, so that the computer approach, making use of Müller-Breslau's principle, is generally most convenient. The distribution properties of the CUBIC space frame considered in Example 8.3 (Figure 8.14) can readily be confirmed by an influence line investigation. To do this, influence lines will be constructed for the axial loads in members CF, BE and AD (Figure 8.17(a)) of the frame. Following the procedure of Example 3.6, influence lines will be generated by applying to the ends of the relevant members forces that will produce an extension of 1 mm. For CF, the forces will be

$$N_{\mathrm{C}} = -N_{\mathrm{F}} = -\frac{EA}{L} \times 1 \times 10^{-3} = \frac{206 \times 10^6 \times 4 \times 10^{-3}}{2} \times 1 \times 10^{-3} = -412\,\mathrm{kN}$$

The forces produced in BE will be similar, and the forces appropriate to AD will be halved due to the halved area of this member that symmetry required. Deflection surfaces resulting from the application of these end forces are shown in Figure 8.17. By Müller-Breslau's principle, the deflection surfaces represent influence lines for the axial loads in the respective members and may be used to check the load distribution properties of the frame. First, comparison of the three surfaces clearly shows that the edge member CF tends to carry the greatest load. From Figure 8.17(a) it may be seen that vertical, downward loading at any node will produce compression in CF and that the greatest compression will naturally be obtained from loading at C. Loading A, B, E, D or F will

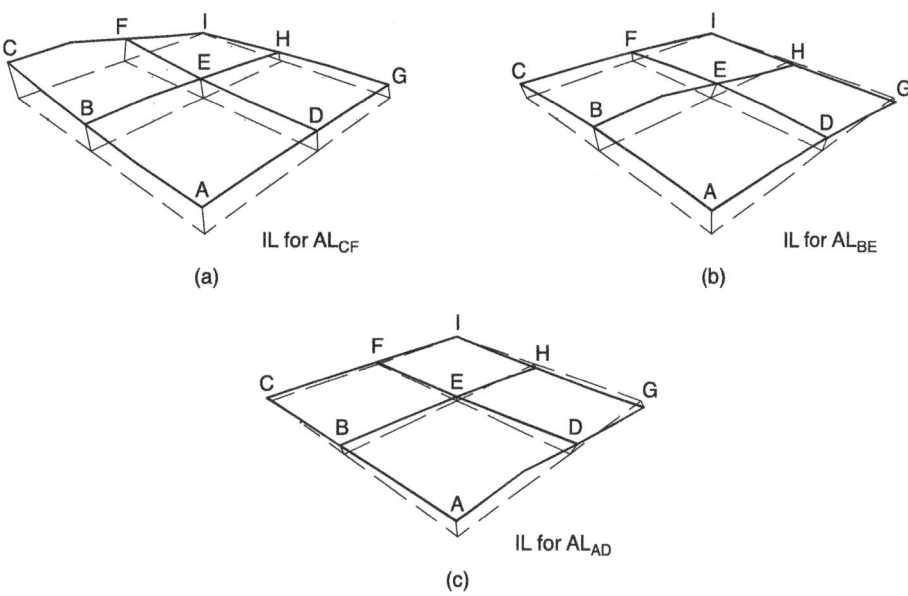

Figure 8.17   CUBIC space frame influence lines: (a) CF, (b) BE, (c) AD

produce rather similar loads in CF. The same observation holds good for BE (Figure 8.17(b)), except that loading H or G will now produce a very small tension in BE. Significant load is only produced in AD (Figure 8.17(c)), however, if node A is loaded. Loading B, C, D, E or F will only result in small compressive loading in AD, simply indicating that loads applied to these nodes follow paths to the corner support that obviously do not pass through the centre position. The same is true of loading applied to nodes G and H, which, as with BE, will produce tension in AD.

8.6   CASE STUDY

C8.1.1   The roofs of the biome greenhouses for the Eden Project (Jones *et al.*, 2001 and Stansfield, 2000) shown in the frontispiece to this chapter were originally conceived as single-layer, hexagon grid geodesic domes. The design evolved, however, into a double layer, hex-tri-hex system (Figure C8.1.1), in which the outer hexagon grid was supplemented with an inner layer comprising a combined grid of hexagons and triangles. This case study aims to explore the comparative properties of these two forms, based on the geometry of a simpler geodesic dome constructed for a Trade and Industries Fair in India (Ramaswamy *et al.*, 2002). Comparisons will also be made with the single-layer triangulated form adopted for the Indian design. This dome has a base radius of 6 m, a rise of 3 m, and its nodes lie on a sphere of radius 7.5 m.

Figure C8.1.1   Eden Project biome – hex-tri-hex system

A single-layer hexagonal grid layout is shown in Figure C8.1.2 and master nodal co-ordinates are given in Table C8.1.1. The remaining co-ordinates may be generated by rotational symmetry.

A double-layer hex-tri-hex system is shown in Figure C8.1.3 and the co-ordinates of the additional master nodes needed to generate this grid are given in Table C8.1.2.

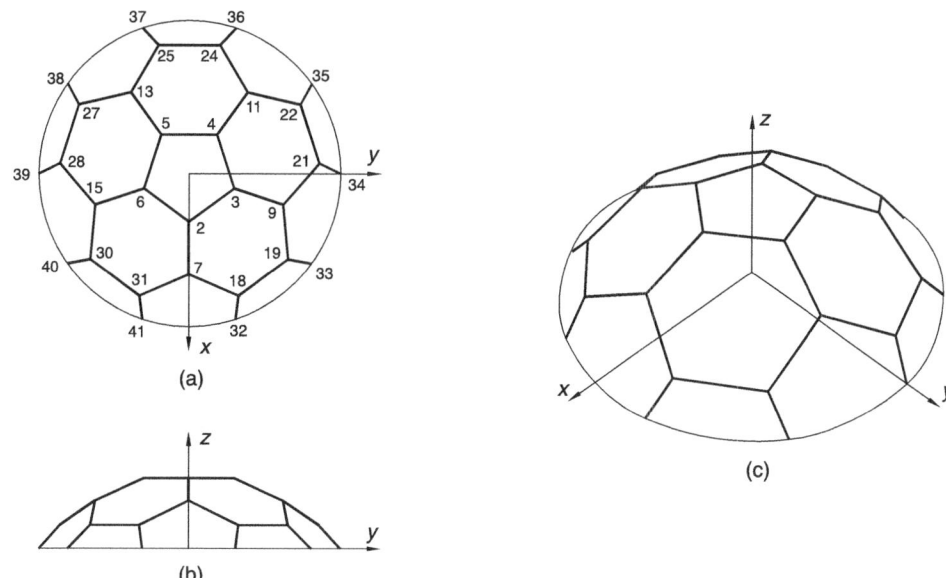

Figure C8.1.2   Hexagonal single-layer dome: (a) plan, (b) elevation, (c) perspective

Finally, a triangular single-layer system is shown in Figure C8.1.4 and the co-ordinates of the additional nodes required to generate the grid are given in Table C8.1.3.

The base nodes (32–41) may be presumed to be free to rotate but fixed in position (ball-joints). Determine the kinematic and statical indeterminacies of each dome based on a space truss assumption (all members pin-jointed). Why is a space truss assumption inappropriate for the single-layer hexagonal design? How can extra members be added to this design to create a determinate space truss?

Analyse each dome for the effects of equal, vertical, downward loads of 10 kN applied to each of the nodes (2–6) of the summit pentagon. The hexagonal layout should be analysed as a rigid-jointed space frame and the other two layouts as pin-jointed space trusses. Analyse the domes, also, for the effects of a uniform temperature rise of 30 °C. Assume that, for each layout, all the members are similar tubes and have the following material properties:

hexagonal: $A = 704\,\text{mm}^2$; $I_v = I_w = 0.277 \times 10^6\,\text{mm}^4$; $J = 0.555 \times 10^6\,\text{mm}^4$;

$r = 30\text{mm}$

hex-tri-hex: $A = 200\,\text{mm}^2$

Table C8.1.1   Master nodal co-ordinates for hexagonal grid

Node	x (m)	y (m)	z (m)
2	1.884	0	2.760
7	3.943	0	1.880
18	4.786	1.971	0.927
19	3.354	3.943	0.927
32	5.706	1.854	0

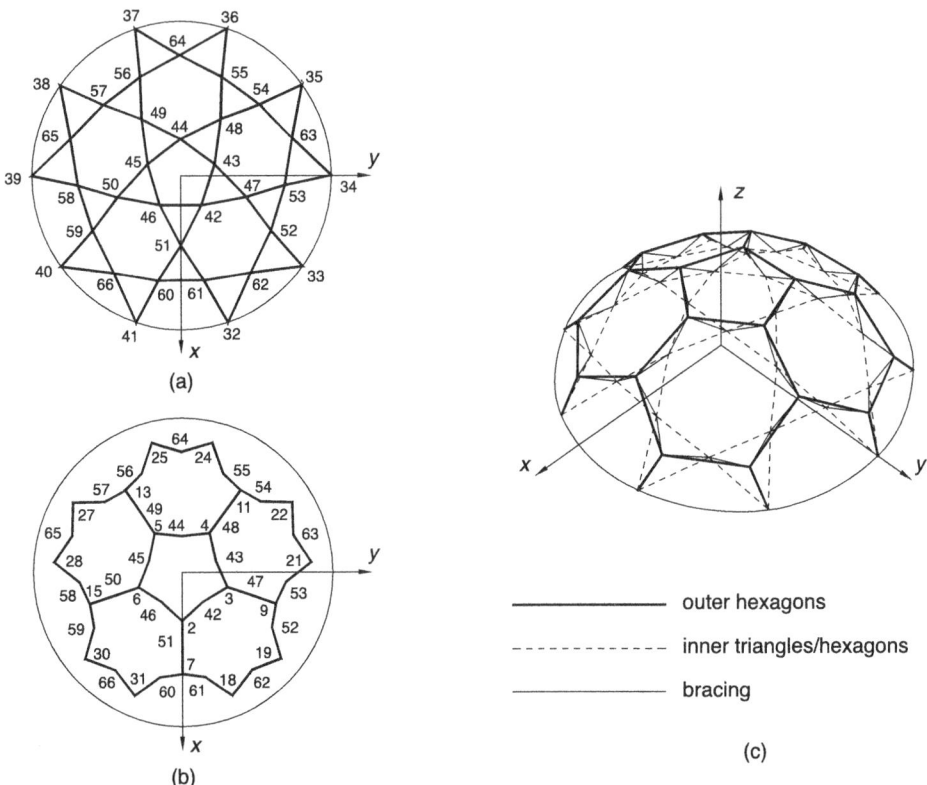

Figure C8.1.3   Hex-tri-hex double-layer dome: (a) plan of inner layer, (b) plan of bracing, (c) perspective

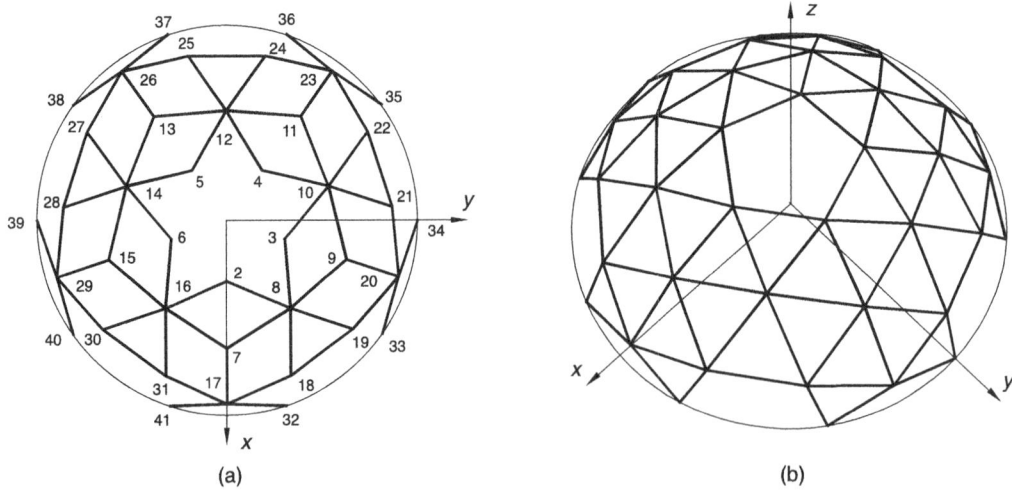

Figure C8.1.4   Triangular grid: (a) additional bembers, (b) perspective

Table C8.1.3   Master nodal co-ordinates for triangular grid

Node	x (m)	y (m)	z (m)
8	2.713	1.971	2.208
9	1.218	3.75	1.880
17	5.650	0	0.431

Table C8.1.2   Master nodal co-ordinates for hex-tri-hex grid

Node	x (m)	y (m)	z (m)
42	1.151	0.836	2.270
47	0.840	2.586	1.859
52	2.134	3.591	1.003
62	3.780	2.760	0.556

triangular: $A = 247\,\text{mm}^2$

all: $E = 206\,\text{kN/mm}^2$; $\alpha = 11.7 \times 10^{-6}/°\text{C}$

The cross-sectional areas quoted are such that each layout has approximately the same self-weight.

Explain the load transmission mechanisms for each of the designs in respect of the summit pentagon loading and explain the nature and magnitudes of the internal force systems induced by the temperature increase. For each loading case, calculate the maximum combined stress (due to axial and bending effects) induced in the members of the hexagon layout and compare these values with the maximum direct stresses induced in the members of the other two designs. Why do you think a hex-tri-hex layout was chosen for the Eden Project biomes?

8.7   REFERENCES

Gere, J. M. and Timoshenko, S. P. (1999) *Mechanics of Materials*. 4th SI edn. London: Stanley Thornes – Section 12.9 of this reference covers principal second moments of area theory.

Jones, A. C., Hamilton, D., Purvis, M. and Jones, M. (2001) 'Eden Project, Cornwall: design, development and construction.' *The Structural Engineer*, **79**(20), 30–36.

Kubik, L. A. (1991) 'The CUBIC space frame in theory and practice'. *Steel Construction Today*, March, 59–63.

Ramaswamy, G. S., Eekhout, M. and Suresh, G. R. (2002) *Analysis, Design and Construction of Steel Space Frames*. London: Thomas Telford – Chapter 8 of this reference describes the analysis of geodesic domes.

Stansfield, K. (2000) 'The biggest greenhouse in the world.' *The Structural Engineer*, **78**(21), 15–19 – progress report on the design and construction of the Eden Project biomes.

9 Answers to problems

9 Answers to problems

9.1 CHAPTER 1 – FUNDAMENTALS

Worksheet 1.1 Kinematic and statical determinacy of trusses

	k	q	Stable
(a)	9	1	Yes
(b)	21	0	No
(c)	10	0	Yes
(d)	21	−1	No
(e)	16	5	Yes
(f)	21	1	Yes
(g)	10	0	No
(h)	17	−1	No
(i)	15	1	Yes
(j)	15	−1	No

Worksheet 1.2 Kinematic and statical determinacy of beams

	k	q	Stable
(a)	0	2	Yes
(b)	6	0	Yes
(c)	3	1	Yes
(d)	5	1	Yes
(e)	3	2	Yes
(f)	2	0	No
(g)	3	3	Yes
(h)	2	0	Yes

Worksheet 1.3 Reaction component influence lines for beams

W1.3.1: $R_B = 115\,\text{kN}$ W1.3.2: $R_A = 50\,\text{kN}$; $M_A = 210\,\text{kN}\,\text{m}$

W1.3.3: $R_A = 240\,\text{kN}$; $R_C = 700\,\text{kN}$ W1.3.4: $R_A = 62\,\text{kN}$; $R_B = 102\,\text{kN}$

9.2   CHAPTER 2 – BEAMS

Worksheet 2.1   Shear force, bending moment and deflection diagrams

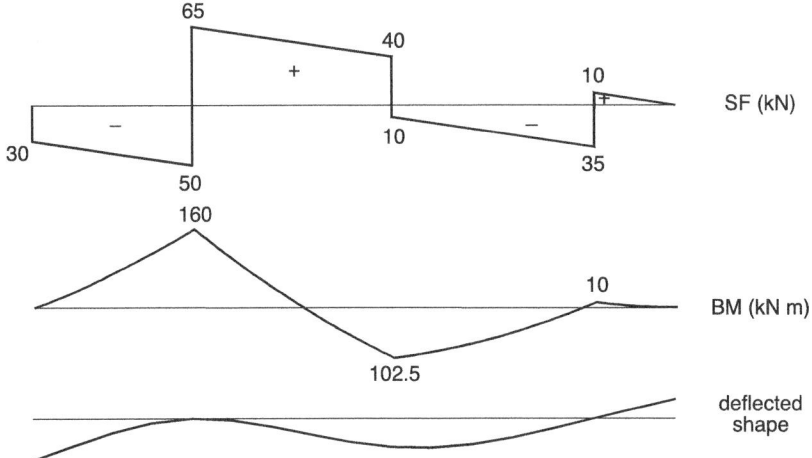

Figure AW2.1.1   Worksheet 2.1.1 answer

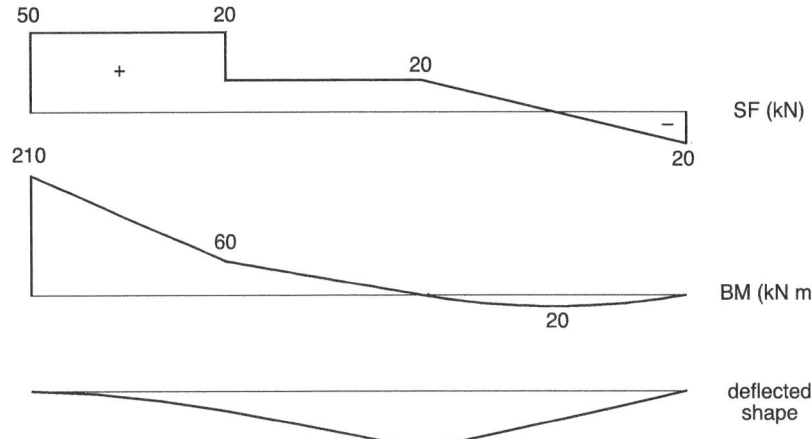

Figure AW2.1.2   Worksheet 2.1.2 answer

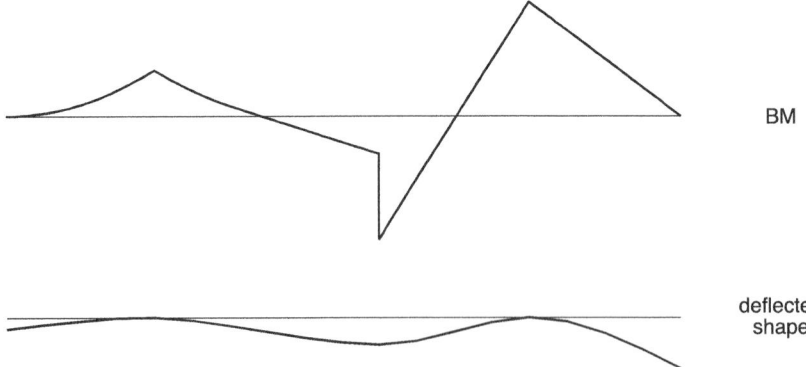

Figure AW2.1.3   Worksheet 2.1.3 answer

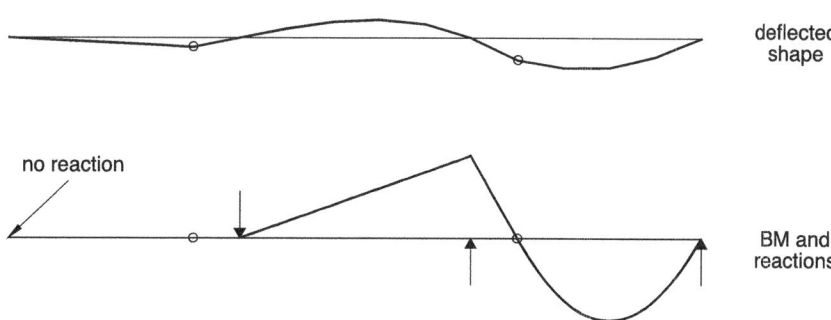

Figure AW2.1.4   Worksheet 2.1.4 answer

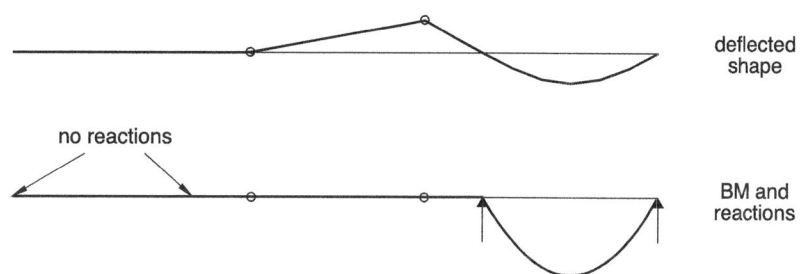

Figure AW2.1.5   Worksheet 2.1.5 answer

Worksheet 2.2 Bending moment and deflection diagrams

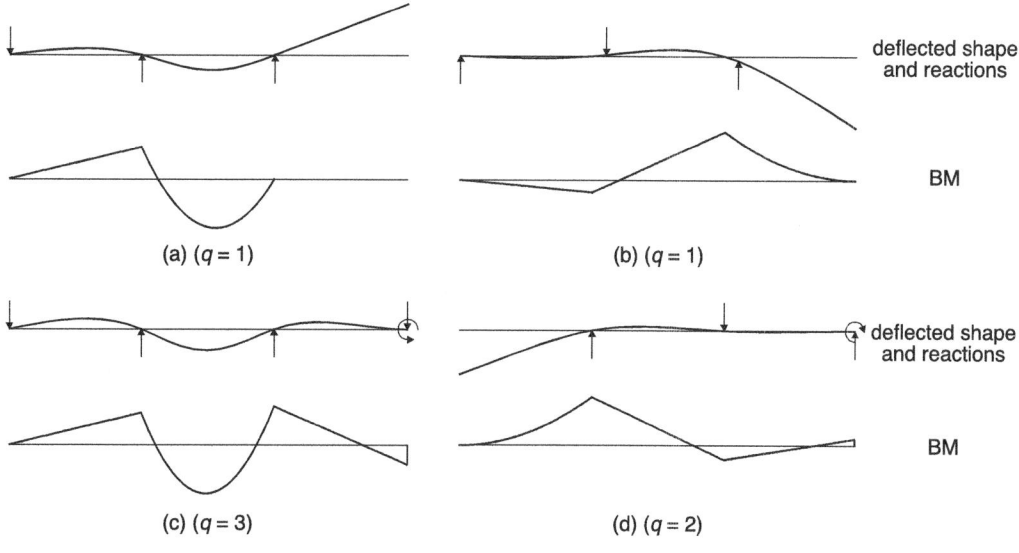

Figure AW2.2.1   Worksheet 2.2.1 answer

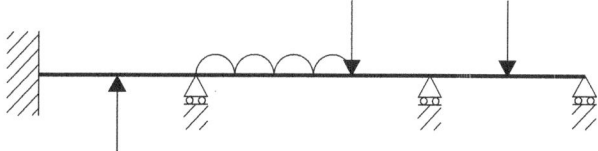

Figure AW2.2.2   Beam analysed

Worksheet 2.3 Computer analysis of beams

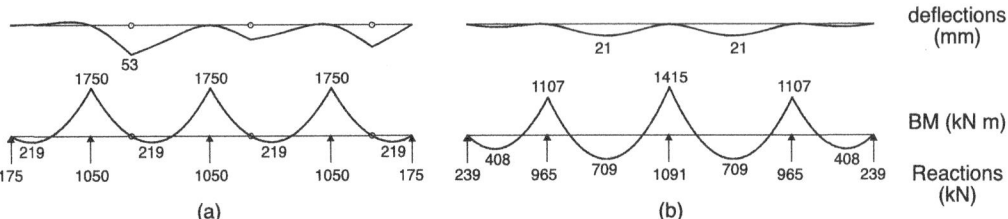

Figure AW2.3.1 (a) Pinned, (b) continuous

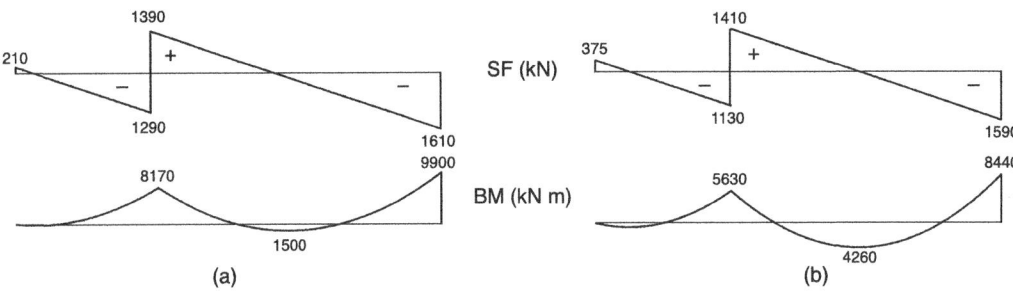

Figure AW2.3.2 (a) Variable section, (b) constant section

Worksheet 2.4 Influence line sketching and calculation

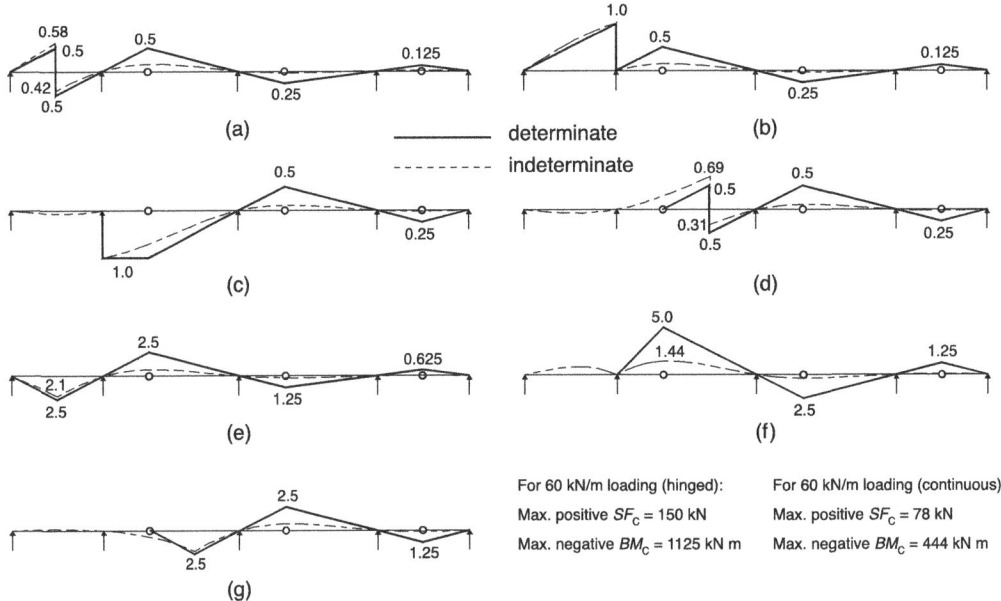

Figure AW2.4.1 Influence lines for SF at (a) A, (b) B⁻, (c) B⁺, (d) C; and BM (m units) at (e) A, (f) B, (g) C

Case studies

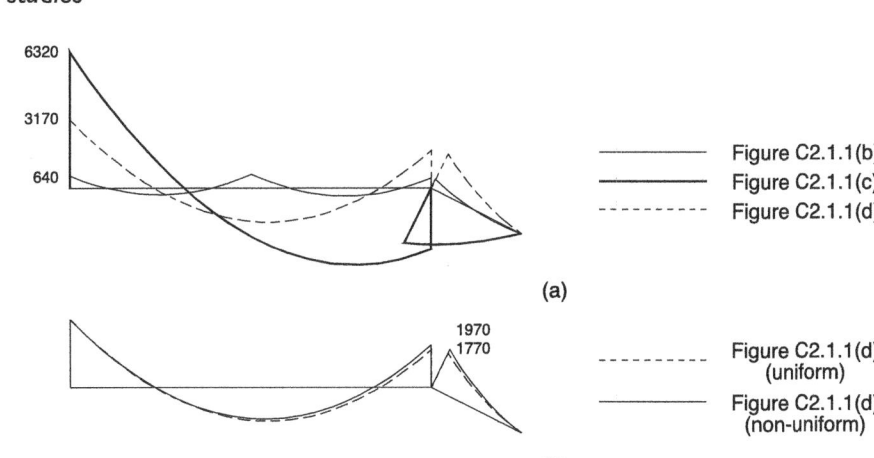

Figure AC2.1.1   Bending moment diagrams (kNm): (a) alternative uniform section designs, (b) uniform and non-uniform section designs

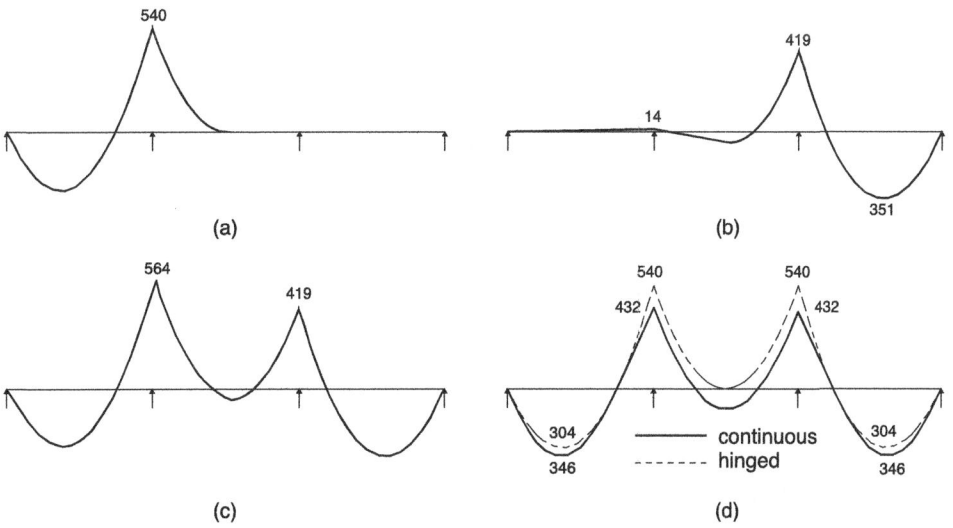

Figure AC2.1.2   (a) left-hand side constructed, (b) effect of right-hand dead load, (c) combined dead load, (d) diagrams for complete construction

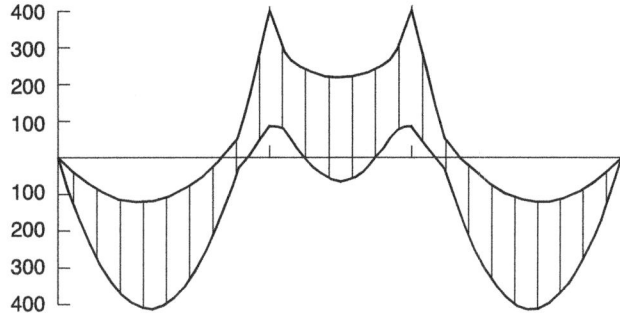

Figure AC2.1.3   Bending moment (kNm) envelope

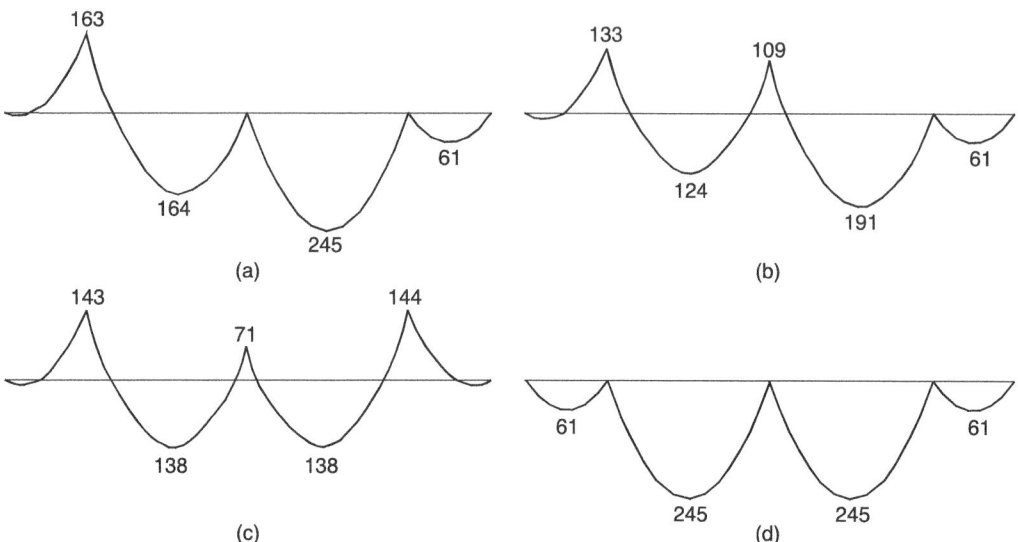

Figure AC2.1.4   Bending moment (mNk) diagrams at (a) stage 1, (b) stage 2, (c) stage 3, (d) simply supported spans

9.3   CHAPTER 3 – PLANE TRUSSES

Worksheet 3.1   Parallel-sided trusses

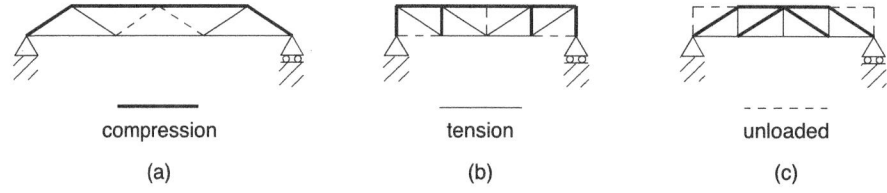

Figure AW3.1.1   Worksheet 3.1.1 answer

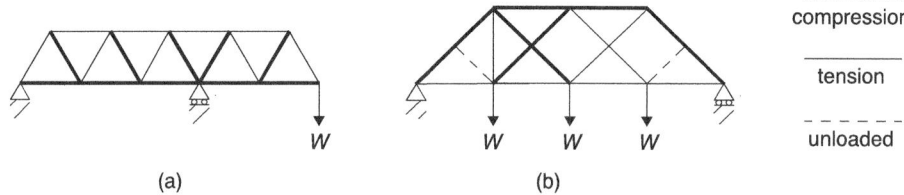

Figure AW3.1.2   Worksheet 3.1.2(a), (b) answers

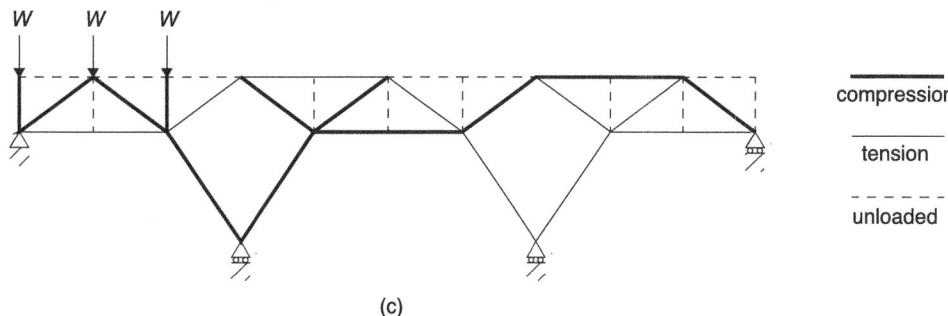

(c)

Figure AW3.1.2 (cont.)   Worksheet 3.1.2(c) answer

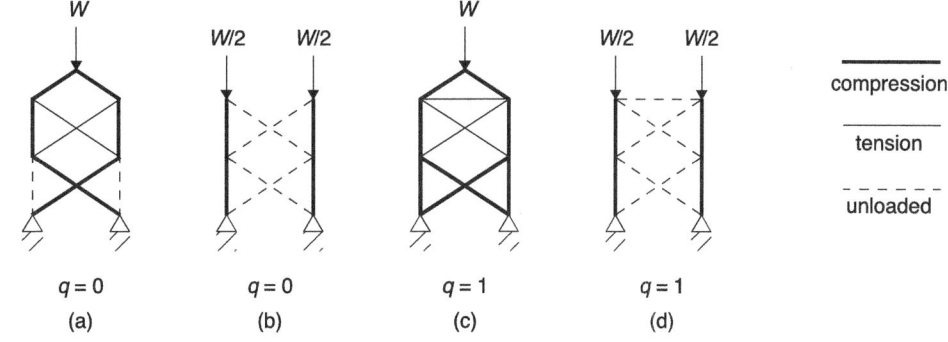

Figure AW3.1.3   Worksheet 3.1.3 answer

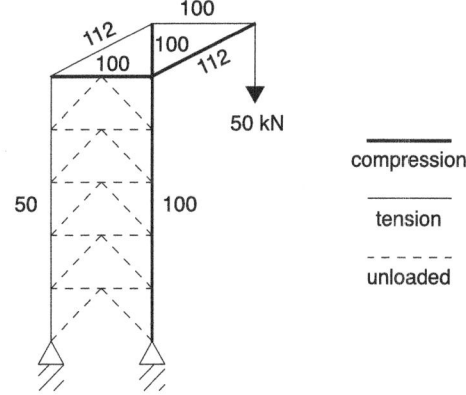

Figure AW3.1.4   Worksheet 3.1.4 answer

Worksheet 3.2 Non-parallel-sided trusses

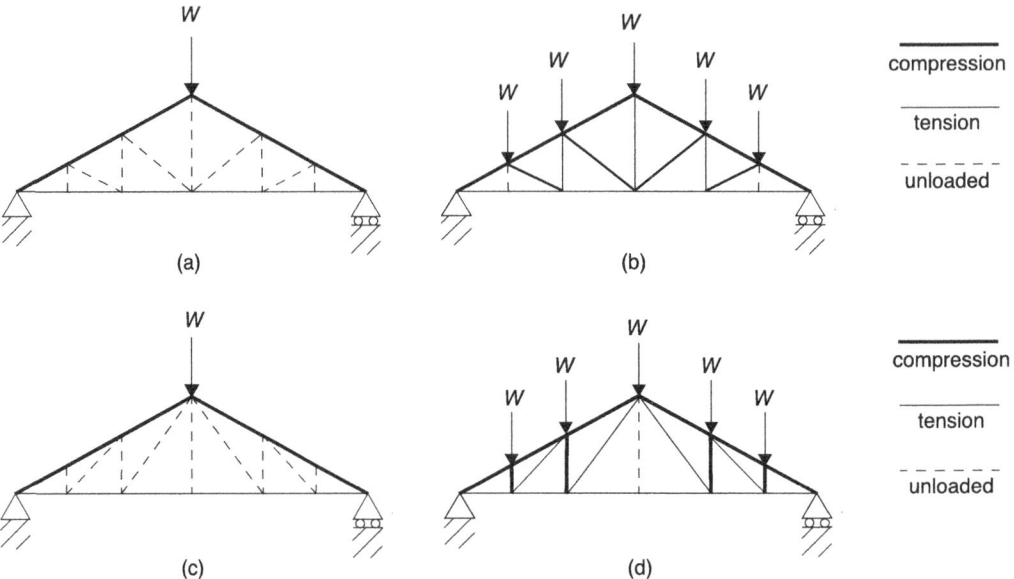

Figure AW3.2.1   Worksheet 3.2.1 answers

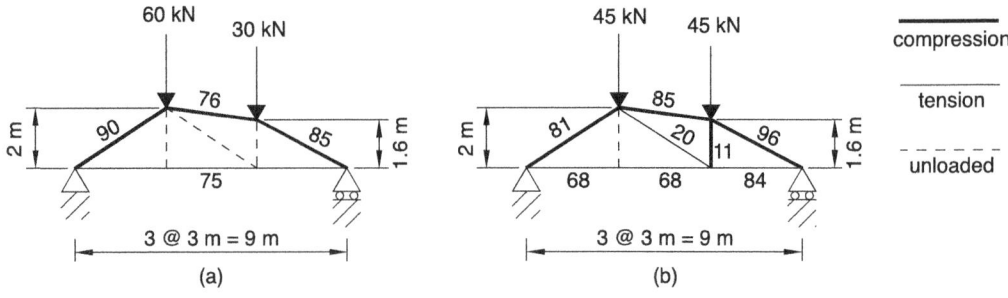

Figure AW3.2.2   Worksheet 3.2.2 answers

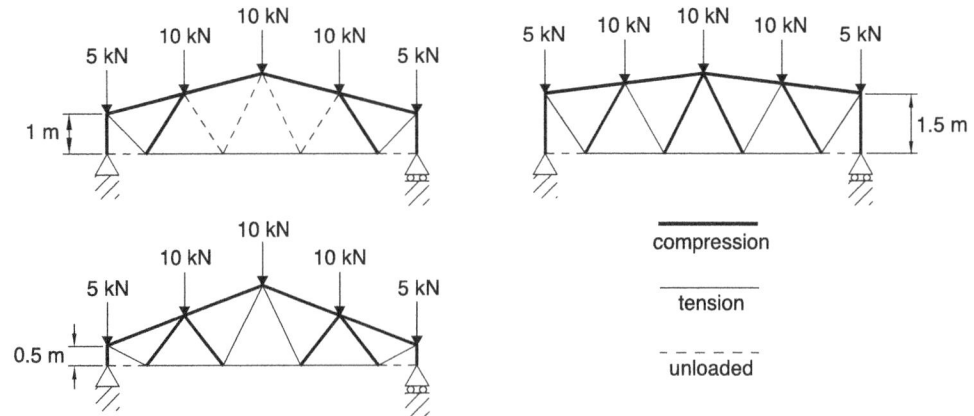

Figure AW3.2.3   Worksheet 3.2.3 answer

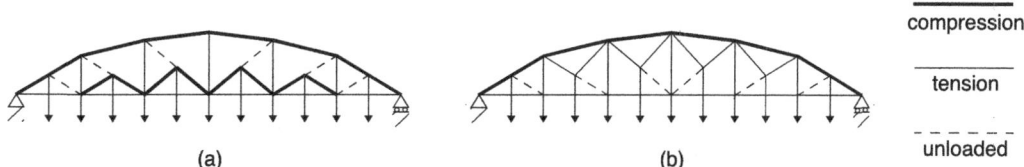

Figure AW3.2.4   Central bay compression higher for (b), but central bay tension higher for (a). Design (b) is probably slightly more economical

Worksheet 3.3   Computer analysis of plane trusses

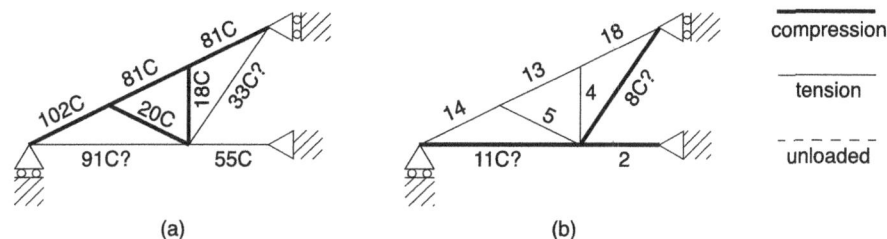

Figure AW3.3.1   Forces (kN) under load cases (a) DOWN, (b) UP – C indicates critical force

W3.3.2   (a) is a mechanism and therefore unstable; (b) is stable.

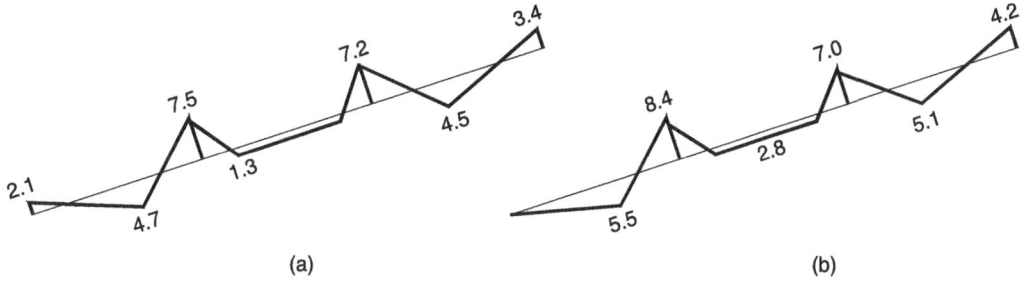

Figure AW3.3.3   (a) Bending moments (kNm) in rafter from plane frame analysis, (b) bending moments in rafter from Figure W3.3.3(c), (d)

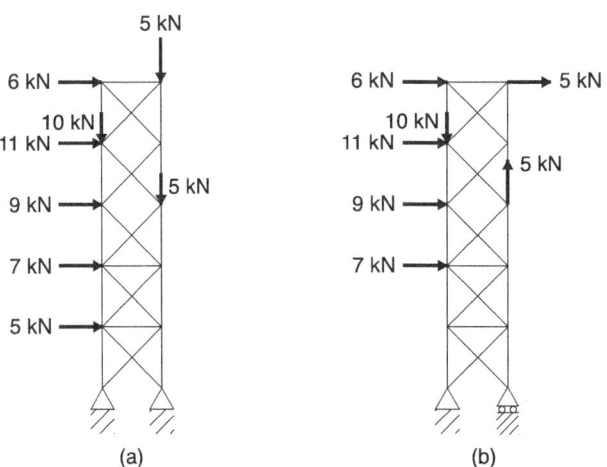

Figure AW3.3.4   (a) Actual, (b) analysed

Worksheet 3.4   Influence line sketching and calculation

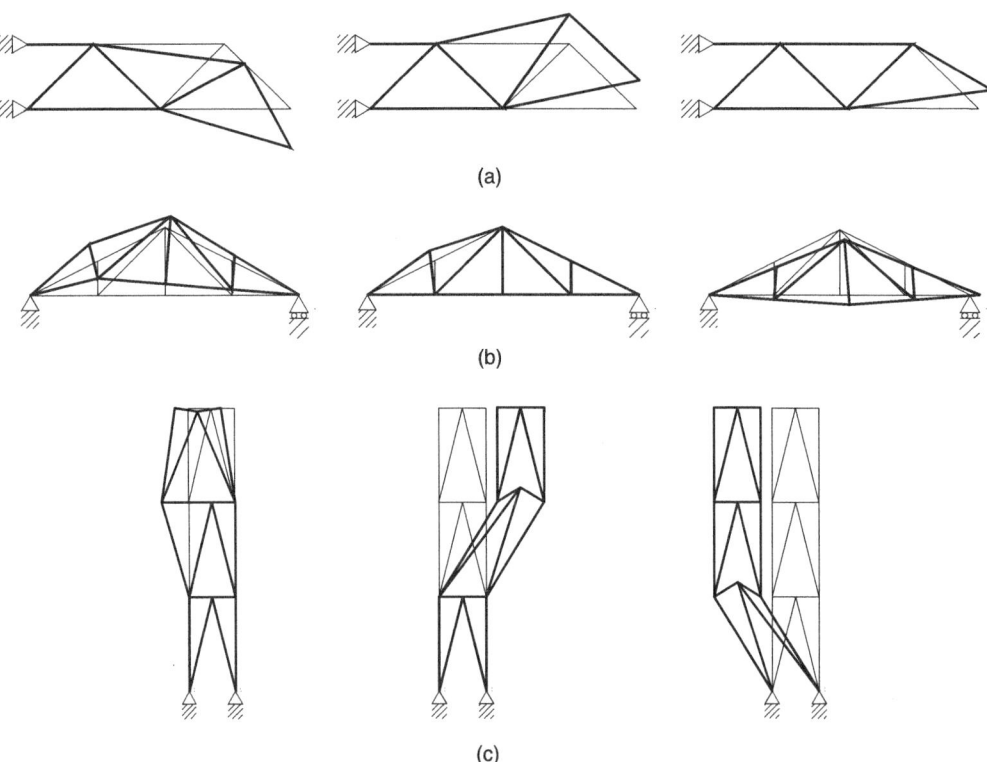

Figure AW3.4.1   (a) Influence lines for Fig. W3.4.1(a); (b) influence lines for Fig. W3.4.1(b); (c) influence lines for Fig. W3.4.1(c)

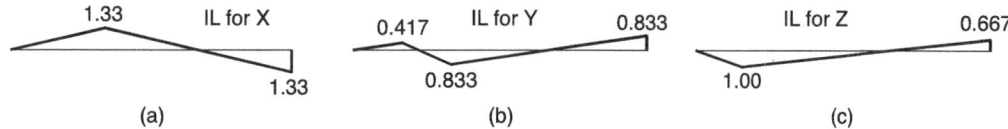

Figure AW3.4.2   Maximum forces in kN are (a) C=T=113 for X, (b) C=T=71 for Y, (c) C=57 and T=90 for Z

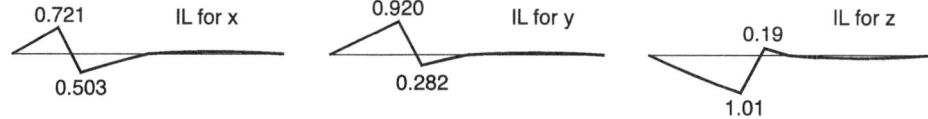

Figure AW3.4.3   Influence lines for x, y, z respectively (x and y would need to be cross-braced, but not z and not others)

Case study

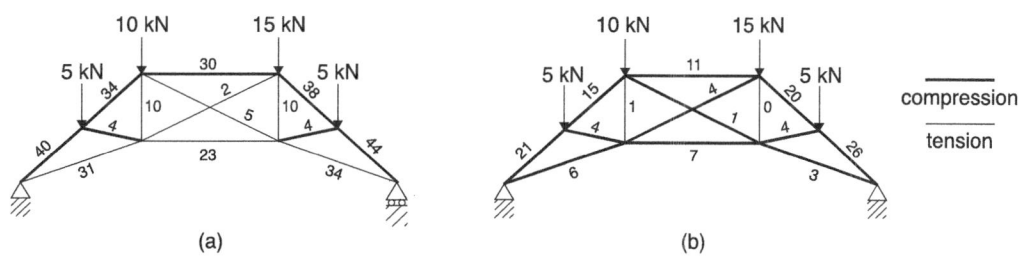

Figure AC3.1.1   Axial loads (kN) for (a) pin/roller supports, (b) pin/pin supports

9.4   CHAPTER 4 – ARCHES

Worksheet 4.1   Lines of thrust and bending moment diagrams for arches

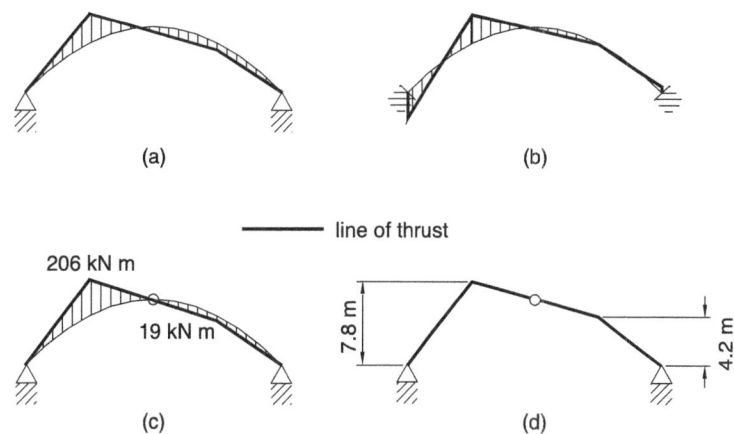

Figure AW4.1.1   (a), (b), (c) Lines of thrust and reversed bending moment diagrams for Figures W4.1.1(a), (b), (c); (d) funicular form for Figure W4.1.1(d)

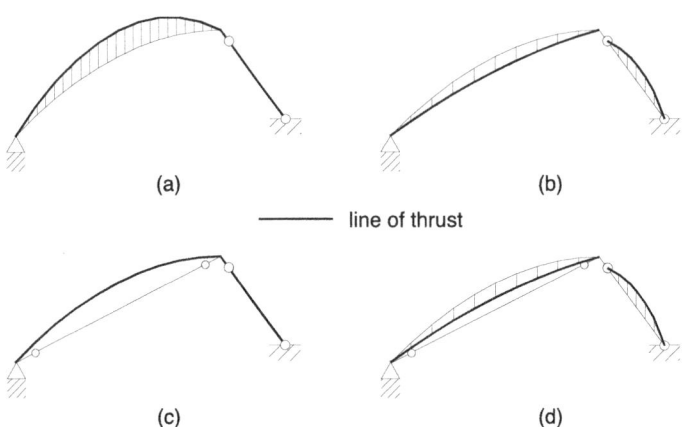

Figure AW4.1.2   Lines of thrust and reversed bending moment diagrams for (a) Figure W4.1.2(a) – l.h.s. loading; (b) Figure W4.1.2(a) – r.h.s. loading; (c) Figure W4.1.2(b) – l.h.s. loading; (d) Figure W4.1.2(b) – r.h.s. loading

Worksheet 4.2   Computer analysis of arches

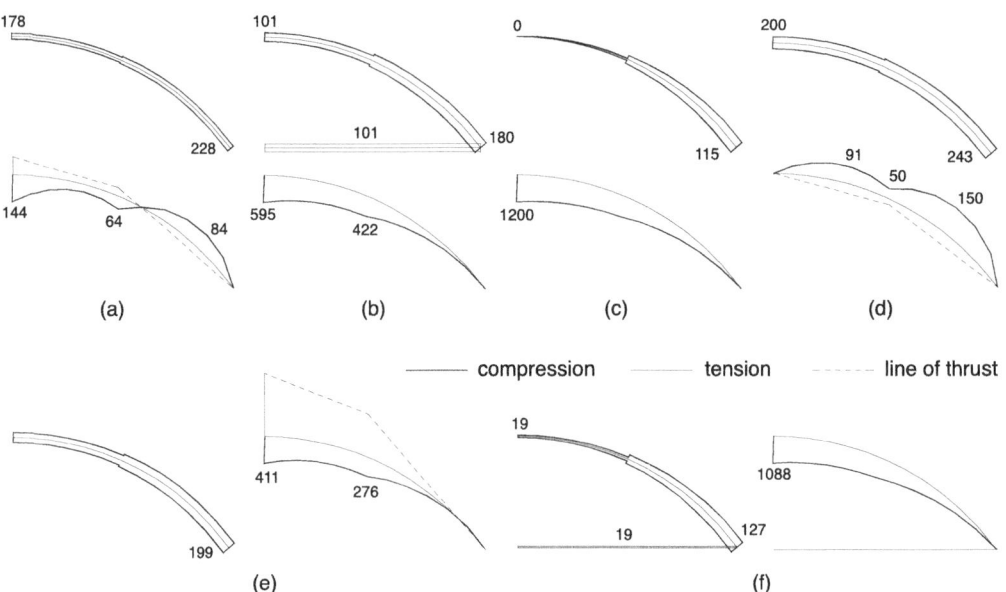

Figure AW4.2.1   (a), (b), (c), (d) Axial load (kN) and bending moment diagrams (kNm) for Figures W4.2.1(a), (b), (c), (d); (e), (f) diagrams for Figure W4.2.1(a), (b) with increased I value for arch

W4.2.2   Maximum positive bending moment at $I = 1050$ kN m due to point load at I and uniformly distributed load applied from $x = 4$ m to $x = 20$ m. Maximum negative bending moment at $I = 900$ kN m due to point load at apex and uniformly distributed load applied from $x = -20$ m to $x = 4$ m.

9.5 CHAPTER 5 – PLANE FRAMES

Worksheet 5.1 Kinematic and statical determinacy of plane frames

	(a)	(b)	(c)	(d)	(e)	(f)	(g)	(h)	(i)	(j)
k	2	10	27	16	16	8	14	10	21	14
q	1	2	9	7	3	7	3	8	9	1

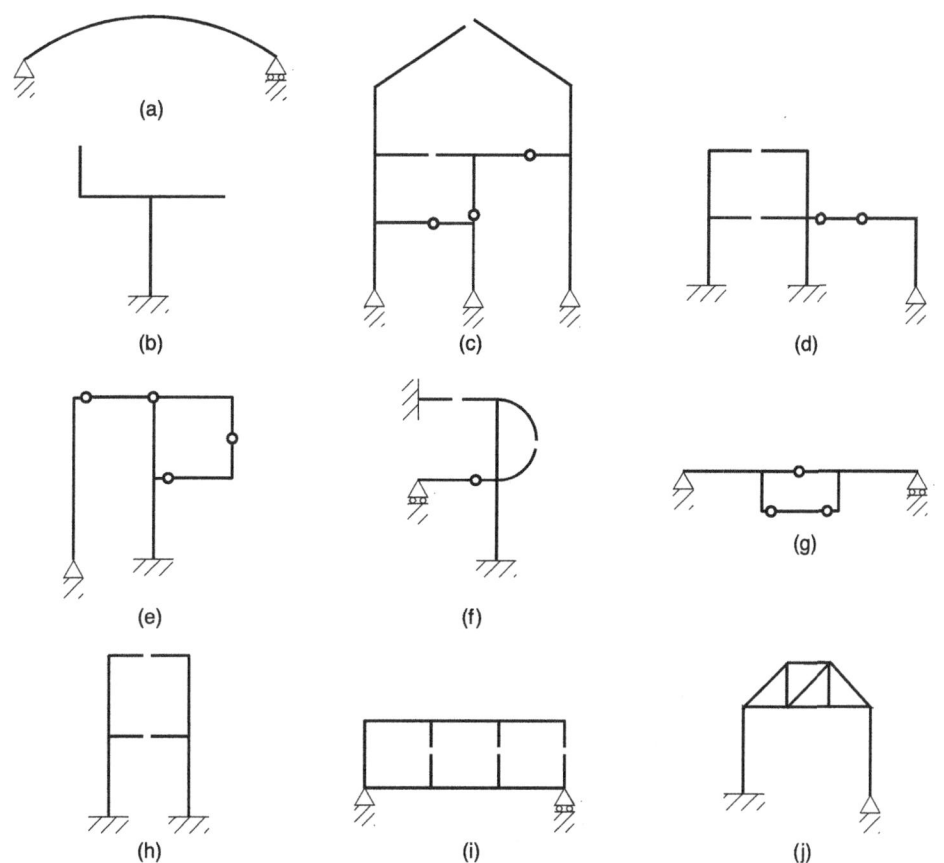

Figure AW5.1.1   Possible release systems

Worksheet 5.2 Stress resultant diagrams for sloping beams

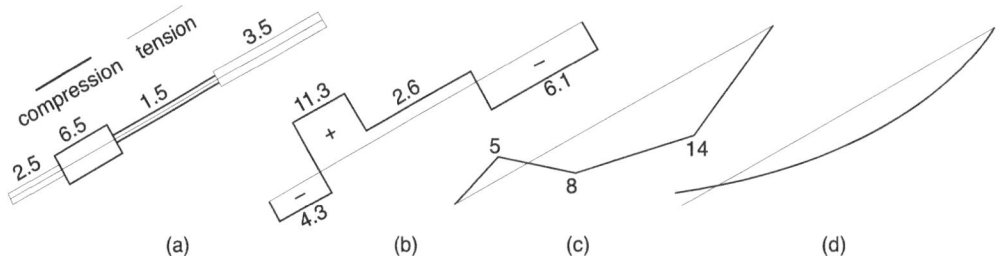

(a) (b) (c) (d)

Figure AW5.2.1 (a) Axial load, (b) shear force (kN), (c) bending moment (kNm), (d) deflected shape sketch, for W5.2.1(a)

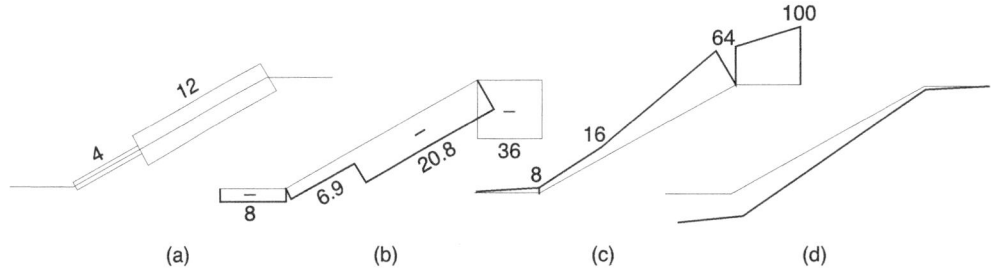

(a) (b) (c) (d)

Figure AW5.2.1 (cont.) (a) Axial load (kN), (b) shear force (kN), (c) bending moment (kNm), (d) deflected shape sketch, for W5.2.1(b)

Worksheet 5.3 Stress resultant diagrams for statically determinate plane frames

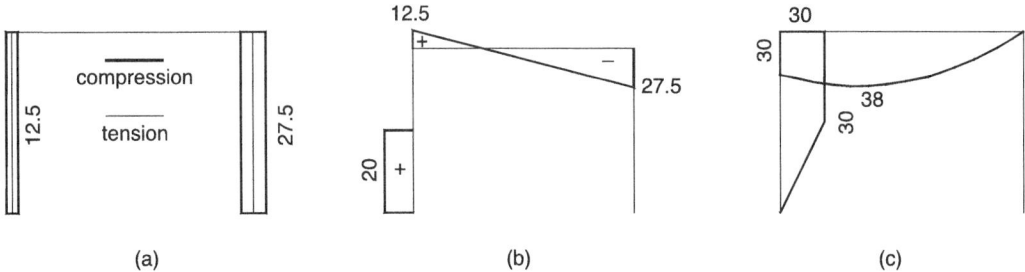

(a) (b) (c)

Figure AW5.3.1 (a) Axial load (kN), (b) shear force (kN), (c) bending moment (kNm)

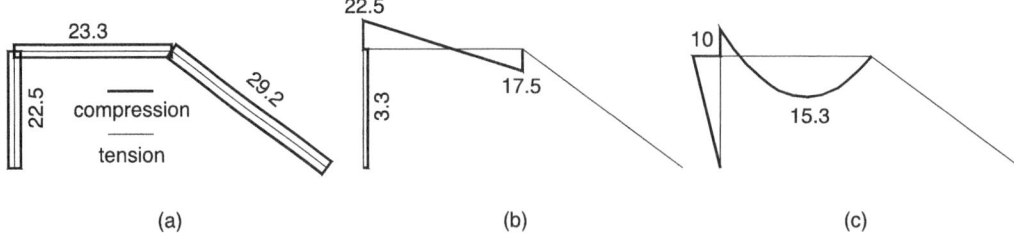

(a) (b) (c)

Figure AW5.3.2 (a) Axial load (kN), (b) shear force (kN), (c) bending moment (kNm)

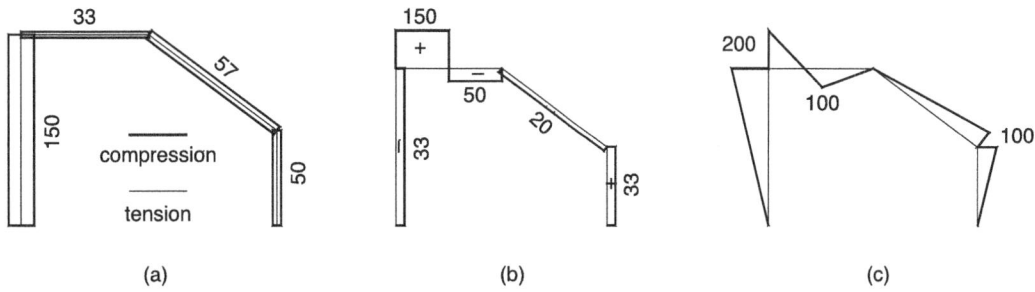

Figure AW5.3.3   (a) Axial load (kN), (b) shear force (kN), (c) bending moment (kNm)

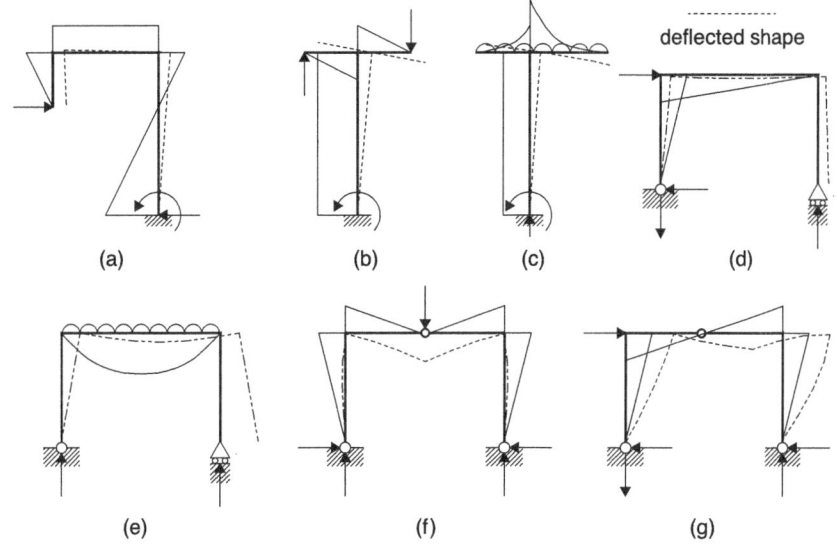

Figure AW5.3.4   Reactions, deflection and bending moment sketches

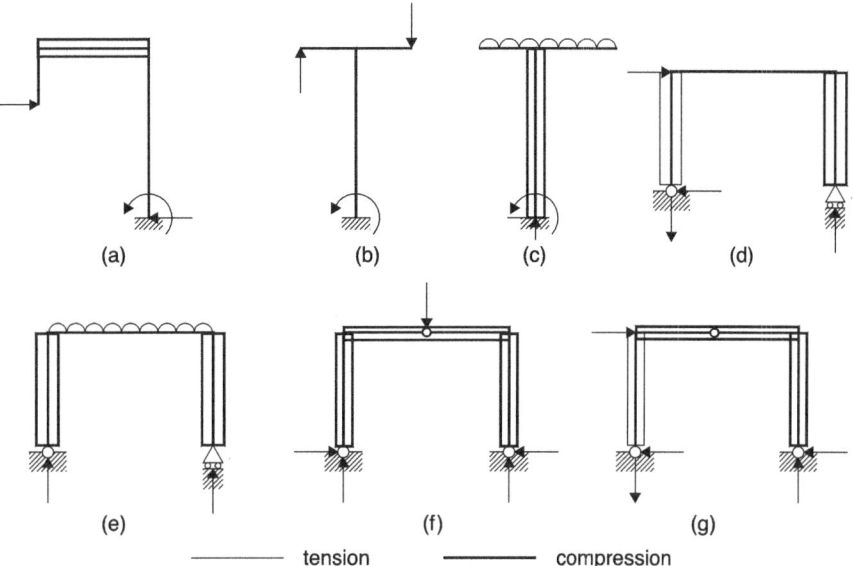

Figure AW5.3.4 (cont.)   Axial load sketches

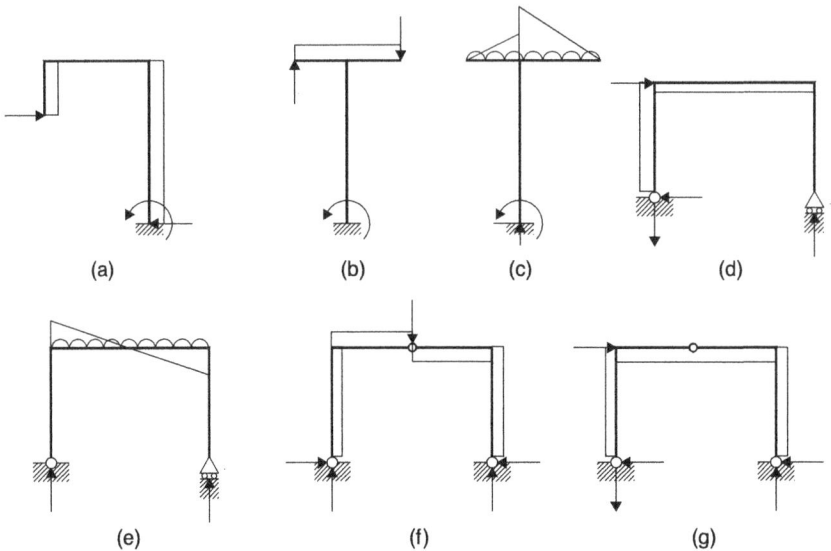

Figure AW5.3.4 (cont.)   Shear force sketches

Worksheet 5.4   Stress resultant diagrams for statically indeterminate plane frames

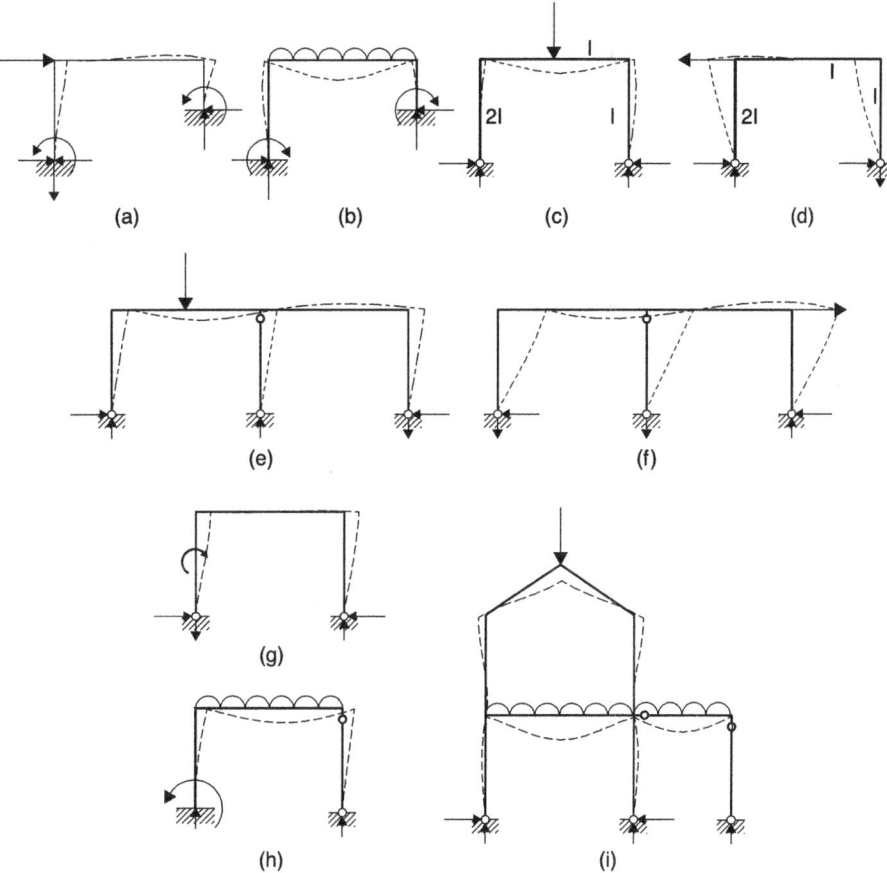

Figure AW5.4.1   Reactions and possible displaced shapes

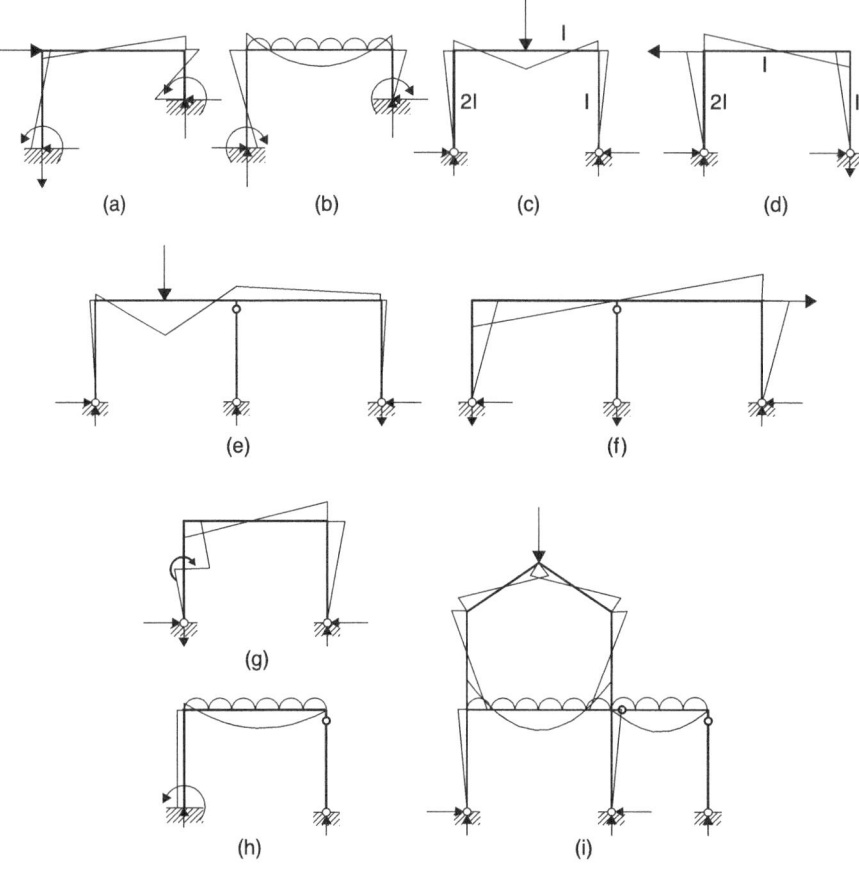

Figure AW5.4.1 (cont.)   Possible bending moment diagrams

Worksheet 5.5   Computer analysis of plane frames

W5.5.1 Maximal rim values based on analyses at 15° interval analyses: bending moment = 38 300 N mm; shear force = 602 N; axial load = 418 N; and displacement = 4 mm.

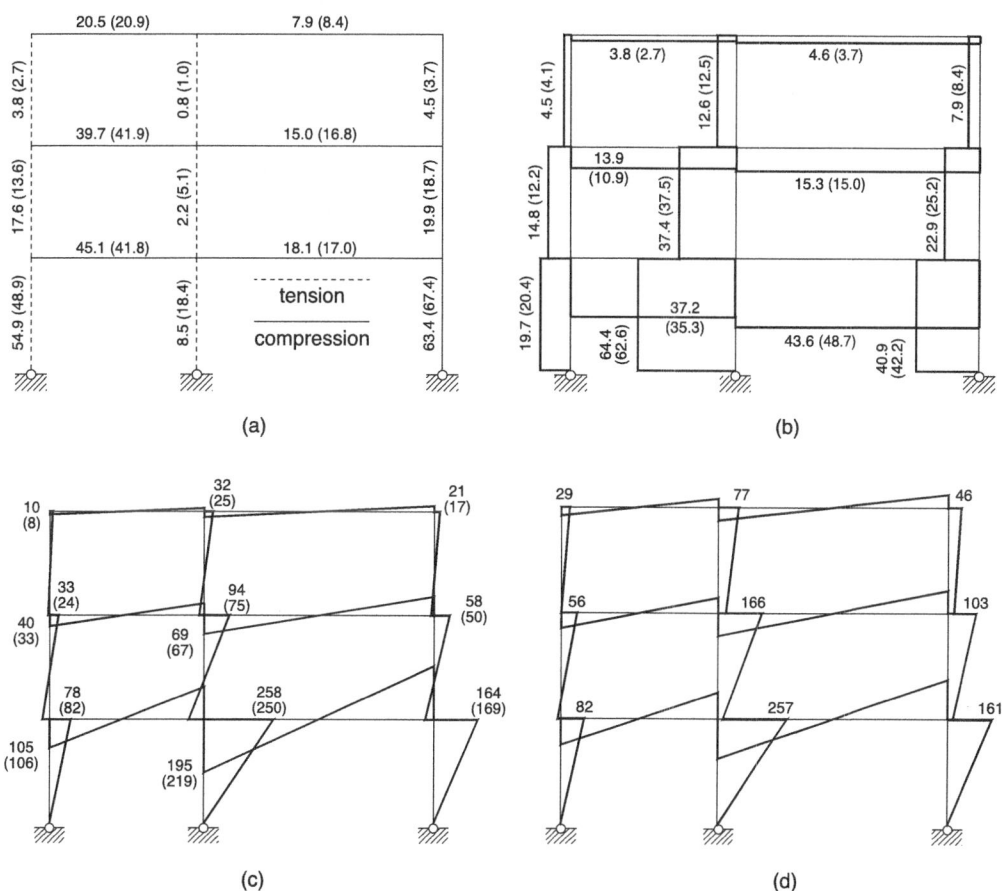

Figure AW5.5.2   Computer (hand) values of (a) axial load (kN), (b) shear force (kN), (c) bending moment (kNm), (d) bending moment (kNm) with reduced beam sections

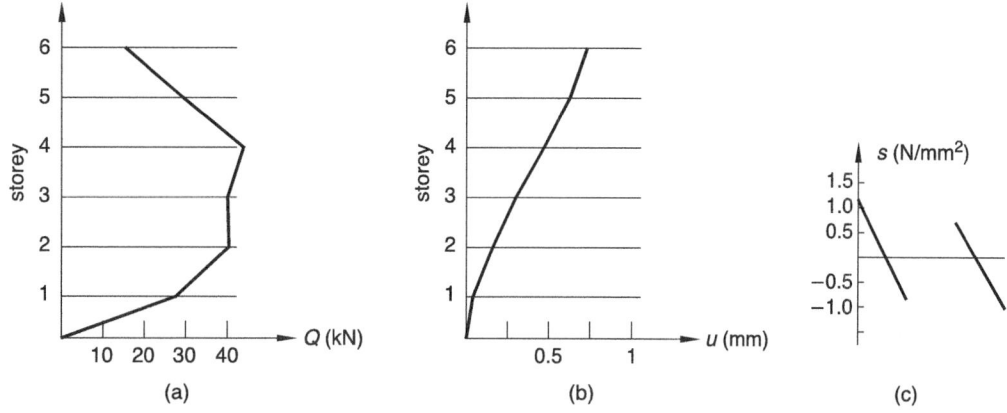

Figure AW5.5.3   (a) Shear force in connecting beams, (b) lateral wall displacements, (c) vertical stresses at wall bases

Worksheet 5.6 Influence lines for plane frames

Table AW5.6.1

Position	Maximum BM (kN m)	BM due full load (kN m)
A	35	26
B	99	65
C	135	133

Table AW5.6.2

Position	Maximum BM (kN m)	BM due full load (kN m)
A	197	197
B	94	85
C	35	35
D	35	35

Case studies

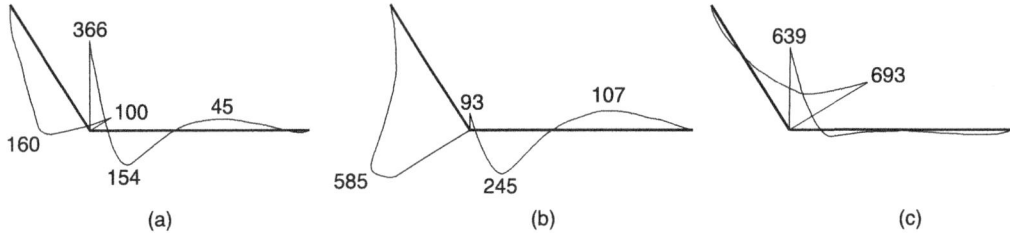

(a)   (b)   (c)

Figure AC5.1.1 Bending moment diagrams for loading cases (a) DL + CL, (b) 0.9DL, (c) 1.1DL + CL

C5.1.3 Maximum bending moment is 142 kN m located just below right-hand eaves position under the action of dead, wind from the right, surcharge and earth loadings.

9.6   CHAPTER 6 – SPACE TRUSSES

Worksheet 6.1   Qualitative analysis of space trusses

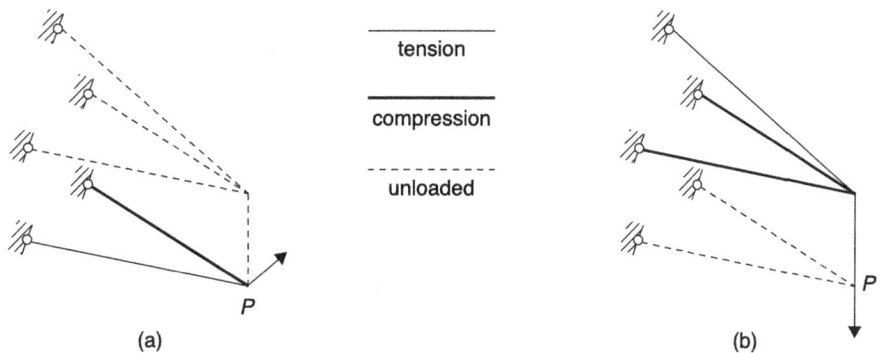

Figure AW6.1.1   Nature of forces due to load at P: (a) in positive y direction, (b) in negative z direction

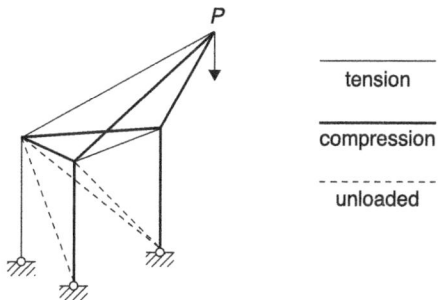

Figure AW6.1.2   Nature of forces due to load at P in negative z direction

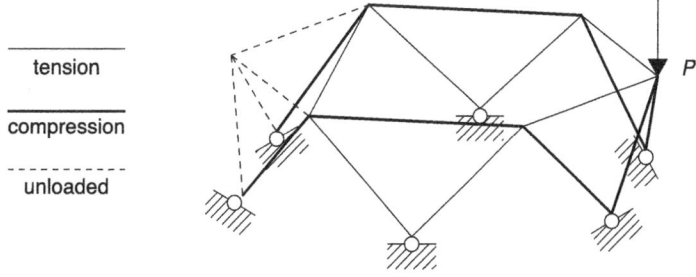

Figure AW6.1.3   Nature of forces due to load at P in negative z direction

Worksheet 6.2 Computer analysis

Table W6.2.1

Element	Plan angle (°)	Tie angle to horizontal (°)	Member force (kN)
Jib	Any	0	−141
Tie	Any	−20	140
Stay	±135	−20	−161
	±45	−20	161
Post	0	−20	−180
	±135	−20	45
Sleeper	±45	−20	−132
	±135	−20	132

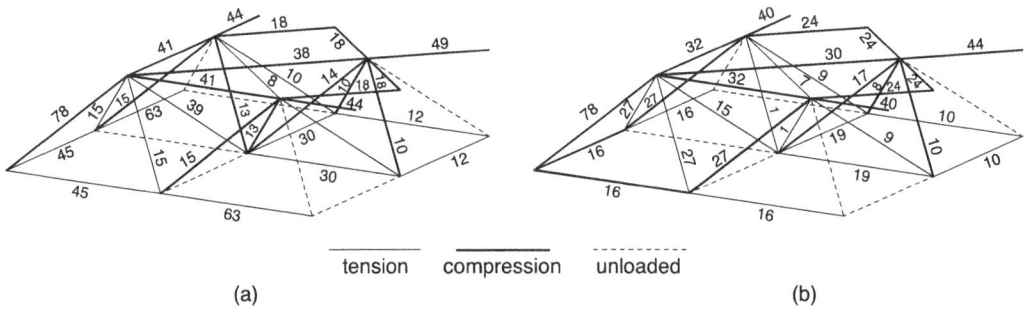

Figure AW6.2.2   Axial loads (kN) in quarter frame: (a) vertical support only, (b) fully-pinned support

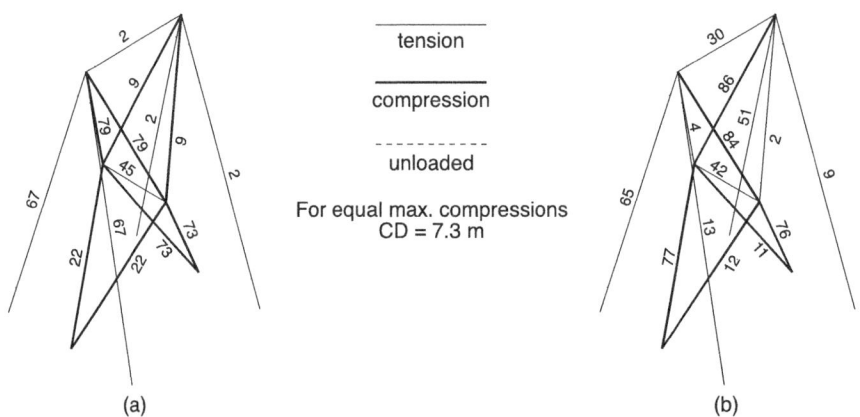

Figure AW6.2.3   Axial loads (kN) due to: (a) prestress and transverse loading, (b) prestress and longitudinal loading

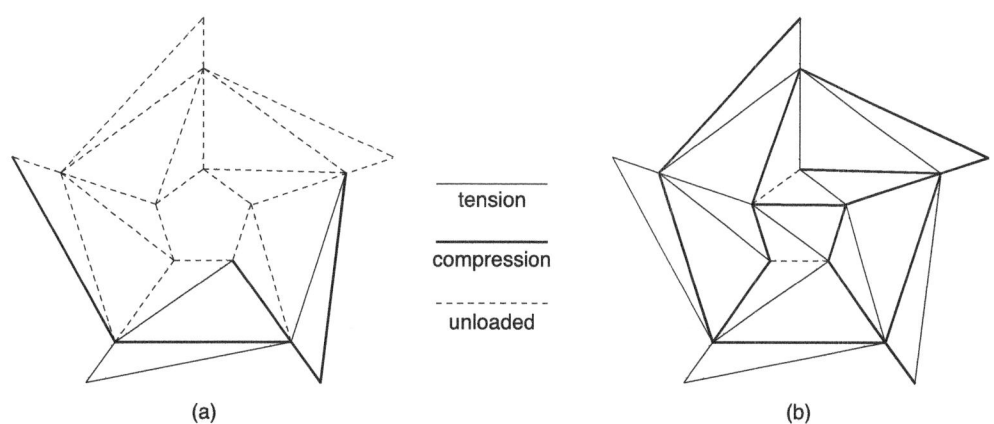

Figure AW6.2.4   Nature of axial loads: (a) unbraced, (b) braced

Worksheet 6.3   Influence lines for space trusses

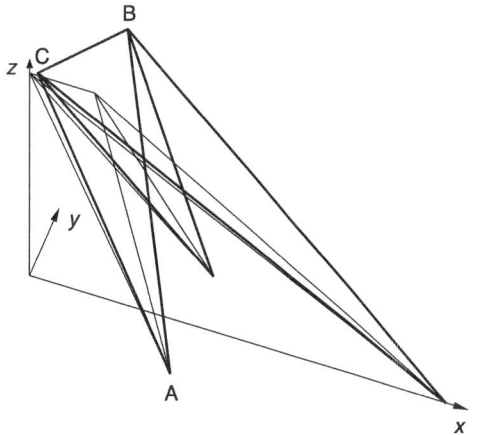

Figure AW6.3.1   Influence line for transverse guy, AB
W6.3.2 Maximum boom compression = 597 kN; maximum inclined bracing compression = 87 kN.

9.7   CHAPTER 7 – GRIDS

Worksheet 7.1   Statically determinate grids

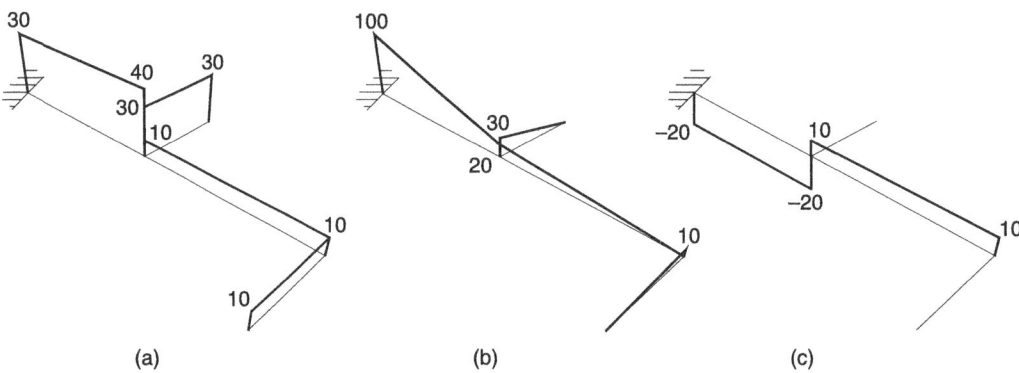

Figure AW7.1.1   (a) Shear force (kN), (b) bending moment (kNm), (c) torque (kNm) for grid (a)

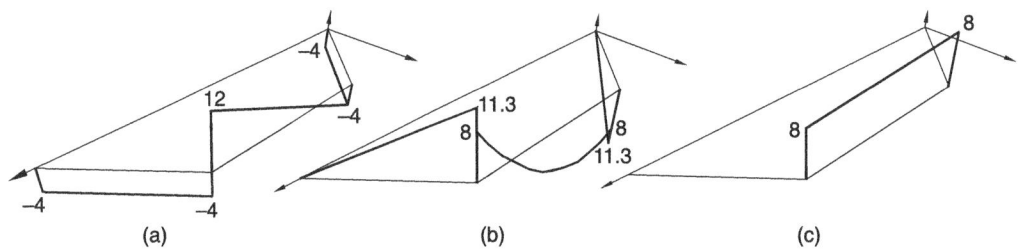

Figure AW7.1.1 (cont.)   (a) Shear force (kN), (b) bending moment (kNm), (c) torque (kNm) for grid (b)

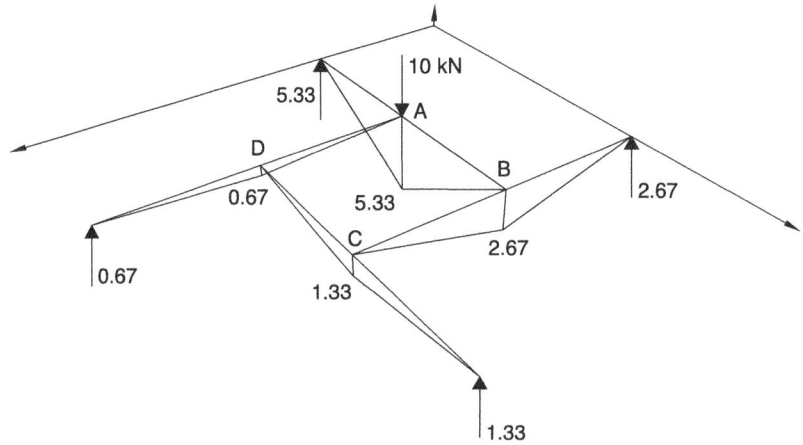

Figure AW7.1.2   Reactions (kN) and bending moment diagram (kNm)

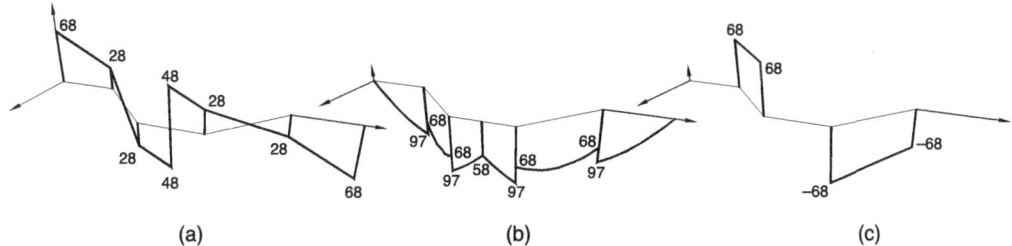

(a) (b) (c)

Figure AW7.1.3 (a) Shear force (kN), (b) bending moment (kNm), (c) torque (kNm)

Worksheet 7.2 Computer analysis of grids

Table W7.2.1

Section	Downwards displacement (mm)	Clockwise rotation (°)
Rectangular hollow	30	−0.05
Channel	40	6.54

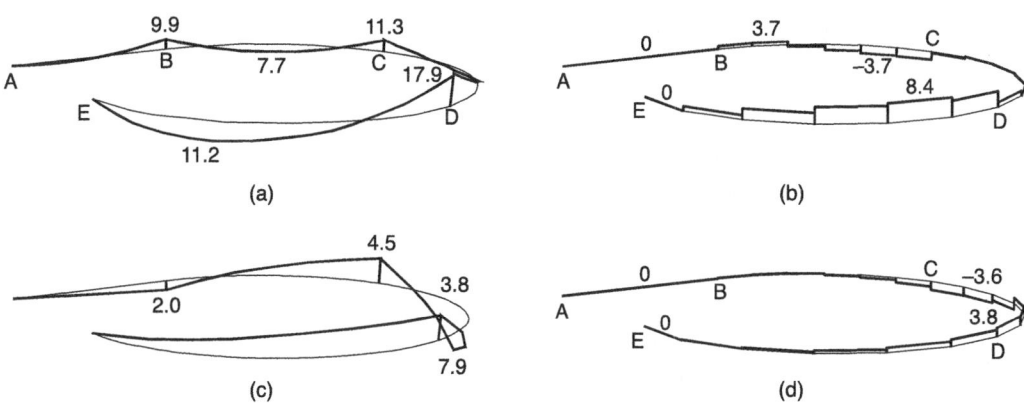

(a) (b)

(c) (d)

Figure AW7.2.2 (a) Bending moments (MN m) due to self weight, (b) torque (MN m) due to self weight, (c) bending moment (MN m) due to offset load, (d) torque (MN m) due to offset load

Worksheet 7.3 Influence lines for grids

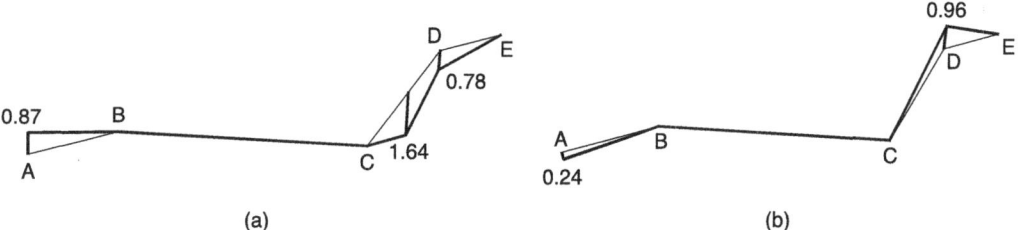

(a) (b)

Figure AW7.3.1 Influence lines for (a) bending moment (m), (b) torque (m) – maximum positive BM = 474 kNm; maximum negative BM = 185 kNm; maximum torque = 252 kNm

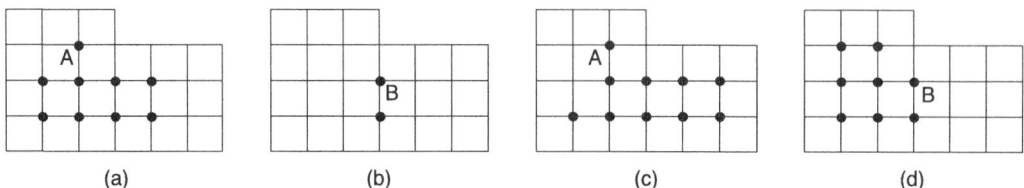

Figure AW7.3.2   Loaded positions for (a) maximum positive bending moment at A, (b) maximum positive bending moment at B, (c) maximum absolute torque at A, (d) maximum absolute torque at B

9.8   CHAPTER 8 – SPACE FRAMES

Worksheet 8.1   Statically determinate space frames

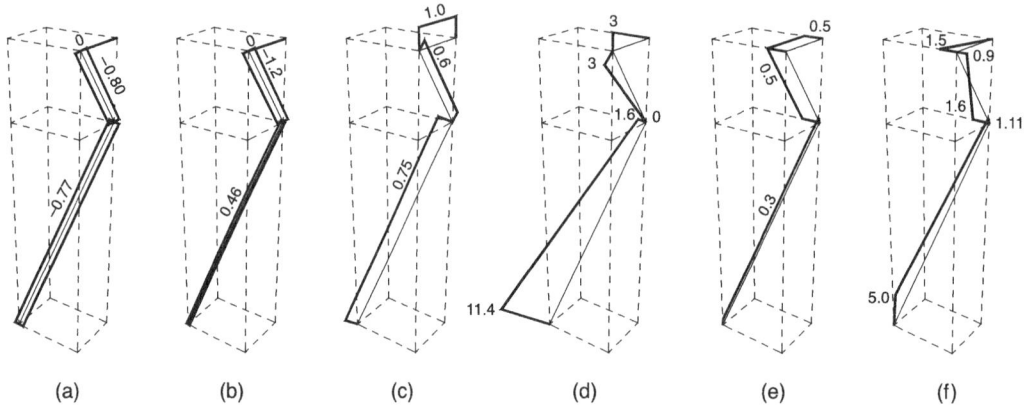

Figure AW8.1.1   (a) Axial load (×W), (b) torque (×Wa), (c) SF_w (×W), (d) BM_v (×Wa), (e) SF_v (×W), (f) BM_w (×Wa)

Designating 1–2 as 2–1 would change signs of all values calculated for the member, but only the SF diagrams would be reversed.

Worksheet 8.2   Computer analysis of space frames

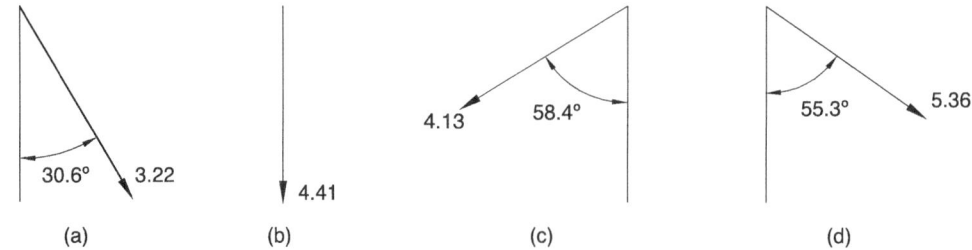

Figure AW8.2.1   Displacements (mm): (a) angle, (b) rotated angle, (c) zed, (d) rotated zed

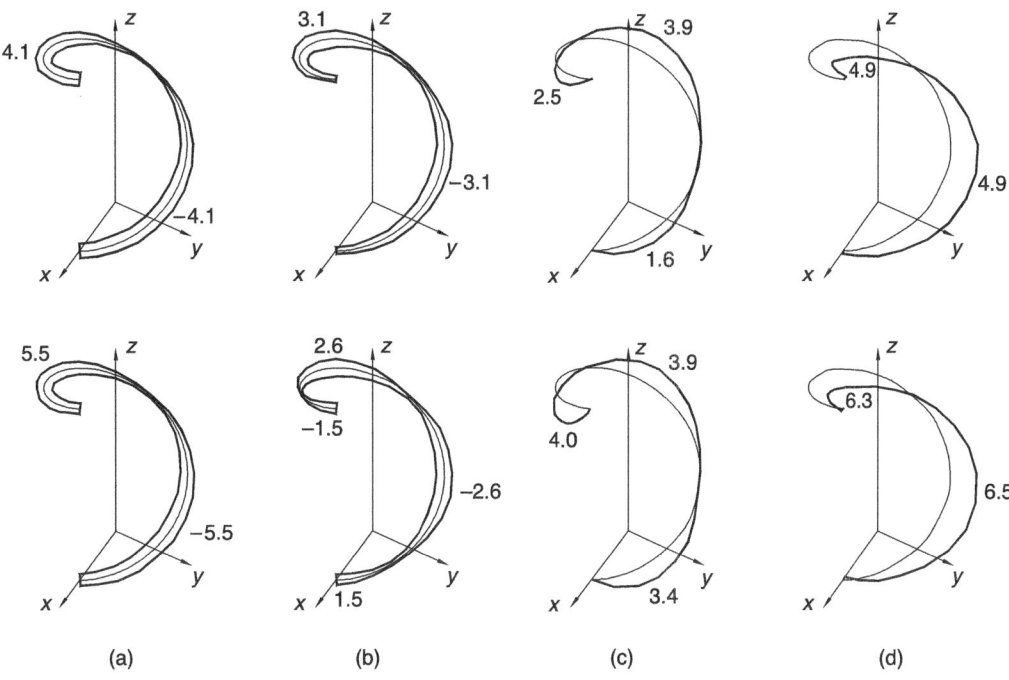

Figure AW8.2.2 Internal forces for box and double channel, respectively: (a) axial load (kN), (b) torque (kN), (c) BM_v (kNm), (d) BM_w (kNm)

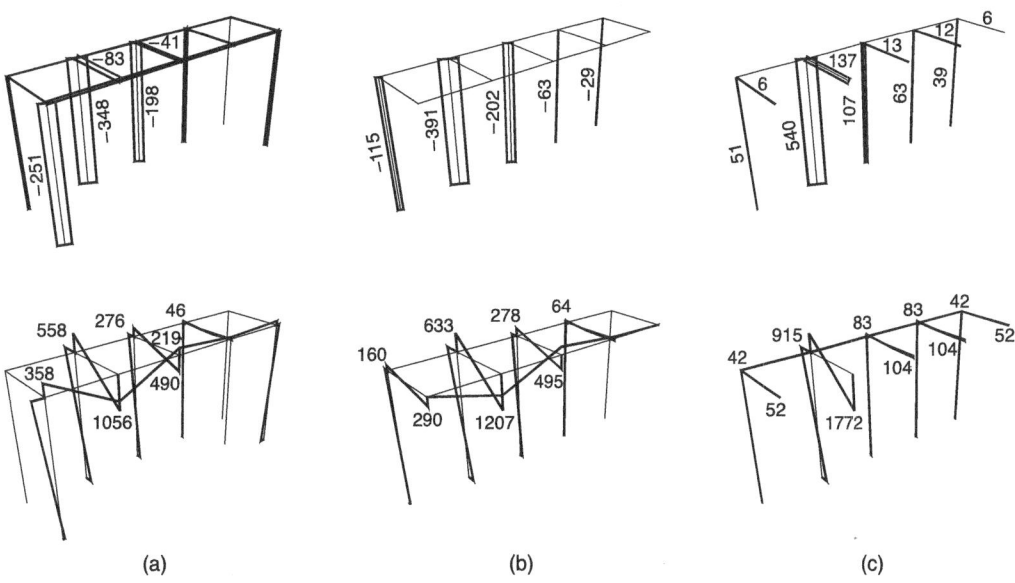

Figure AW8.2.3 Axial force (kN) and BM_v (kNm) for half of (a) basic frame, (b) central end columns omitted, (c) central longitudinal beams and central end columns omitted

Case study

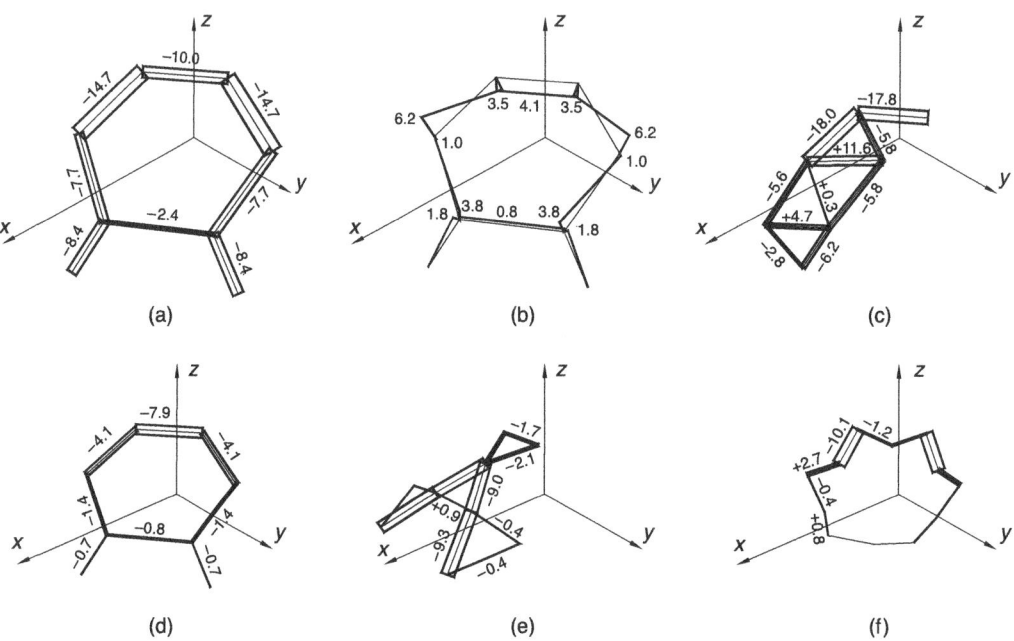

(a)

(b)

(c)

(d)

(e)

(f)

Figure AC8.1.1   Hexagon dome under summit loading: (a) axial loads (kN), (b) BM_v (kNm); triangular dome, (c) axial loads; hex-tri-hex dome axial loads (kN), (d) outer members, (e) inner members, (f) bracing members (symmetric portions only shown in all cases)

Index

Page numbers in *italics* denote illustrations/photographs.

The manufacturer's authorised representative in the EU for product safety is Easy Access System Europe – Mustamäe tee 50, 10621 Tallinn, Estonia, easproject.com, +372 5696 8939, gpsr.requests@easproject.com

Printed and bound by CPI Group (UK) Ltd, Croydon, CR0 4YY

14/05/2026

02111093-0001